Evolutionary Biology

PRINCIPLES OF MEDICAL BIOLOGY
A Multi-Volume Work, Volume 1B

Editors: **E. EDWARD BITTAR,** *Department of Physiology,*
University of Wisconsin, Madison
NEVILLE BITTAR, *Department of Medicine,*
University of Wisconsin, Madison

Principles of Medical Biology

A Multi-Volume Work

Edited by **E. Edward Bittar**, *Department of Physiology, University of Wisconsin, Madison and*
Neville Bittar, *Department of Medicine, University of Wisconsin, Madison*

This work in 25 modules provides:

- A holistic treatment of the main medical disciplines. The basic sciences including most of the achievements in cell and molecular biology have been blended with pathology and clinical medicine. Thus, a special feature is that departmental barriers have been overcome.
- The subject matter covered in preclinical and clinical courses has been reduced by almost one-third without sacrificing any of the essentials of a sound medical education. This information base thus represents an integrated core curriculum.
- The movement towards reform in medical teaching calls for the adoption of an integrated core curriculum involving small-group teaching and the recognition of the student as an 'active learner.'
- There are increasing indications that the traditional education system in which the teacher plays the role of expert and the student that of a passive learner is undergoing reform in many medical schools. The trend can only grow.
- Medical biology as the new profession has the power to simplify the problem of 'reductionism.'
- Over 700 internationally acclaimed medical scientists, pathologists, clinical investigators, clinicians, and bioethicists are participants in this undertaking.

The first four to six physical volumes are planned for Fall 1994.

Evolutionary Biology

Edited by **E. EDWARD BITTAR**
Department of Physiology
University of Wisconsin
Madison, Wisconsin

NEVILLE BITTAR
Department of Medicine
University of Wisconsin
Madison, Wisconsin

JAI PRESS INC.

Greenwich, Connecticut *London, England*

Library of Congress Cataloging-in-Publication Data
Evolutionary biology / edited by E. Edward Bittar, Neville Bittar.
p. cm. -- (Principles of medical biology ; v. 1B)
Includes bibliographical references and index.
ISBN 1-55938-802-1: $90.25
1. Molecular evolution. 2. Cells -- Evolution. 3. Human evolution.
I. Bittar, E. Edward. II. Bittar, Neville. III. Series.
QH371.E9273 1994
575--dc20 94-11633
CIP

55 Old Post Road, No. 2
Greenwich, Connecticut 06836

JAI PRESS LTD.
The Courtyard
29 High Street
Hampton Hill, Middlesex TW12 1PD
England

ISBN: 1-55938-802-1
ISBN: 1-55938-779-3 (Set)
ISBN: 1-55938-800-5 (Treatise)

Library of Congress Catalog Card No.: 94-11633

Transferred to Digital Printing 2006

CONTENTS
Volume 1B: Evolutionary Biology

CONTENTS
Volume 1A: Bioethics

LIST OF CONTRIBUTORS

Peter R. Bahn — Bahn Technology Company, Woodlawn, Illinois

Laurel A. Burton — Department of Religion and Health and Human Values, Rush Medical College, Chicago, Illinois

Eric Chivian — Massachusetts Institute of Technology, Cambridge, Massachusetts

David C. Clark — Department of Psychiatry, Rush Medical College, Chicago, Illinois

Colleen D. Clements — Department of Psychiatry, University of Rochester School of Medicine, Rochester, New York

James F. Crow — Department of Genetics, University of Wisconsin, Madison, Wisconsin

Michael H. Day — Department of Palaeontology, The Natural History Museum, London, England

Rem Blanchard Edwards — Department of Philosophy, The University of Tennessee, Knoxville, Tennessee

H. Tristram Engelhardt, Jr.
Center for Ethics, Medicine, and Public Issues
Baylor College of Medicine
Houston, Texas

Linda A. Fothergill-Gilmore
Department of Biochemistry
University of Edinburgh
Edinburgh, Scotland

Sidney W. Fox
Coastal Research and Development Institute
University of South Alabama
Mobile, Alabama

M. Gomendio
Museo Nacional de Ciencias Naturales
Madrid, Spain

Glenn Campbell Graber
Department of Philosophy
The University of Tennessee
Knoxville, Tennessee

Jennifer A. Marshall Graves
Department of Genetics and Human Variation
La Trobe University
Bundoora, Australia

Matti Häyry
Department of Philosophy
University of Helsinki
Helsinki, Finland

Austin L. Hughes
Department of Biology and Institute of Molecular Evolutionary Genetics
The Pennsylvania State University
University Park, Pennsylvania

Motoo Kimura
National Institute of Genetics
Mishima, Japan

Helga Kuhse
Center for Human Bioethics
Monash University
Clayton, Victoria, Australia

John Last — Department of Epidemiology and Community Medicine
University of Ottawa
Ottawa, Canada

Robert J. Levine — Department of Medicine
Yale University School of Medicine
New Haven, Connecticut

A. Lima-de-Faria — Institute of Genetics
University of Lund
Lund, Sweden

Michael McCally — Chicago Community Trust
Chicago, Illinois

Peter Nicholls — Department of Biological Sciences
Brock University
St. Catharines, Ontario, Canada

Simon Potter — Department of Biochemistry
University of British Columbia
Vancouver, Canada

E. R. S. Roldan — Department of Development and Signalling
AFRC Babraham Institute
Cambridge, England

Bernard E. Rollin — Department of Philosophy
Colorado State University
Fort Collins, Colorado

Cecilia Saccone — Dipartimento di Biochimica e Biologia Molecolare
Università di Bari
Bari, Italy

Elisabetta Sbisà — Centro di Studio sui Mitocondri e Metabolismo Energetico
CNR
Bari, Italy

Naoyuki Takahata
National Institute of Genetics
Mishima, Japan

Jaclyn M. Watson
Department of Genetics and Human Variation
La Trobe University
Bundoora, Australia

Daniel Wikler
Department of the History of Medicine and Ethics
University of Wisconsin
Madison, Wisconsin

Kevin Wm. Wildes
Kennedy Institute of Ethics
Philosophy Department
Georgetown University
Washington, D.C.

PREFACE

The object of Volume 1B is to provide a general introduction to the study of evolutionary biology and, in particular, to give due place to several new concepts that have arisen in this field. The volume ends with a chapter on the evolution of man by Michael Day.

Our most grateful thanks are due to the contributors who have made this volume possible, and to the staff members of JAI Press for their courtesy and skill.

E. EDWARD BITTAR
NEVILLE BITTAR

Chapter 1

The Evolution of RNA and Protein as Catalysts

LINDA A. FOTHERGILL-GILMORE and SIMON POTTER

Principles of Medical Biology, Volume 1B
Evolutionary Biology, pages 1–38

ISBN: 1-55938-802-1

INTRODUCTION

Biocatalysts are Essential for Life

Living organisms depend fundamentally on thousands of chemical reactions. Biopolymers such as proteins must be assembled from amino acids, nucleic acids from nucleotides, and carbohydrates from monosaccharides. Metabolic compounds must be synthesized and broken down. Chemicals must be transported into and out of cells. Without these and a host of other chemical reactions, no living organism would exist. In most organisms these reactions take place at neutral pH, at mild temperatures, and at atmospheric pressure. Without assistance, they would require minutes, hours, or even days to complete—an impossible situation for any living organism. Living systems have, therefore, evolved such that almost every chemical reaction has an attendant catalyst. These biocatalysts (or *enzymes*) enable the chemistry of life to proceed rapidly and reproducibly; it is common for an enzyme to accelerate a chemical reaction by over a million times.

What are Biocatalysts?

What types of molecules are enzymes, and how do they differ from the catalysts used in nonbiological (predominantly inorganic) chemistries? In nearly all cases, enzymes are proteins. In fact, until quite recently there were no known non-proteinaceous enzymes. However, studies on the mechanisms of RNA processing by Thomas Cech and Sidney Altman about a decade ago led to the unexpected discovery that some RNA molecules could catalyze their own splicing reactions (a process which removes unwanted sequences of nucleotides or *introns*) in a highly specific manner (reviewed by Cech and Bass, 1986). The number of different types of RNA enzymes (or *ribozymes*) is very small in comparison with the vast array of protein enzymes, although they have proved to be more widely distributed than was first predicted. Nevertheless, their discovery has had a profound impact on thinking about how living systems first arose and subsequently evolved.

Enzymes are typical catalysts in that: (1) they accelerate a chemical reaction by lowering the activation energy barrier between a substrate and its product (Figure 1); (2) they do not alter the equilibrium of the reaction, but simply allow the equilibrium to be attained more rapidly; and (3) they are not affected by the

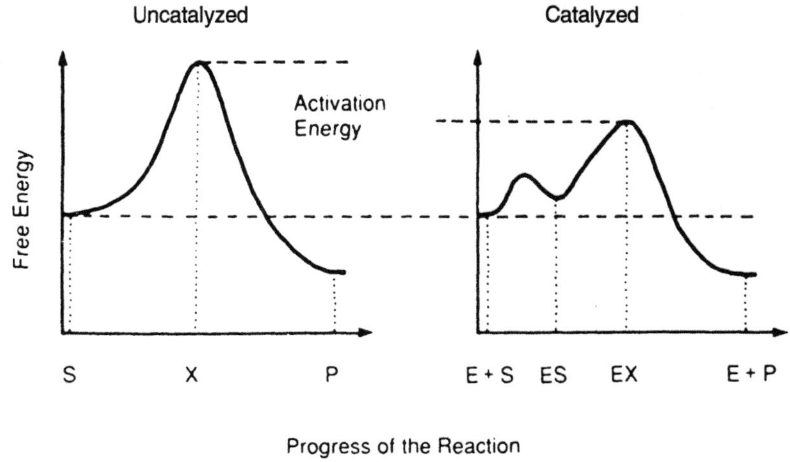

Figure 1. Catalysis accelerates a reaction by lowering the activation energy barrier. Abbreviations: S, substrate; X, transition state; P, product; E, enzyme; ES, enzyme substrate complex; EX, transition state complex.

reactions which they catalyze and are, therefore, reusable. Enzymes also have properties which make them the most superior catalysts known. First, they are very powerful catalysts and are, therefore, capable of reducing the activation energy of a reaction substantially. For example, the production of ammonia industrially from nitrogen using an iron catalyst requires temperatures of 500 °C and pressures of up to 200 atmospheres. The same reaction can occur in bacterial cells at room temperature and pressure with the assistance of the nitrogenase enzyme complex. Second, enzymes are extremely specific, in contrast to inorganic catalysts. This is advantageous in that few side products are formed from the reaction. Finally, enzymes can be precisely regulated and hence reactions can effectively be switched on and off in response to a variety of biological signals. The manner in which enzymes have evolved to acquire these superior properties is the main concern of this chapter.

The Origins of Biocatalysts

In the same way that present-day living organisms depend upon enzymes, so the early stages in the development of life must have been inextricably tied up with the origins of biocatalysts. One aspect of early evolution has seemed particularly baffling: How could the system of information flow from DNA to RNA to protein have developed when the whole process depends on protein catalysts? How did nucleic acid molecules replicate without enzymes? The discovery of ribozymes provides a clue and a possible solution to the puzzle.

There are many theories which have been proposed to explain the origins of life on Earth. One of the most plausible is the "RNA world" hypothesis of Gilbert (1986). In this world, the first self-reproducing (replicating) molecules were RNAs. The hypothesis was suggested by the discovery by Cech (1986) and others that certain species of RNA in the ciliate protozoan *Tetrahymena* were capable of catalyzing self-splicing reactions (a process by which introns are removed from gene sequences) without the assistance of any protein molecules. It was then suggested that the primitive RNAs subsequently developed the ability to synthesize proteins which replaced the ribozymes because of their superior catalytic versatility (we shall see later that ribozymes are not capable of the same catalytic versatility as protein catalysts). In the next step, DNA was formed by reverse transcription of RNA and eventually replaced RNA as the ultimate genetic material because of the enhanced stability of the double helix.

Recently, the "RNA world" hypothesis has been given a double boost; first by the discovery that self-splicing introns are present in the transfer-RNA genes of widely divergent bacteria (Reinhold-Hurek and Shub, 1992), and second by the discovery that ribozymes can almost certainly catalyze the formation of peptide bonds during protein synthesis. The latter work was performed in the Noller laboratory in Santa Cruz on ribosomes isolated from the thermophilic bacterium *Thermus aquaticus*. Ribosomes are composed of both RNA and protein constituents, and are responsible for translation of genetic information carried by messenger RNA into protein. In essence, Noller and his co-workers removed most of the protein component and surprisingly found that the activity of the ribosome was not affected (Noller et al., 1992). If, however, they disrupted the RNA component, leaving the protein apparently untouched, protein synthesis activity was abolished. The results are not entirely conclusive as 5% of the protein proved to be impossible to remove from the RNA tertiary structure, but they do provide strong evidence that RNA catalysts are capable of a wider repertoire of reactions than was previously thought. Hence, while the first observation suggests that ribozymes are widespread, the second confirms a central tenet of the "RNA world" hypothesis; namely, that ribozymes were capable of directed protein synthesis (i.e., following a template to produce a polypeptide sequence).

Alternatives to the "RNA world" hypothesis do exist. It is possible that the original replicating molecules (and catalysts) were neither RNA nor protein in nature. In 1990, Julius Rebek's group designed a synthetic, self-replicating molecule (amino-adenosine triacid ester) that can catalyze the synthesis of copies of itself, some of which contain mistakes—a prerequisite for subsequent evolution (Tjivikua et al., 1990). The process of evolution requires that mistakes occur in the copying of biopolymers and that some of these changes become fixed (i.e., widespread throughout descendant populations). The implication of this work is that the original biocatalysts may not have been any of the commonly considered candidates. They may have produced RNA and protein and then have been totally replaced by them, much as in the "RNA world" protein catalysts supplanted the original ribozymes.

Other theorists have suggested that the original biocatalysts were charged surfaces. In his book, *Blueprint for a Cell*, de Duve (1991) proposed that thioesters, which arose from the condensation of carboxylic acids and thiols in the prebiotic soup, acted as an energy source for the formation of oligomeric "proto-enzymes" (also formed from thioesters) which then catalyzed a cascade of chemical reactions akin to those occurring in modern metabolism. These reactions would eventually have formed RNA molecules and gave rise to the "RNA world." Other possible catalytic surfaces have been suggested such as iron pyrites (a metallic mineral containing iron and sulfur), a possible precursor of the FeS complexes found in metabolic processes today, or even crystalline clays. The latter alternative offers the intriguing possibility that the original replicating "cell" was in fact a clay crystal; structures which have been shown to be capable of self-replication in a life-like fashion.

How Do Biocatalysts Evolve?

The very early stages of evolution mentioned in the previous section are so distant as to be essentially inaccessible except to theorizing or simulations. In particular, the details of the origins of the first biocatalysts are difficult to imagine, whether they are proteins or RNA molecules. A study of present day organisms, however, can reveal or suggest a great deal about evolution since the origin of multicellular organisms about 1,500 million years ago. This chapter will thus be mainly concerned with the evolution of enzymes during these more recent phases of evolution; a period that is nevertheless very long, and has seen the origin and demise of an extraordinary variety of living organisms.

It has become apparent that enzymes have evolved by a great many different processes. Single base changes alone in the DNA have given rise to enzymes with altered properties. Somewhat more extensive DNA changes, involving insertions and/or deletions of several codons have occurred. In many cases whole genes have duplicated, thus enabling one copy of the gene to acquire new properties by releasing it from its previous functional constraints. The presence of intervening (intron) sequences in the genes of eukaryotic organisms has favored the origin of new enzymes from portions of existing genes. Sometimes the fusion of whole genes (or portions of genes) has created enzymes with extended repertoires of properties. Occasionally, genes are transferred from one organism to another (lateral gene transfer) by a plasmid- or viral-mediated process. Such genes then encounter a new range of evolutionary constraints and may give rise to new gene products such as enzymes with altered properties.

Molecular Evolution

Prior to the 1970s, taxonomic studies of organisms and their phylogenetic interrelationships were based exclusively upon complex morphologies and large numbers of phenotypic characters (e.g., presence of vertebrae, pentadactyl limbs, chlorophyll etc.). The evolutionary relationships between organisms were then extrapolated from these taxonomic studies using evidence garnered from the fossil record as a guide. Studies of this type led to the acceptance of the five-kingdom phylogeny proposed by Whittaker (1959), which was subsequently developed and enlarged upon by other groups (for a review, see Margulis and Guerrero, 1991). In this proposal, organisms were split into five equally ranked groups or "kingdoms": animals, plants, fungi, protists, and the bacteria (the prokaryotes). A major problem with analyses of this kind, however, is that microbial morphologies are too simple and/or uninterpretable to provide a valid taxonomy and, therefore, evolutionary considerations were necessarily confined to complex organisms. The histories of such higher organisms at best cover only the most recent 40% of evolutionary time, and therefore little evolutionary information could be deduced for the previous 60% (a period of time amounting to nearly 2,000 million years).

In the 1970s, however, the sequencing revolution (the recent large increase in the number of biopolymer sequences available for study on computer databases as a direct result of the development of new molecular biology techniques) led to the established dogma being challenged. The basis for the challenge was a series of comparative studies of the protein and DNA sequences of conserved macromolecules isolated from various prokaryotic and eukaryotic sources. These molecules became known as phylogenetic markers or molecular chronometers and were required to fulfill the following three criteria: ubiquity, ease of isolation, and constancy of function (which would ensure that any mutations would be selectively neutral). In a pioneering study of this series, Carl Woese of the University of Illinois chose the 16S/18S ribosomal RNA molecule as a phylogenetic marker because, in addition to satisfying the requirements of such a marker molecule, the rRNA is thought to possess regions which evolve at different rates, thereby giving the molecular chronometer both a "minute hand" (the faster mutating regions) for detecting close taxonomic relationships (e.g., within phyla) and an 'hour hand' (the slower mutating regions) for detecting relationships between organisms which diverged from one another much earlier.

Phylogenetic trees are diagrams which illustrate the taxonomic relationships between organisms where any two organisms are separated by a line, the length of which is determined by the perceived difference between them. To produce such trees using the 16S/18S rRNAs as marker molecules, Woese originally used a process known as oligonucleotide cataloging (Woese and Fox, 1977). 16S/18S rRNA molecules were digested with ribonuclease T1 and the resultant oligonucleotides were separated electrophoretically and sequenced. This generated a catalog of information which could then be used to determine the number of nucleotide differences between the rRNA sequences, and from that information trees could be constructed. This approach has since been superseded by complete rRNA sequencing studies.

Trees of this type showed up the expected groupings of the eubacteria, or true bacteria, and the eukaryotes, which could be further divided into the four kingdoms of higher organisms: animals, plants, protists, and fungi (Whittaker, 1959). In 1978, Schwartz and Dayhoff produced a tree in which the eukaryotic lineage appeared to branch off from within the bacteria. As the studies continued, however, adding more and more rRNA sequences isolated from further organisms, it was noted that, in addition to the above groupings, there was a third, large cluster developing and that all three clusters diverged from one another at about the same time. All the species in the third cluster were prokaryotic in nature and yet their rRNA sequences, while being similar to one another, were as different to those of other prokaryotes as they were to those of eukaryotic species. In 1977, Woese proposed that this new, third grouping of organisms should be treated as a distinct taxonomic kingdom and christened them the archaebacteria. A more recently derived phylogenetic tree illustrating the three major groupings of organisms is shown in Figure 2.

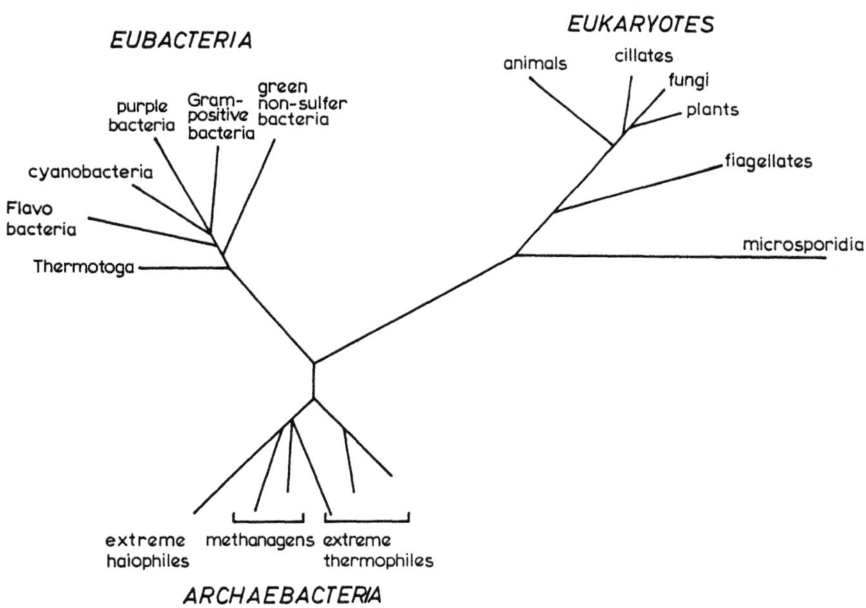

Figure 2. Universal phylogenetic tree (the archaebacteria tree) showing the three domains of organisms defined according to Woese et al. (1990). Branching order and branch lengths are based upon rRNA sequence comparisons.

THE EVOLUTION OF PROTEIN CATALYSTS

The evolution of proteins can be considered to be occurring at various levels from single amino acid replacements to extensive changes involving the entire protein. We will now consider these levels of evolutionary change, and a number of specific examples will be given. Several points should be kept in mind. One is that the processes described probably apply to all enzymes, despite the fact that the majority of the examples are derived from one particular subset of well studied enzymes. Another is that studies on enzymes are very heavily biased towards enzymes isolated from three groups of organisms: mammals, baker's yeast (*Saccharomyces cerevisiae*), and bacteria (particularly *Escherichia coli*). This bias exists for the simple reason that these organisms are most often the subject of biochemical investigations. It is possible that higher plants or parasitic animals or protists have exploited and developed distinct evolutionary mechanisms of their own which have yet to be described. Finally, it is likely that most enzymes have evolved not just by one process, but by a combination of many of the processes described.

Amino Acid Changes

This section is concerned with sequence comparisons of homologous enzymes. In this chapter, a precise definition of the word homologous is employed: "derived from a common ancestor" (Reeck et al., 1987). Thus, homologous proteins are those that have sufficiently similar amino acid sequences that it is possible to deduce that they are likely to have evolved from a common ancestral protein. Two proteins are either homologous or they are not; it is, therefore, not meaningful to use the term "percentage homology" (in the same way that a woman cannot be 50% pregnant!). It is, however, both legitimate and interesting to give a measure of how closely related a group of homologous proteins are. For this purpose the percentage of identical or similar residues between pairs of sequences is informative.

The DNA of an organism is constantly subjected to a variety of mutagenic agents, such as ultraviolet radiation. Chemical changes to the DNA can be caused by these agents and the resulting mutations are often manifested as amino acid changes in the proteins encoded by the DNA. A substantial proportion of mutations are deleterious to the function of a protein, and thus do not persist (become fixed) because they are lethal or put the organism at a selective disadvantage. For this reason, most of the amino acid changes observed when comparing homologous proteins have had relatively small effects on the properties of the protein (this is known as the "neutral theory of evolution"). Much more rarely, the mutations have conferred a new property which is a selective advantage.

The phosphoglycerate mutase family of enzymes provides suitable examples of the many ways in which enzymes can evolve and will be mentioned in several sections of this chapter. Phosphoglycerate mutases (reviewed by Fothergill-Gilmore and Watson, 1989) catalyze reactions involving the transfer of phosphogroups among the three carbon atoms of phosphoglycerates. These reactions are essential components in the metabolism of glucose and/or 2,3-bisphosphoglycerate. A major branch of the mutase family has two members that are clearly homologous, but are readily distinguished by the fact that the main reaction each catalyzes is different. Monophosphoglycerate mutase primarily catalyzes the interconversion of 2- and 3-phosphoglycerates, and requires the cofactor 2,3-bisphosphoglycerate. Bisphosphoglycerate mutase, on the other hand, is mainly active in catalyzing the synthesis of 2,3-bisphosphoglycerate from 1,3-bisphosphoglycerate. In fact, both enzymes are rather versatile, and can catalyze the same three reactions, albeit at substantially different relative rates. This branch of the phosphoglycerate mutase family is well characterized structurally. The high resolution structure of yeast monophosphoglycerate mutase has been determined (Winn et al., 1981) and six complete amino acid sequences are shown in Figure 3. The sequence of a seventh enzyme that catalyzes a quite different reaction from the mutases, but is clearly homologous to them, is also included in the figure and will be discussed in the section on gene fusions.

The crystal structure of phosphoglycerate mutase (Figure 4) enables those amino acid residues likely to be important for activity to be identified. Thus, for example, histidine-8 and histidine-181 are thought to participate in the phospho transfer reaction, and several residues such as serine-11, threonine-20, and arginine-59 contribute to ligand binding. In contrast, residues 125–150 form an extended loop separate from the active site and are unlikely to be particularly important for activity.

The gene encoding phosphoglycerate mutase has undergone at least two gene duplication events (see later) sometime during vertebrate evolution to give rise to the three isoenzymes that exist in mammals: the muscle, brain, and erythrocyte forms. The erythrocyte isoenzyme has high bisphosphoglycerate mutase activity and is the main enzyme for controlling the concentration of 2,3-bisphosphoglycerate in erythrocytes, and hence for controlling hemoglobin oxygen affinity. The muscle and brain isoenzymes have high monophosphoglycerate mutase activity.

What general points concerning the evolution of enzymes emerge from a comparison of phosphoglycerate mutase sequences? First, it can be seen that a substantial proportion of the enzyme structure is conserved, even through very long periods of evolution. The approximate time of divergence of yeast and mammals is about 1,000 million years ago, but nevertheless about 50% of phosphoglycerate mutase has remained unaltered during this time (Table 1). Moreover, it is apparent that the degree of conservation is not evenly distributed. For example, residues 82–90 and 180–200 are relatively invariant, whereas residues 128–148 are particularly variable. The overall rate of change is such that about 5% of the residues are changing every 100 million years. Thus, mammals and birds that diverged about 300 million years ago would be expected to be about 15% different in amino acid sequence. This rate of change is typical of glycolytic enzymes and other intracellular enzymes, but is slower than that of extracellular enzymes such as trypsin.

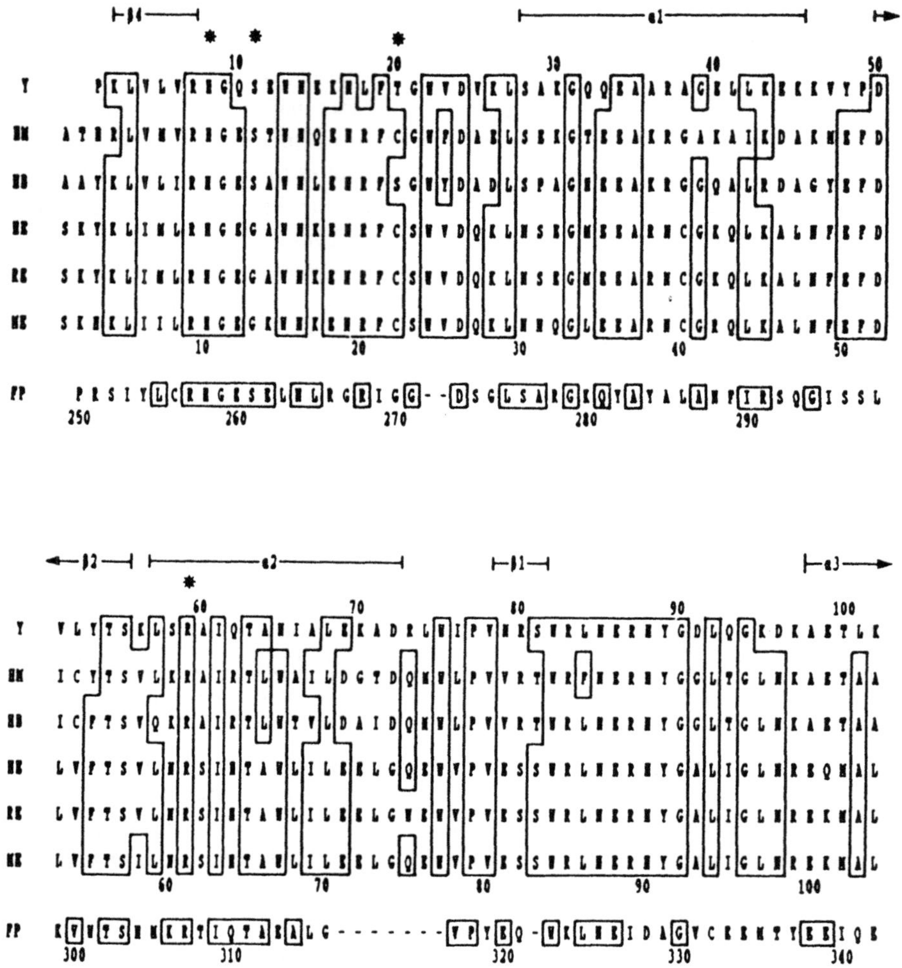

Figure 3. Comparison of the sequences of phosphoglycerate mutases. Residues identical in four or more sequences are boxed. Elements of regular secondary structure observed in the yeast enzyme are shown, and those residues implicated as being important for activity are indicated by an asterisk. Gaps have been introduced to maximize sequence similarities. The sequences are from the following references: Y, yeast monophosphoglycerate mutase (White and Fothergill-Gilmore, 1988); HM, human muscle monophosphoglycerate mutase (Tsujino et al., 1989); HB, human brain monophosphoglycerate mutase, (Blouquit et al., 1988); HE, human erythrocyte bisphosphoglycerate mutase (Joulin et al., 1986); RE, rabbit erythrocyte bisphosphoglycerate mutase (Yanagawa et al., 1986); and ME, mouse erythrocyte bisphosphoglycerate mutase (Le Boulch et al., 1988). The sequence of the phosphatase domain of FP, 6-phosphofructo-2-kinase/fructose-2,6- bisphosphatase (Lively et al., 1988) is also included, and residues identical to any of the mutase sequences are boxed.

(*continued*)

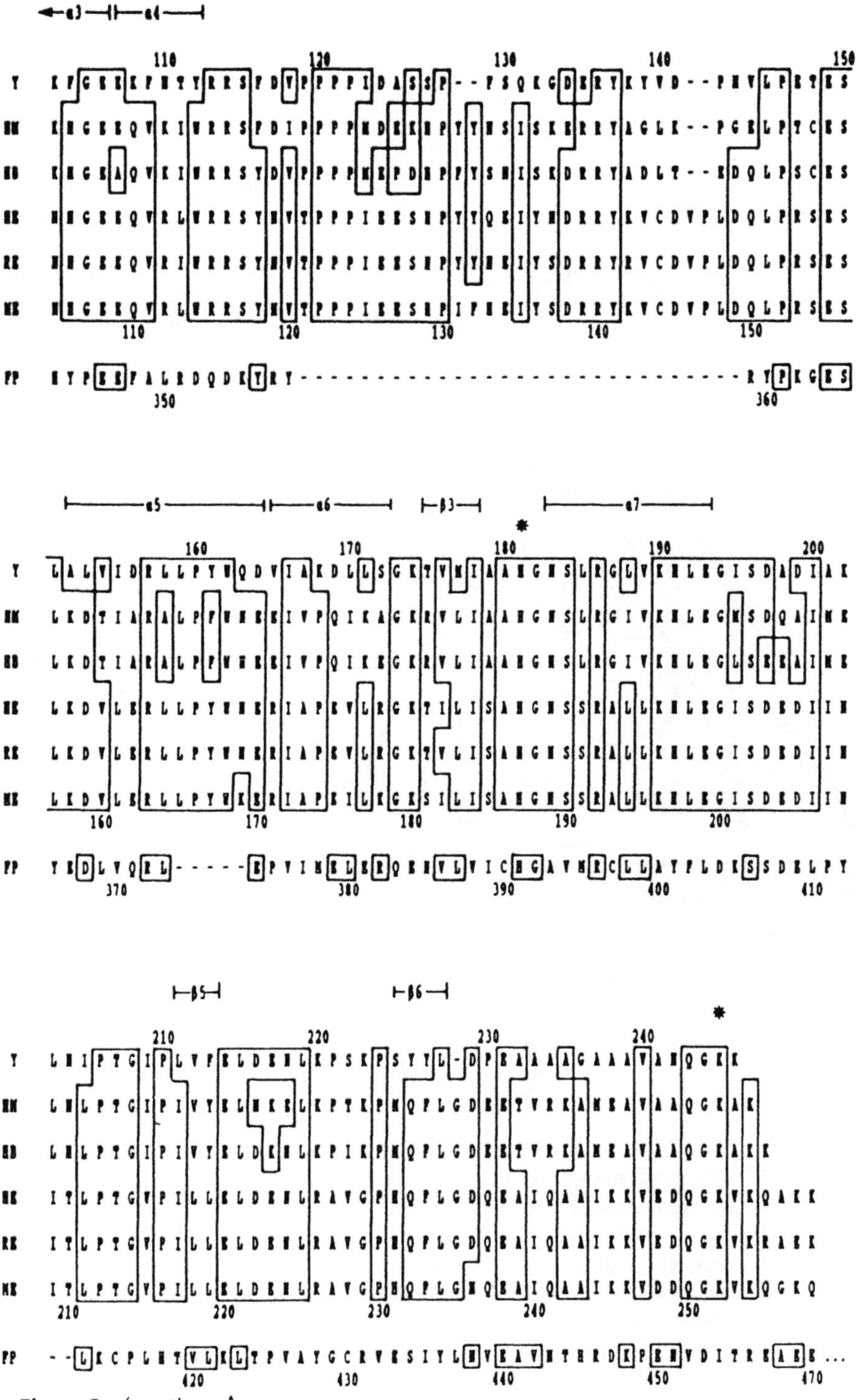

Figure 3. (*continued*)

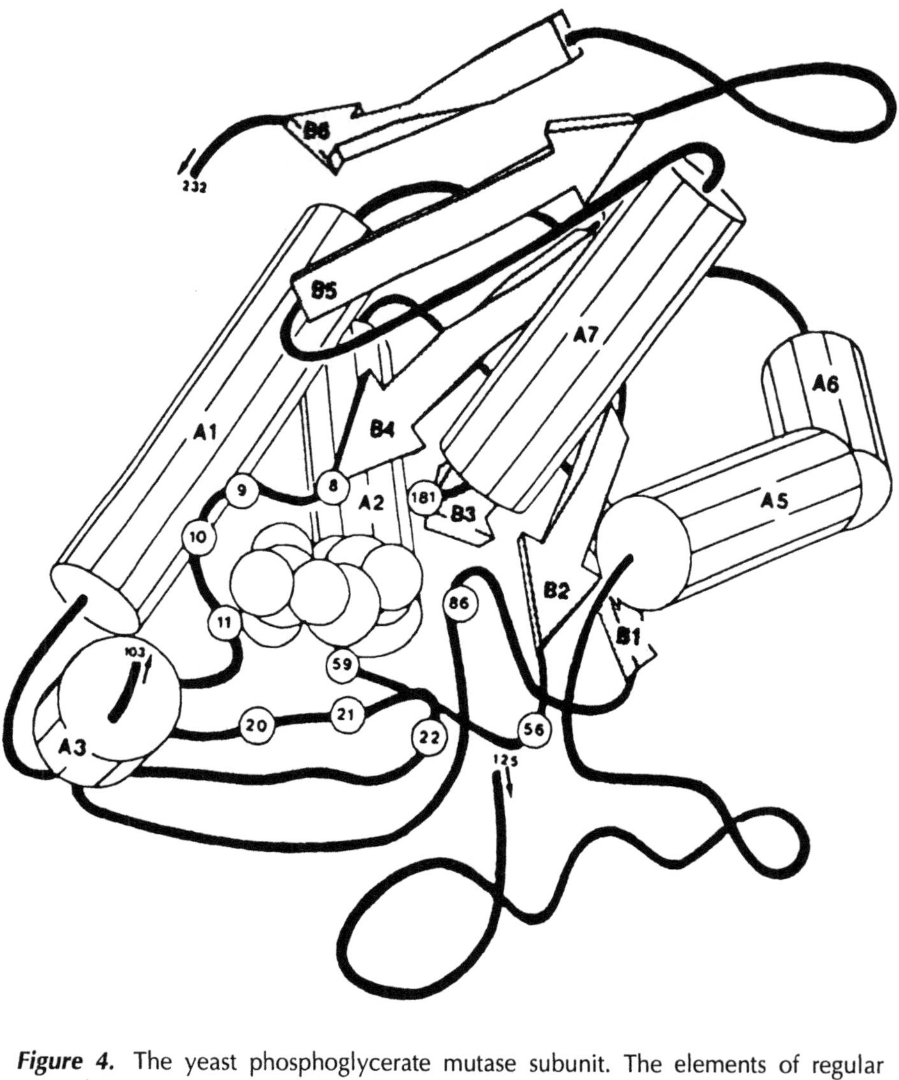

Figure 4. The yeast phosphoglycerate mutase subunit. The elements of regular secondary structure are labeled A1–A3 and A5–A7 for the α-helices (drawn as cylinders), and B1–B6 for the β-strands (drawn as large arrows). The general conformation of the polypeptide backbone connecting the elements of secondary structure is indicated in dark stringlike shapes. A portion of the polypeptide chain including helix 4 has been removed to reveal the α-carbon positions of residues at the active site, as well as the position occupied by bound ligands as indicated by the space-filling drawing of 3-phosphoglycerate.

Table 1. Overall Sequence Comparisons of Phosphoglycerate Mutases

Sequences Compared	*Number of Residues Compared*	*Number of Identities*	*Identities (%)*
Y-HM[a]	246	128	52
Y-HB	246	125	51
Y-HE	246	120	49
HE-RE	258	249	97
HE-ME	258	237	92
ME-RE	258	234	91
HM-HB	252	203	81
HB-HE	253	135	53
HM-HE	252	130	52
Y-FP	220	48	22
HM-FP	221	47	21
HB-FP	221	49	22
HE-FP	221	53	24

[a]The abbreviations for the sequences are as given in Figure 3.

The reasons why the rates of amino acid sequence variation are not constant over the entire sequence of an enzyme are quite readily understood in terms of its structure and function. For example, in the case of phosphoglycerate mutase, the highly conserved residues 180–200 are in the core of the enzyme and also contribute a crucial active site histidine (Figure 4). Very few changes could be tolerated in this region without disruption of function. In contrast, residues 128–148 in the extended loop are unlikely to be essential for the correct functioning of the enzyme.

A comparison of the monophosphoglycerate mutase sequences with the bisphosphoglycerate mutase sequence (Figure 3) shows how an enzyme catalyzing one reaction may evolve into one catalyzing a different reaction by a series of amino acid replacements. Of particular interest are the residues of the active site. Most of these residues are conserved in both enzymes (e.g., His-8, His-181, Arg-7, and Arg-59), but residue 11 is an exception. In monophosphoglycerate mutases this residue is a serine, and it is in a suitable position to provide a ligand for the phospho group at the active site. In bisphosphoglycerate mutases, residue 11 is a glycine, and these enzymes thus have one fewer phospho ligand. This is probably a major factor in ensuring that the product of the erythrocyte enzyme, 2,3-bisphosphoglycerate, dissociates relatively easily from bisphosphoglycerate mutase, but is tightly bound as a cofactor and intermediate by monophosphoglycerate mutase. The serine–glycine replacement at residue 11 is not sufficient to specify the different activities of the two enzymes as has been confirmed by site-directed mutagenesis studies (White and Fothergill-Gilmore, 1990). It is quite likely that a number of amino acid replacements have been required.

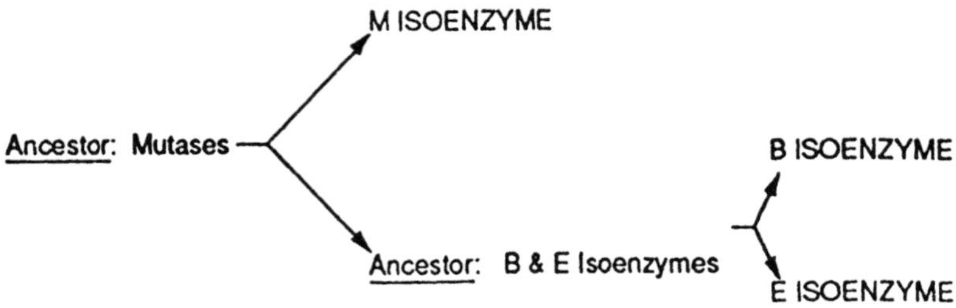

Figure 5. Proposed evolutionary pathway for phosphoglycerate mutases.

Another property that varies among the phosphoglycerate mutases is quaternary structure. The enzymes exist in different organisms as either tetramers, dimers, or monomers as a consequence of amino acid sequence differences. The yeast enzyme is a tetramer, whereas all mammalian enzymes are dimers. A consideration of the structure of the tetrameric yeast enzyme shows that α-helix 6 (residues 165–171) contributes to one set of intersubunit contacts. The sequence in this region in all the mammalian enzymes contains a proline residue that probably disrupts the helix and thus precludes intersubunit contacts. It is likely that this single amino acid change would be sufficient to specify the difference in quaternary structure, a suggestion that has recently been confirmed by site-directed mutagenesis; a technique by which specific amino acid residues in a protein can be altered (White et al., 1993).

Sequence comparisons can also help to understand a likely pathway of evolution for an enzyme, although other evidence must usually be taken into consideration, and often a number of assumptions have to be made. Figure 5 shows a possible pathway of evolution for the phosphoglycerate mutases. A comparison of the M- and B-types of phosphoglycerate mutase shows that they are about 80% identical in sequence (Table 1). These two enzymes catalyze the same reaction and have similar properties, and it is reasonable to assume that they have been evolving at about the same rate of 5% of the residues changing per 100 million years. This means that these two enzymes are likely to have diverged after a gene duplication event about 400 million years ago. This time corresponds to the divergence of mammals and fish early in vertebrate evolution, and is consistent with the occurrence of phosphoglycerate mutase isoenzymes in all vertebrates examined (mammals, birds, reptiles, amphibians, and fish; Mezquita and Carreras, 1981). The time of the divergence of bisphosphoglycerate mutase from monophosphoglycerate mutase probably occurred more recently. It is known that amphibia use bisphosphoglycerate as a regulator of hemoglobin oxygen affinity, but fish do not (reviewed by Coates, 1975). It is therefore probable that the gene duplication event that gave rise to bisphosphoglycerate mutase occurred after the mammal–fish divergence, but before the mammal–amphibia divergence about 350 million years ago. An indication that the erythrocyte isoenzyme has diverged from the brain isoenzyme rather than the muscle isoenzyme comes from the structures of the genes. The upstream flanking region of the gene encoding the brain isoenzyme is more similar to the corresponding region of the erythrocyte gene than it is to the muscle gene (Sakoda et al., 1988).

Insertions and Deletions

Proteins evolve as a consequence of changes in the structures of the genes encoding them. The previous section was concerned with relatively small-scale changes leading to single amino acid replacements. Mutations can also occur that involve the insertion or deletion of nucleotides. The mechanisms of such mutations are not always well understood, but may result from mistakes during the process of DNA replication or from unequal DNA recombination events. As in the case of single base mutations, a large proportion of mutations resulting from insertions or deletions are lethal. For example, if the insertion or deletion involves anything other than a multiple of three nucleotides, then the reading frame of the gene will be disrupted and any part of the sequence downstream of the mutation will be nonsense.

Almost without exception, fixed insertions and deletions of amino acids occur in surface loops of proteins where they can be tolerated without disastrous disruption of structure or loss of function. The highly conserved phosphoglycerate mutases show only a few such mutations, and these are mostly located in the extended loops described by residues 128–148. This loop is also particularly variable in sequence. Other less well-conserved enzymes such as extracellular serine proteases can show very extensive variations in the lengths of surface loops. For example, a comparison of the crystal structures of bovine chymotrypsin (Birktoft and Blow, 1972) and the bacterial protease B from *Streptomyces griseus* (Delbaere, 1975) shows deletions of up to eight amino acid residues in 14 different surface loops as well as five insertions.

In addition to variations in the length of external loops, it can be seen from a comparison of the sequences in Figure 3 that there are variations in length at the two ends of the proteins. This is a common observation and many examples could be cited. Why should the terminal regions tolerate variations in length? A possible explanation may relate to the fact that the terminal regions are usually not buried and thus can vary in structure the same way as surface loops. In some cases, variation of the terminal regions can be correlated with the "acquisition" of additional properties. The glycolytic enzyme pyruvate kinase provides a particularly striking example. Two of the four mammalian isoenzymes of pyruvate kinase can be regulated by phosphorylation, and it is quite clear that this property correlates with a longer N-terminal sequence (Muirhead et al., 1986). This enzyme will be described in greater detail in the next section.

Introns and Exons

The protein-encoding portions of genes in bacteria have continuous, uninterrupted stretches of triplet codons. It was assumed for many years that genes in higher organisms would be the same, and it was, therefore, a considerable surprise when it was discovered in 1977 that the mouse β-globin gene is discontinuous (Tilghman et al., 1978). It is now known that many genes of eukaryotes are divided into coding and noncoding regions, termed exons and introns. Indeed, most genes of higher eukaryotes, such as vertebrates, have introns. Some genes are very large and complex having up to 30 introns. During the process of gene expression, it is necessary to remove the introns before the encoded protein can be assembled on the ribosome. This intron-removal, or splicing, process is done at the level of the primary RNA transcript of the gene and will be discussed in more detail in the section on the evolution of RNA catalysts.

The fact that introns are not present in the genes of all organisms raises intriguing questions concerning their evolution. Did introns arise very early in the evolution of living organisms only to be lost from bacterial and yeast genomes? Or are introns more recently developed entities which have been inserted into eukaryotic genomes? The comparisons of the structures of homologous genes give strong indications that introns are primitive features which have probably been present since the start of evolution. For example, the genes encoding maize and chicken triose phosphate isomerase (Marchionni and Gilbert, 1986) are shown aligned in Figure 6. The seven exons of the chicken gene correspond very closely with the nine plant exons. It can be seen that the two terminal introns in the chicken gene have been lost, and that there is a small difference in position at the boundary of one of the introns. The close correspondence of the two genes is excellent evidence that they have evolved from a common ancestral gene containing at least nine exons. This observation shows that the presence of introns predates the divergence of plants and animals, which occurred about 1,000 million years ago. Evidence that

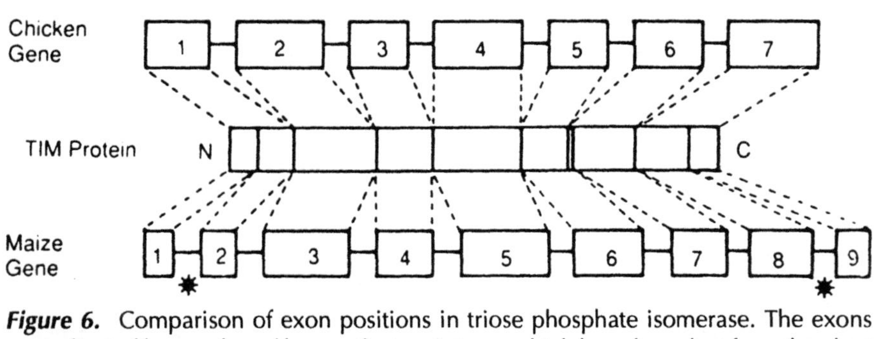

Figure 6. Comparison of exon positions in triose phosphate isomerase. The exons are indicated by numbered boxes; the two introns which have been lost from the plant gene are shown by asterisks. The sizes of the exons and introns are not drawn to scale. Abbreviation: TIM, triose phosphate isomerase.

introns are even older, and predate the divergence of prokaryotes and eukaryotes comes from comparisons of glyceraldehyde-3-phosphate dehydrogenase genes from plants, birds, nematodes, and bacteria (Quigley et al., 1988). Thus it may be concluded that "introns are as old as the genes themselves" (Marchionni and Gilbert, 1986).

Introns, therefore, seem to have been present since the start of evolution. What is their function? Why have they been so well conserved in eukaryotes, but lost from prokaryotes? It has been noted that exons have a restricted size range, encoding on average about 40 to 45 amino acid residues (Blake, 1983). Very frequently, these blocks of amino acids correspond to structural units in proteins (Go, 1981). It seems very likely that proteins have arisen and evolved by the assembly and reassortment of exons. The presence of introns with their noncoding nucleotide sequences enables the exons they flank to be "shuffled" around without loss of function. Any mutations arising in the introns would not be manifested in the encoded protein. Another possible selective advantage of the presence of exons is that they enable a family of related proteins to be synthesized from a single gene by alternative RNA splicing. In fact, this turns out to be quite a common mechanism, and has been described for systems as diverse as antibodies (IgM), muscle proteins (myosin), cell-surface proteins (fibronectin), blood coagulation factors (kininogen), neurotransmitters (substance P), and enzymes (pyruvate kinase).

Exon shuffling is such a useful evolutionary trick that it led Gilbert (1978) to propose that introns were purposely introduced into eukaryotic genomes after the divergence of eukaryotes and prokaryotes in order to speed evolution by assembling new genes from parts of old ones: the "introns-late" argument. Other workers in the field of molecular evolution (Darnell and Doolittle, 1986) have since suggested that introns have indeed been present since before the divergence of prokaryotes and eukaryotes, and that they are simply leftovers from some precellular stage in the assembly of genes. They go on to propose that introns were present in the genomes of the last common prokaryote–eukaryote ancestor and were subsequently lost from prokaryotic lineages through selection for efficiency in replication and gene expression; a process known as "genomic streamlining." This "introns-early" school of thought only makes sense when taken together with the new phylogeny of Woese and Fox (1977), which suggests that prokaryotes and eukaryotes are equally ancient lineages, and which has been previously alluded to (see Figure 2). The new evolutionary trees based on rRNA sequence comparisons suggest that prokaryotes diverged from eukaryotes at the dawn of cellular life. Proof for this theory would require the discovery of introns within protein-encoding genes of prokaryotes. Introns have since been found in the genes of some archaebacterial species; for example, the DNA polymerase gene of *Thermococcus litoralis* (Perler et al., 1992). This gene is split by two introns, neither of which is similar to previously described introns, but which have open reading frames which resemble those of mobile group I introns. This discovery may not be relevant to the introns-early argument as it is thought that protein rather than mRNA splicing is

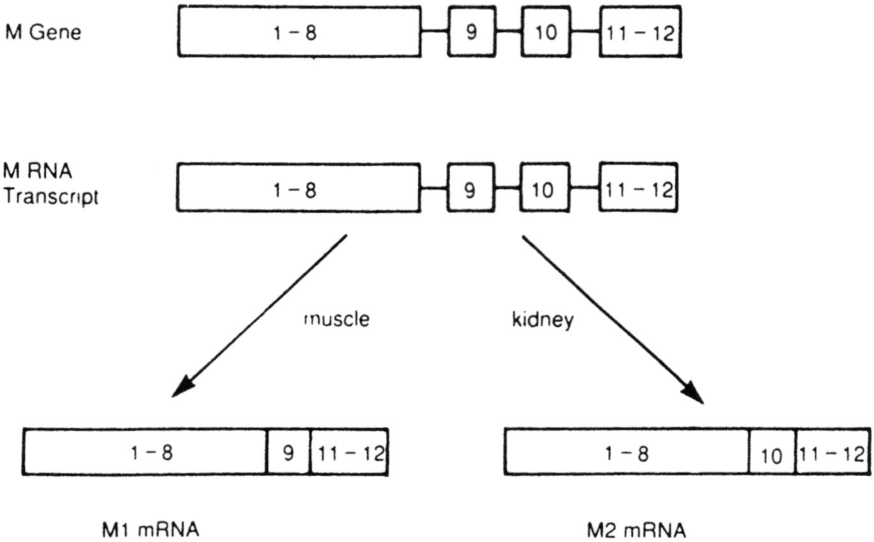

Figure 7. Alternative splicing in different tissues of RNA transcribed from the pyruvate kinase M gene. The gene has 12 exons of which exon 9 is specific to the M1 isoenzyme, and exon 10 specific to the M2 isoenzyme. The sizes of the exons and introns are not drawn to scale.

responsible for production of the mature polymerase. The controversy between the introns-late and introns-early schools of thought has yet to be resolved satisfactorily. The resolution may only come from further studies on the genes of primitive eukaryotes, such as the parasite *Giardia lamblia*, and on those of archaebacterial species.

The glycolytic enzyme pyruvate kinase provides a good example to illustrate the roles that exons, whatever their origins, may play in the evolution of enzymes and other proteins. Pyruvate kinase is found in all cells, including archaebacteria, and catalyzes an essentially irreversible reaction which yields ATP and pyruvate. The enzyme can be isolated from mammalian tissues as four different isoenzymes that have different properties, reflecting the different metabolic requirements of the tissues. Three of the isoenzymes (M2 in kidney, L in liver, and R in erythrocytes) are allosterically regulated such that the pyruvate kinase activity is inhibited when the tissues are carrying out gluconeogenesis. This control is necessary to prevent energetically wasteful futile cycling between glycolysis and gluconeogenesis. The Ml isoenzyme is found in skeletal muscle (a non-gluconeogenic tissue), and in contrast to the other isoenzymes is not subject to allosteric control. It was originally expected that the four pyruvate kinase isoenzymes would be encoded by four different genes. It is now known, however, that there are only two different genes: one encoding the L and R isoenzymes, and one for the M1 and M2 isoenzymes. These two genes have recently been isolated and sequenced (Noguchi et al., 1987;

Takenaka et al., 1989), and it is clear that the L and R isoenzymes arise by the use of different promoters, whereas the M1 and M2 isoenzymes are produced by alternative RNA splicing. Figure 7 shows how the M gene can yield two types of mRNA molecule by the alternative use of exons 9 and 10. These two exons are homologous, and have probably arisen from the duplication of an ancestral exon followed by sequence divergence, such that 21 of the 55 amino acids encoded by exons 9 and 10 now differ. These are the only sequence differences between the M1 and M2 isoenzymes, which are otherwise identical over the remaining 509 amino acids. An indication of how these relatively small structural variations can give such an important difference in properties comes from a consideration of the three-dimensional structure of pyruvate kinase (Muirhead et al., 1986). The amino acid residues encoded by the alternative exons fold into two α-helical regions that constitute a major subunit interface. Apparently the contacts across this intersubunit boundary determine whether or not the isoenzyme can be allosterically regulated.

Gene Duplication and Gene Fusion

The single most important factor in enzyme evolution is probably gene (or exon) duplication. Once a gene has duplicated, then one copy is released from functional constraints and can subsequently acquire a new repertoire of properties by mutations in its DNA sequence. The process of gene duplication has occurred frequently during evolution, and many examples can be cited. All isoenzymes such as the phosphoglycerate mutases and pyruvate kinases mentioned previously have, at some stage, arisen by gene duplication. The molecular basis of gene duplication is not understood, but may involve unequal crossovers between sister chromosomes during meiosis. Almost always the observed duplications are of complete genes or complete exons, with the cross-overs having occurred in noncoding DNA sequences such as introns or intergenic regions. Presumably partial duplications do occur, but they are usually without function and thus do not persist.

Occasionally, duplicated genes or exons fuse to give rise to proteins with new properties. An example of a gene with a duplicated exon is the M gene of pyruvate kinase described in the previous section. Much larger fusions that extend to whole genes have been recognized in a wide range of enzymes and other proteins. Some fusions involve homologous genes only, whereas others can be mixtures of nonhomologous genes, and can give rise to complex assemblies of protein domains with different functions.

The evolution of the glycolytic enzyme phosphofructokinase is typical of the process of duplication followed by fusion of the duplicated genes. The enzyme from bacteria is about half the size of the enzyme from lower and higher eukaryotes, including fungi, nematodes, and mammals. It is likely that the duplication and fusion events occurred sometime early in evolution, not long after the divergence of prokaryotes and eukaryotes. Phosphofructokinase catalyzes the synthesis of fructose-1,6-bisphosphate from fructose-6-phosphate and ATP, and is an important control point in metabolism. The activity of the enzyme is regulated by a number of different molecules, and it is striking that the bacterial phosphofructokinase responds to a much smaller range of effector molecules than does the enzyme from some mammalian tissues, particularly liver. Bacterial phosphofructokinase has one type of regulatory site that binds ATP and AMP which enables the enzyme to respond to the ATP/AMP ratio in the cell. The enzyme is inhibited at a high ATP/AMP ratio, and this inhibition is released when the ATP/AMP ratio is low. Glycolysis is thus turned down when the concentration of ATP is high but is turned back up again when the ATP level is low. Phosphofructokinase in mammalian liver is sensitive to a number of additional effector molecules, especially fructose-2,6-bisphosphate as an activator, and citrate as an inhibitor. The mammalian enzyme thus appears to have more than one type of effector site, and this can be understood in terms of the structures of the prokaryotic and eukaryotic enzymes as indicated in Figure 8. The three-dimensional structure of bacterial phosphofructokinase shows that each subunit has two domains with the active site located in a cleft between the domains (Evans et al., 1981). The effector sites are shared between the adjacent subunits in the tetramer, and are identical. Determination of the amino acid sequence of rabbit phosphofructokinase showed that it consists of two halves which are homologous to each other and to the bacterial enzyme (Poorman et al., 1984). One consequence of the gene duplication/fusion process was that one of the duplicated active sites was released to develop into a new class of effector site, because it was no longer required for catalysis. This has had the critically important effect of enabling glycolysis to be activated or inhibited as appropriate, in response to a range of different signals. In this way, the overall balance of metabolism of an organism can be made responsive to changing circumstances with the minimum loss of efficiency.

The metabolite fructose-2,6-bisphosphate plays a particularly important role in metabolic control because it coordinates glycolysis and gluconeogenesis. The enzyme responsible for synthesizing and degrading fructose-2,6-bisphosphate has arisen from the fusion of two quite different genes. This enzyme has two distinct activities that were originally thought to correspond to two different enzyme molecules but which are now known to reside on a single protein. The N-terminal half of the bifunctional protein possesses the kinase activity for fructose-2,6-bisphosphate synthesis, whereas the C-terminal half has fructose bisphosphatase activity. The two activities are reciprocally controlled by phosphorylation of a single serine residue. Quite unexpectedly, the determination of the amino acid sequence showed that the bisphosphatase portion is homologous to phosphoglycerate mutase (Lively et al., 1988). Both enzymes have active site histidine residues that become phosphorylated during catalysis, and it is apparent from the sequences aligned in Figure 3 that the bisphosphatase is particularly similar to the mutases in the region of the active site. The major differences are in the extended surface loop (residues 125–150) that is anyway poorly conserved in the mutases. The bisphos-

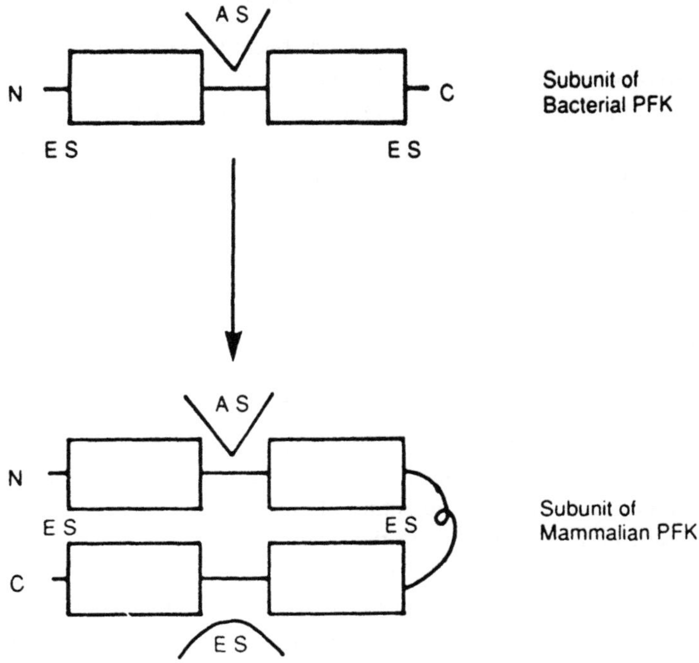

Figure 8. Evolution of phosphofructokinase (PFK) by gene duplication and fusion. The diagrams represent the domain structure of the enzyme with the active site (AS) located in a cleft between domains. Phosphofructokinase is active as a tetramer in both bacteria and mammals, although for simplicity only a single subunit is represented. The effector site (ES) in the bacterial enzyme is shared between two adjacent subunits in an arrangement similar to that proposed for the subunit of the mammalian enzyme. The effector site shown at the bottom of the mammalian subunit is homologous to the active site.

phatase sequence is slightly more similar to the erythrocyte isoenzyme of phosphoglycerate mutase than to the other isoenzymes (see Table 1), which suggests that it may have diverged from this branch of the family. Clearly, the bifunctional enzyme 6-phosphofructo-2-kinase/fructose-2,6-bisphosphatase has evolved by the fusion of a kinase gene with that of a phosphoglycerate mutase gene (see Bazan et al., 1989). A major selective advantage in favor of the bifunctional enzyme over separate enzymes is that the two activities of the fused enzyme can be coordinately regulated, thus ensuring that the concentration of fructose-2,6-bisphosphate is efficiently and responsively controlled.

Lateral Gene Transfer

Lateral gene transfer is the movement of genetic information from one organism to another. Examples of this process would include the transfer of plasmids (nonchromosomal DNA molecules) from one bacterium to another, or the movement of DNA between eukaryotes and symbiotic bacteria. Lateral gene transfer is of concern in studies of molecular evolution because it disrupts existing evolutionary rates. The transferred gene is quite suddenly placed in a new environment where it may experience a completely new range of selective pressures. A main consequence of lateral gene transfer is that whenever it occurs it becomes problematical to devise taxonomic relationships based on the structural similarities of the affected homologous proteins or genes.

Striking instances of rapid evolution in pathogenic bacteria can occur when antibiotics are used indiscriminately. Under these conditions, the bacteria can suddenly manifest resistance to a range of antibiotics such as chloramphenicol, streptomycin, sulfanilamide, and tetracycline. The genes conferring resistance have been transferred on a plasmid from a resistant bacterium to a sensitive one. The resistance plasmids also often contain genes for conjugation between bacterial cells, for DNA replication, and for integration of DNA into the chromosome (Clewell and Gawron-Burke, 1986), and are thus very efficient systems for transferring genetic information between bacteria.

The existence of symbiotic bacteria within eukaryotic organisms seems to have provided opportunities for the transfer of DNA between the two organisms. It is, for example, quite plausible that mitochondria and chloroplasts are descendants of symbiotic bacteria. Evidence in support of this hypothesis comes from an examination of the sequences of the cytosolic and chloroplast isoenzymes of glyceraldehyde-3-phosphate dehydrogenase from tobacco (Shih et al., 1986). The evolutionary tree derived from the sequences suggests that the chloroplast and cytosolic enzymes diverged about 1,700 million years ago, long before the divergence of plants and animals about 1,000 million years ago. These observations imply that the chloroplast lineage was introduced into plants probably by lateral gene transfer from bacteria at some time after plants and animals diverged.

The examples of lateral gene transfer considered above have all been in the direction of transfer from prokaryote to eukaryote. Gene transfer in the opposite direction appears to be possible also. One of the best documented examples concerns the presence of a copper/zinc superoxide dismutase in the bacterium *Photobacterium leignathi* (Bannister and Parker, 1985). Superoxide dismutase is a widely distributed enzyme responsible for scavenging toxic superoxide anion radicals. The enzyme containing copper and zinc is normally found only in eukaryotes, whereas the bacterial enzymes contain manganese or iron. The anomalous presence of the copper/zinc form of the enzyme can be explained by considering the life style of *P. leignathi.* The bacterium can exist either in a free-living mode, or as a symbiont of ponyfish. A comparison of the structural features of superoxide dismutase from *P. leignathi* with the enzymes from eukaryotic species strongly suggest that it has descended from the ponyfish enzyme. Another example is the discovery that the *Escherichia coli* glyceraldehyde-3-phosphate dehydrogenase A enzyme clusters in phylogenetic trees with those isolated from such eukaryotes as yeast and trypanosomes, in contrast with the *E. coli* B enzyme which clusters with prokaryotic glyceraldehyde-3-phosphate dehydrogenases.

THE DIVERSITY OF PROTEIN CATALYSTS

The cases of enzymic evolution which we have so far discussed have been exclusively single proteins treated as separate entities in isolation. It is important in any discussion of the evolution of protein catalysts to note that enzymes do not, in fact, function in isolation within living systems. Enzymes most often perform steps along metabolic pathways, the other steps of which are also enzyme catalyzed. Many enzymes, furthermore, exist complexed with other proteins in multienzyme complexes to increase the efficiency of these pathways. The pathways in which the enzymes operate must also themselves evolve in order to cope with the new environments into which organisms move as they colonize different ecological niches. Protein catalysts have, therefore, become widely diversified and can be found operating in every living organism, no matter how extreme its living conditions.

Multienzyme Complexes

Multienzyme complexes are assemblies of different enzyme proteins which catalyze two or more steps in a metabolic sequence. The complexes that have been most thoroughly studied are composed of two to six functionally related enzymes and have molecular weights ranging from hundreds of thousands to a few million. The complex particles contain fixed amounts of each of the enzymes which are organized in a specific and regular manner.

Physiologically, such complexes are thought to possess the following advantages. First, they allow enhancement of intrinsic catalytic activity. This can be brought about by mutual restriction of the conformations of the interacting polypeptide chains and/or by close spacial location of the catalytic sites. Secondly, complexes can sequester reactive intermediates. This is considered to be the major function of many multienzyme complexes, as confining intermediates to the complex prevent their loss via chemical reactions which compete with the desired reaction. In some cases this causes an increase in the specificity of catalysis. This effect is known as "substrate channeling." Channeling also leads to locally elevated concentrations of substrates which may prove to be advantageous if the combination of enzyme and substrate is the rate-limiting step in catalysis. Channeling may possibly be brought about by a number of mechanisms. There may be diffusion of substrate between spacially adjacent sites; there may be a shared site on the two enzymes; there may be a "swinging arm" which transfers the substrate between sites; or there may be conjunction between the sites via a conformational change (Hammes, 1981).

Both enhancement of catalytic activity and substrate channeling are mechanisms for regulating metabolic flux. Further to these, many complexes are regulated by classical allosteric mechanisms (such as feedback inhibition and antagonism by specific metabolites).

The Pyruvate Dehydrogenase Complex (PDC)

To examine the physiological role of multienzyme complexes more closely, it is informative to look at a specific example. The complex chosen, the pyruvate dehydrogenase complex (PDC), is one of the best studied. In the prokaryote *Escherichia coli* the PDC has a molecular weight of 5 million and contains three types of enzyme which function successively to catalyze the oxidative decarboxylation of pyruvate to acetyl CoA. It therefore lies metabolically between glycolysis and the citric acid cycle. The functional core of the complex is composed of 24 copies of dihydrolipoamide acetyltransferase (designated E_2) which are assembled with octahedral symmetry. To this core are bound multiple copies of pyruvate decarboxylase (E_1), and dihydrolipoamide dehydrogenase (E_3). The reactions catalyzed by the complex are shown in Figure 9, while a schematic of its quaternary structure is shown in Figure 10.

Substrate channeling (and, therefore, reaction mechanism) in the PDC is mediated by the cofactor lipoic acid, two molecules of which are associated with each molecule of E_2. The classical mechanism proposed suggested that the long, flexible lipoic acid molecules acted as "swinging arms" which transferred the reaction intermediates by rotating between the three catalytic sites of E_1, E_2, and E_3. This is now regarded to be unlikely since a single lipoic acid arm could only traverse a distance of 30 Å, and other studies have shown the catalytic sites are in fact 50 Å or more apart. Current theories suggest that the reaction intermediates may be transferred between lipoic acid arms and thence to the catalytic sites.

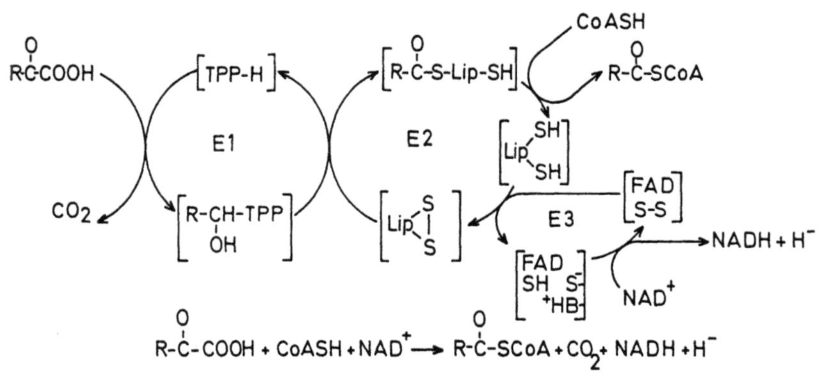

Figure 9. The pyruvate dehydrogenase multienzyme complex of eubacteria and eukaryotes (E1, pyruvate decarboxylase; E2, dihydrolipoyl acetyltransferase; E3, dihydrolipoamide dehydrogenase). Abbreviations used: R,-CH_3; TPP-H, thiamine pyrophosphate; Lip, lipoic acid; B, an amino acid base on the E3 component.

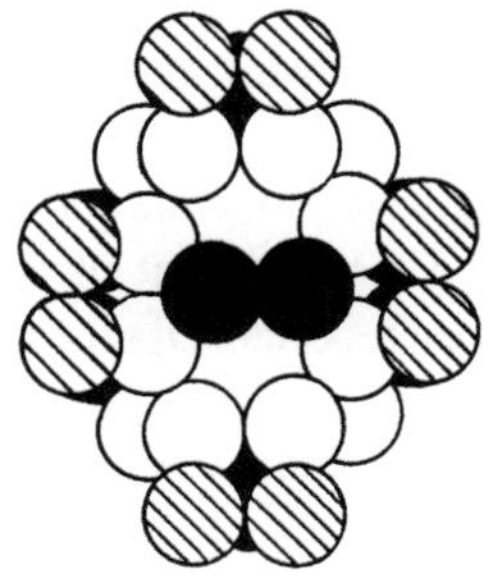

Figure 10. A model of the pyruvate dehydrogenase multienzyme complex from *E. coli*. The view is one face of the octahedral structure. The white spheres are the lipoate acetyltransferase with three polypeptide chains on each corner of a cube. The dark spheres on the faces of the cube are dimers of lipoamide dehydrogenase. A striped sphere representing pyruvate decarboxylase should be on the edge of each lipoate acetyltransferase, but not all of these are shown so as not to obscure the model.

The regulation of pyruvate and acetyl CoA concentrations is of the utmost importance in metabolism. It is not surprising, therefore, that the PDC is subject to multiple metabolic regulations. E_1, which is the subject of most controlling mechanisms, is allosteric (i.e., can be activated or inhibited by molecules which interact with the enzyme at binding sites other than its active site). It is activated by phosphoenolpyruvate and oxaloacetate, and is inhibited by acetyl CoA and GTP. The mammalian PDC is subject to further control by phosphorylation–dephosphorylation of E_1 by a kinase/phosphatase system which responds to changes in the nutritional and hormonal state of the organism.

From 50 to 80% of the energy sources in the diets of omnivores is channeled through the PDC system. For a human adult this corresponds to about 250–400 g of glucose per day. Since the reactions of the PDC are essentially irreversible (committing pyruvate to oxidation or channeling into lipids or steroids), continued flux through this step during periods of carbohydrate deprivation could lead to depletion of the body's reserves of glucose. Erythrocytes and the renal medulla have an absolute requirement for glucose (until blood levels of ketone bodies rise sufficiently to reduce the requirement). Hence, it is no surprise that hormones are involved in the regulatory mechanism of the PDC. Insulin activates the PDC at physiological concentrations and does so rapidly (in minutes). It activates the complex indirectly by activating the PDC phosphatase which requires the presence of a metabolizable sugar. The phosphatase is subject to inhibition by an inhibitory protein and by nucleoside di- and triphosphates. This form of PDC control has been described for white and brown adipocytes, fibroblasts, myocytes, and lactating mammary gland, but has not been found in liver cells.

In humans, PDC deficiency is a potentially severe inborn error of oxidative metabolism. More than 100 cases of human PDC deficiency have been identified by measurements of PDC activity and, while the clinical presentation of the patients varied greatly, there were common factors. These included lactic acidosis and neurological disabilities ranging from moderate ataxia to lethal brain stem dysfunction. Most defects involve the E_1 component of the complex, although defects involving E_2 and E_3 have also been reported. Diagnosis of specific PDC deficiencies can be laborious and, as yet, no treatment has been proven to demonstrate long-term benefits, although many dietary and pharmacological approaches are being tried.

For a comprehensive review of the regulation of PDC and a clinical appraisal of current approaches to treatment and diagnosis see *Annals of the New York Academy of Sciences* 573 (1989).

Enzymes in Extreme Environments

Most of the examples of protein catalysts dealt with in this discussion so far have been mesophilic in nature. That is to say, they operate most efficiently in largely physiological conditions (approximately neutral pH, temperatures of around 30–37 °C etc.). Enzymes are not, however, simply confined to such environments. Recently, enzymes have been discovered which display spectacular levels of tolerance to what are ordinarily thought of as extreme conditions which would denature irreversibly most mesophilic proteins. A good source of such "extremophilic" proteins are the archaebacteria, the phylogenetically distinct group of prokaryotes which tend to exist preferably in extreme environmental niches.

The archaebacteria fall largely into three groups. The methanogens are obligate anaerobes which reduce carbon dioxide to methane and which exist in swampy areas and ruminant gut regions. The halophiles have an absolute requirement for high salt levels, some even flourishing in saturated salt (5.2 M) and exist in salt flats (they also spoil salted fish, a property which the Thais have exploited to generate the condiment Nam Pla, or "fish sauce"). The thermophiles have an absolute requirement for high temperatures for growth—the current world record for life is an organism which grows best at 113 °C!—and exist in boiling water pools and hydrothermal vents beneath the oceans. Detailed study of the proteins of these unusual organisms has shown that they have used a number of techniques to adapt to functioning in extreme conditions.

Halophilicity

High concentrations of salt in solution denature proteins by withdrawing water from them, causing increased strength of hydrophobic interactions within the molecule. These strong interactions cause the polypeptide chains to aggregate and collapse. Halophilic proteins appear to have overcome this problem by having an increased percentage, relative to their mesophilic counterparts, of polar amino acid residues in their structures. These residues compete with the salt for the available water and thereby maintain a hydration shell around the protein even at high ionic strengths. The levels of the acidic residues, glutamate and aspartate, are particularly elevated as they have the highest hydration capacities (6.0–7.5 moles of water/mole of residue). This high frequency of polar amino acids, and the consequent low occurrence of nonpolar residues, produces the disadvantage that hydrophobic interactions in the protein are weak and, therefore, the high-ambient salt concentration is actually required for the protein to maintain its structure. In the absence of salt, the hydrophobic interactions collapse and the protein will unfold. Primary structure is not the only adaptation to halophilicity. Special secondary and tertiary structural elements of the protein may contribute to its binding of water. For example, the malate dehydrogenase of *Halobacterium marismortui* has an active core comprising 80% of the protein, which shows a high degree of similarity to the entire eukaryotic enzyme (from pig heart). The remaining 20% of the archaebacterial enzyme is a large interface with the solvent, and has a highly involuted surface.

Thermophilicity

Thermostable enzymes do not appear to exhibit pronounced differences to their heat-labile counterparts in mesophilic organisms. From a comparison of the sequences and three-dimensional structures of hemoglobins from thermophiles and mesophiles, Perutz (1978) has suggested that the source of a protein's greater heat stability is an increase in the number of salt bridges on the surface of the protein. It has been shown with the enzyme kanamycin nucleotidyl transferase (Matsumura et al., 1984) that one extra salt bridge can significantly increase the heat stability of the protein molecule. There is no associated change in the three-dimensional structure, and only one base change in the coding DNA is required to confer the extra thermostability.

Catalytic Pathways

As we have seen previously, a great deal of information about how individual enzymes evolve has been gathered over the past two decades or so, largely from comparisons of protein and gene structures. Such comparisons are facilitated using sequence databases which, to date, hold upwards of 100 million nucleotides of information. It is now understood, for example, that enzyme evolution occurs on various levels, ranging from single amino acid changes through insertions, deletions, and exon-shuffling, to gene duplication/fusion and lateral gene transfer events. Unfortunately, much less is known about the overall evolution of the metabolic pathways of which these enzymes are often a part.

A good case in point is the glycolytic pathway. This sequence of metabolic reactions is catalyzed by 10 enzymes and is unique in terms of the amount of information available about it. Sequences of all 10 enzymes are available from a variety of sources as are their crystal structures (reviewed by Fothergill-Gilmore and Michels, 1992). This information has revealed some interesting interrelationships between the enzymes which may begin to shed some light on how the glycolytic pathway as a whole may have evolved. Some of the most interesting of these insights have arisen as a result of studies on the glycolytic enzymes of the archaebacteria: the methanogens, halophiles, and the thermophiles. While they are all prokaryotic in nature and organization, they possess macromolecular constituents which more closely resemble their eukaryal counterparts. In fact, they are now recognized as constituting a third taxonomic superkingdom or "domain" of equal ranking with the bacteria and the eukaryotes.

Use of Computers

Before discussing the information gleaned from comparative studies of archaebacterial glycolytic enzymes and their bacterial and eukaryotic counterparts, it is necessary to briefly review the theoretical background to the studies. If two proteins are homologous (i.e., derived from a common ancestor), when their sequences are aligned they should be over 25% identical. The sequences may be aligned using

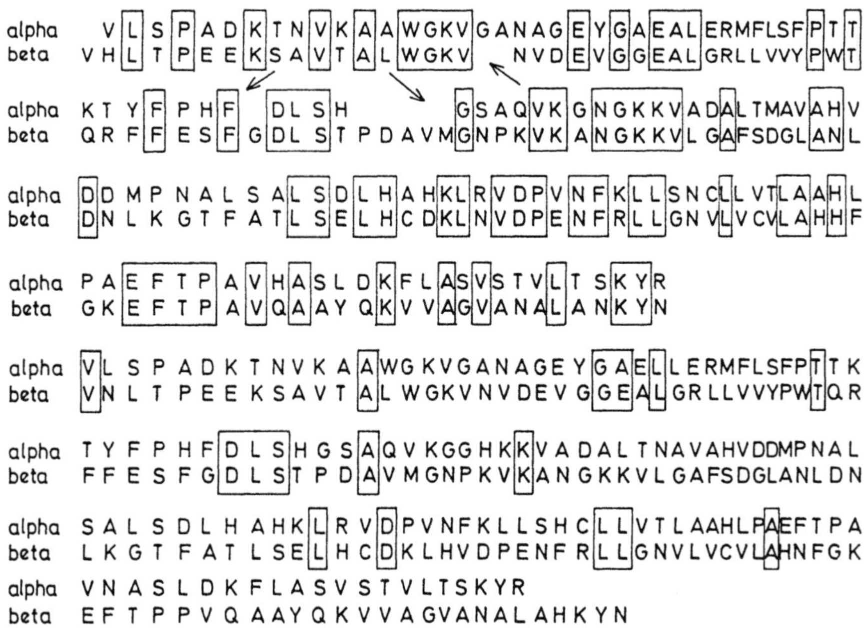

Figure 11. Alignments of α and β haemoglobins with introduced gaps (*upper panel*) or with direct sequence alignment without gaps (*lower panel*). The gaps in the upper alignment were inserted in three places (indicated by arrows) to maintain obviously similar sequences in register.

computer programs such as CLUSTAL (Higgins and Sharp, 1988) which produce maximum identity between sequences by introducing small gaps into them. For an example of the use of gapping in sequence alignment, see Figure 11 which shows two alignments of the α and β hemoglobins, one with gaps and the other without. The threshold value of 25% identity is an average of the values that would be expected were two unrelated protein sequences to be aligned with gaps. The reason for this is that alignment programs allow gaps to be incorporated into either sequence if, by doing so, the alignment is significantly improved. In this regard, an appropriate gap penalty is imposed in order that gapping not be unrestrained. For example, gaps are seldom introduced into regions of probable secondary structure, such as α-helices. If gapping were not allowed, two random sequences would be about 5 to 6% identical. If two sequences show identity above the threshold value, they can be said to be homologous. If they show identity around the threshold value or below it, it is probable that they are not homologous (although it is impossible to say that with certainty as they may be distantly related to one another, retaining only vital structural and functional regions).

In the case of multiple sequence alignments, which can be performed by programs such as CLUSTAL, it is possible to construct a matrix of percentage sequence identities (or differences) which relates each protein to its counterparts from other organisms. Although more and more proteins are being subjected to this type of analysis, the pioneering work of Carl Woese and his co-workers in the 1970s was performed using alignments of RNA sequences. A matrix of rRNA differences can be used to generate a phylogenetic tree of the type shown in Figure 2. In this diagram, the branch lengths of the tree are related to the percent differences between the RNA molecules of the various organisms being compared.

The Application of Sequence Alignments to the Enzymes of Glycolysis

Studies on the 3-phosphoglycerate kinases of methanogenic archaebacteria have shown them to be 30 to 36% identical to their bacterial and eukaryotic counterparts (Fabry et al., 1990). It is probable, therefore, that phosphoglycerate kinases are all descended from a common ancestor. This is complemented by a preliminary study of the pyruvate kinase of a thermophilic archaebacterium (Potter and Fothergill-Gilmore, 1992) which also shows considerable identity to regions of other pyruvate kinases. These data suggest a phylogenetic tree similar to that shown in Figure 2. In contrast, it appears that the archaebacterial glyceraldehyde-3-phosphate dehydrogenases are not homologous with other glyceraldehyde-3-phosphate dehydrogenases, as they exhibit only 15% identity to them (Hensel et al., 1989). This suggests that the archaebacterial enzyme is descended from an entirely different protein than that which gave rise to the bacterial/eukaryotic glyceraldehyde-3-phosphate dehydrogenases. It is, therefore, possible that glycolysis arose as an association of enzymes which worked on substrates of similar structure. Some of the enzymes therefore share common ancestors, and some others are descended from different "parent" molecules.

Glycolysis in the Thermophilic Archaebacteria

Another important insight into the origins of glycolysis has come from the discovery that it is not ubiquitous, as had once been thought. In fact, in some thermophilic archaebacteria, only two of the common 10 enzymes are present: enolase and pyruvate kinase (Danson, 1988). This may suggest that the glycolytic pathway may have evolved from the "bottom up" (i.e., in the gluconeogenic sense) starting with enolase and pyruvate kinase, the only two parts of the pathway still thought to be ubiquitous. As time progressed, other enzymes joined the pathway, some of which became common to all organisms. The final event in the assembly of the eukaryotic–glycolytic pathway was presumably the incorporation of 6-phosphofructokinase, the point at which bacterial glycolysis is blocked. Some of the thermophilic archaebacteria seem to have been left behind by this process, and hence may provide a glimpse into the distant evolutionary past of glycolysis.

Glycolysis may not be the only pathway to have its distant origins in thermophilic archaebacteria. *Archaeglobus fulgidus*, a sulphate reducer, has recently been shown to possess elements of the methanogenerative pathway which was once thought to be unique to the methanogens (Moller-Zinkham et al., 1989). In conclusion, our opinions about the evolution of metabolic pathways are still largely based on speculation but, with the recognition of the archaebacteria and their unusual biochemistry, a powerful window onto the past has been opened by the molecular evolutionist.

Catalytic Cascades

A rather less common type of enzymatic, catalytic pathway encountered in living systems is the catalytic cascade. This involves a series of enzymes which precipitate a chain reaction to produce a large biochemical response to a tiny signal. Two examples of such cascades can be found in human cells. The first is the complement system which operates as a part of the immune response. The complement cascade serves to amplify the plasma levels of several proteins, one of which (C9) is a membrane-perforating agent which lyses invading bacteria. The second is the blood clotting system. This is a multicomponent enzyme system in which inactive zymogens are converted to active proteolytic enzymes by an amplification cascade. A striking feature of many of the blood clotting enzymes is that they are composed of varying arrangements of a number of basic domain types. These enzymes have probably arisen by a series of gene and exon fusion events (Pathy, 1985). As an example, Hageman factor (also known as factor XII) has seven domains as indicated in Figure 12 (McMullen and Fujikawa, 1985). Similar assemblies of domains are also found in other complex proteolytic enzymes such as tissue

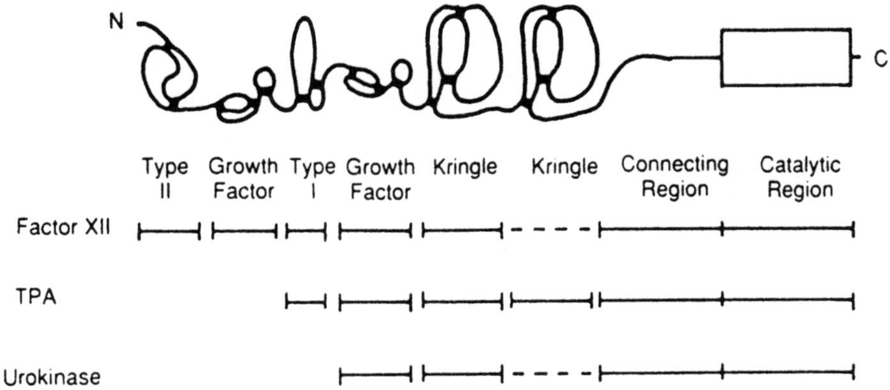

Figure 12. Domain organization in factor XII of the blood clotting cascade compared with the domain structures of tissue plasminogen activator (TPA) and urokinase. The domain types are: finger domains (Types I and II) found in fibronectin; growth factor domain found in proteins that stimulate cell proliferation; kringle domain (named after a Scandinavian pastry because of its shape) found in prothrombin; and a catalytic domain which in these examples is a serine protease homologous to trypsin. Areas shown by dashes are absent in factor XII and urokinase.

plasminogen activator and urokinase. The catalytic regions provide the appropriate enzyme activities, and the other domains apparently mediate the binding of the proteases and their zymogens to other macromolecules. It is these specific macromolecular interactions that regulate the cascades of blood clotting and fibrinolysis.

THE EVOLUTION OF RNA CATALYSTS

A wealth of information is available concerning the structure, function, and evolution of protein catalysts. By contrast, the study of RNA catalysts is in its infancy. The fact that RNA molecules could be catalytically active has been recognized only in the last decade, and the reactions that they catalyze are components in processes as yet not fully understood. The observations which can be made concerning the evolution of RNA catalysts are, therefore, necessarily limited.

RNA enzymes (see reviews by Cech, 1986; Cech and Bass, 1986) catalyze a narrow range of reactions, involving either the cleavage and formation of phosphodiester bonds in RNA molecules, or peptidyl transfers as part of the ribosomal protein synthesis complex; that is, nucleophilic additions to either carbonyl or phosphate groups.

Ribonuclease P is an enzyme with both RNA and protein subunits which catalyzes the maturation of tRNA molecules. The enzyme has a strict requirement for divalent metal ions such as Mg^{2+}. Ribonuclease P excises tRNA molecules from RNA primary transcripts containing multiple tRNA copies, and is also responsible for the maturation of the 5′ ends of the tRNA molecules. Both of these reactions involve the cleavage of phosphodiester bonds. The catalytic activity of ribonuclease P resides in the RNA subunit, with the protein subunit apparently playing a role in shielding parts of the highly charged RNA molecule. The protein subunit may also be important for mediating conformational changes during catalysis. The nucleotide sequence of this RNA enzyme appears to be very variable, as shown by comparisons of the sequences of the RNA subunits from the two bacteria *Escherichia coli* and *Bacillus subtilis*. However, some aspects must be similar because the RNA subunits from the two bacterial enzymes can form active complexes with each others protein subunit. The fact that the RNA sequences are quite different means that it is difficult to make any observations about whether the RNA enzymes may have diverged from a common ancestor, or whether they may have converged from separate origins toward a common function.

Most other RNA enzymes discovered so far are concerned with RNA splicing. Splicing is the process whereby noncoding sequences or introns are removed from precursor RNA molecules. Splicing has stringent requirements for specificity, especially in mRNA precursors. If there was an error in a splice site of even one nucleotide, then the reading frame downstream of the splicing would be shifted to give a completely different amino acid sequence. Splicing enzymes can be grouped into two broad categories depending on whether they occur as introns that catalyze their own removal (self-splicing) or whether they catalyze intron removal from other distinct RNA molecules.

Self-splicing reactions are typically encountered in the processing of rRNA precursors. In fact, it was the study of rRNA processing in *Tetrahymena* (a ciliated protozoan) that led to the discovery of RNA catalysts in 1981 (Kruger et al., 1982). There appear to be two main types of self-splicing RNA: Group I enzymes require a guanosine or guanyl nucleotide as a cofactor, whereas Group II enzymes must have an adenosine residue in a specific position in the intron to be excised (Figure 13). Group I introns have been found in many organisms in genes encoding nuclear rRNA, mitochondrial mRNA, mitochondrial rRNA, and chloroplast tRNA. A number of common features are present that can be correlated with the catalytic activity, and a consensus sequence can be written for the regions necessary for cyclization, guanosine binding, and catalysis. However, the introns are otherwise variable in sequence. In a similar way, limited consensus sequences can be described for Group II introns. These observations are difficult to interpret in evolutionary terms because they are compatible with two alternative theories. One possibility is that two ancestral forms of self-splicing introns arose early in evolution, and only those features necessary for each type of catalysis have been preserved. An alternative theory is that self-splicing introns have arisen many different times, and these have converged to one or other of the two different types of splicing mechanisms.

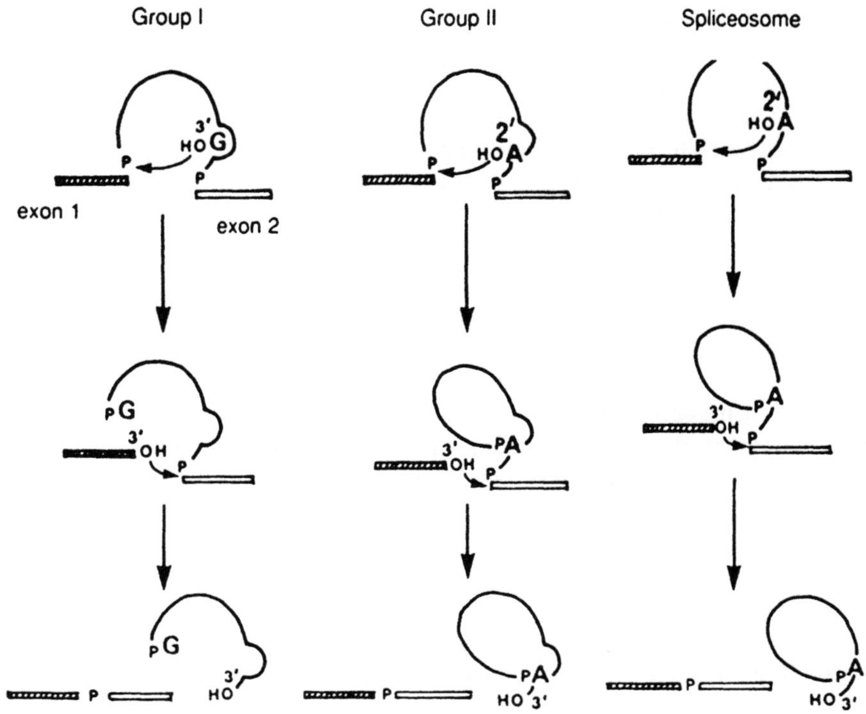

Figure 13. Comparison of reactions catalyzed by RNA enzymes. The Group I and Group II reactions are self-splicing, and therefore occur independently of protein molecules. The splicing of precursor mRNA molecules, however, is much more complicated and is done by a spliceosome, a complex assembly of proteins and RNA molecules. The reactions catalyzed by the spliceosome have some features in common with those catalyzed by Group II self-splicing enzymes.

Are such Group I introns ancient? This question has a direct bearing on the introns-late against introns-early controversy that has been referred to previously. Many mitochondria have intron-rich genomes and were probably derived from a subgroup of the purple bacteria or *Proteobacteria*. Recently, two small self-splicing Group I introns were discovered in members of two proteobacterial subgroups, *Agrobacterium tumefaciens* and *Azoarcus* species (Reinhold-Hurek and Shub, 1992). The introns were present in two tRNA genes. Their occurrence in these ancient bacteria and also in some phylogenetically diverse cyanobacterial species indicates that Group I introns used to have a widespread distribution amongst prokaryotes (before genomic streamlining). As more bacterial introns are discovered and sequenced, a comparison of the sequences of introns in homologous genes (e.g., for the same tRNA) will provide conclusive evidence concerning the antiquity of Group I introns.

A remarkable aspect of the self-splicing of one of the rRNA precursors from *Tetrahymena* is that the Group I intron released has its own catalytic activities. This RNA molecule is capable of binding the oligonucleotide pentacytidylate (C_5), and then either extending or shortening the oligonucleotide. In other words, this RNA catalyst can act either as a ribonuclease or an RNA polymerase. The physiological significance of these reactions is not yet clear, but it is interesting to note that plant RNA viroids resemble Group I introns. These viroids are the smallest infectious agents known, and can cause devastating diseases such as cadang-cadang, which has killed vast tracts of coconut palms. It is possible that the life cycle of the viroid includes reactions like those catalyzed by the *Tetrahymena* intron.

The processing of precursors of mRNA is much more complicated, presumably because of the stringent requirements for precision in splicing, and the greater variability in intron structure. The splicing reactions are now known to be catalyzed by extraordinarily complex assemblies of protein and RNA, which have been called spliceosomes. Spliceosomes have only relatively recently been discovered, and are particles of the same order of size as ribosomes. It is not surprising that many aspects of their structure and function are not yet well understood. For example, it is not even known whether the splicing reactions are catalyzed by the protein or the RNA components of the spliceosomes. However, some details of the splicing reaction are well characterized, and show important similarities to Group I and Group II self-splicing. In particular, the first cleavage reaction is done by the 2′OH of an adenosine near the 5′-end of the intron as is the case for Group II splicing (Figure 13). It is tempting to speculate that the spliceosome-catalyzed splicing has evolved from RNA-catalyzed self-splicing (as proposed by Sharp, 1987). Group II splicing may represent an intermediate between Group I splicing and that occurring in spliceosomes.

The recent discoveries of RNA catalysts have provoked a great deal of interest in the early evolution of living systems. Intriguing and fundamental questions are being asked about the first catalytic molecules. Minds have been opened by the unexpected discoveries of introns and ribozymal-directed protein synthesis. There is no doubt that this is an exciting time for those interested in enzymology and evolution. Indeed, such studies into the dark past of biological catalysis will certainly help to direct the future evolution of protein and RNA catalysts.

The techniques of molecular biology are being increasingly employed to engineer enzymes for therapeutic and biotechnological purposes. Site-directed mutagenesis techniques not only allow insights into structure–function relationships of existing enzymes but they can also be used to alter the properties of existing ones. Work is in progress to produce enzymes of commercial importance which have enhanced thermostability or halophilicity by such procedures. Fusion proteins can also be artificially produced by more drastic genetic manipulations. A recent example is the development of the Antibody Directed Enzyme Prodrug Therapy treatment for metastatic tumors (the ADEPT system). This treatment is based on an artificial fusion protein containing two proteins: an antibody directed against antigens specific to the tumors, and an enzyme which catalyzes the transformation of an inactive prodrug into an active drug. The chemotherapy is, therefore, directed specifically to the affected tissues, and general (and debilitating) chemotherapy is not necessary.

SUMMARY AND CONCLUSIONS

Biocatalysts (or *enzymes*) are essential for life. They enable the thousands of chemical reactions of living systems to proceed rapidly and efficiently even under mild conditions. Enzymes are superior catalysts that are extremely specific and can be readily controlled with great precision. Most enzymes are proteins, although certain RNA molecules can catalyze a limited repertoire of reactions. It is an attractive hypothesis that RNA molecules were the first catalytic molecules to arise at the beginnings of life, and were then almost entirely replaced by the more versatile protein enzymes.

Much of our understanding of enzyme evolution is derived from comparisons of homologous amino acid sequences. The gradual accumulation of amino acid change through single point mutations is the most common way in which the properties of enzymes have altered during evolution. Often a rapid burst of change has occurred after a gene duplication event. In these circumstances one copy of the gene is released from functional constraints and can subsequently acquire a new repertoire of properties. Other ways in which enzyme properties can evolve are from rearrangements of DNA sequences by insertions, deletions, fusions, or the shuffling of exons. In some cases the transfer of genetic information from one organism to another, such as in plasmid-transmitted antibiotic resistance, has given rise to an abrupt evolutionary change.

The majority of enzymes are embedded in metabolic pathways or multienzyme complexes, or other multicomponent systems that have their own collective properties. Any consideration of evolution must take into account the ways in which pathways and other collective systems adapt to cope with new environments encountered as organisms colonize different ecological niches. Enzymes have thus become widely diversified and can be found in all living organisms no matter how extreme their living conditions. Enzymes can function not only in the mild conditions of most living systems, but also in spectacularly extreme environments such as those encountered by archaebacteria that flourish at 110 °C or in saturated salt solutions.

REFERENCES

Bannister, J. V. & Parker, M. W. (1985). The presence of copper/zinc superoxide dismutase in the bacterium *Photobacterium leiognathi:* A likely case of gene transfer from eukaryotes to prokaryotes. Proc Natl. Acad. Sci. USA 82, 149–152.

Bazan, J. F., Fletterick, R. J., & Pilkis, S. J. (1989). Evolution of a bifunctional enzyme: 6-Phosphofructo-2-kinase/fructose-2,6-bisphosphatase. Proc. Natl. Acad. Sci. USA 86, 9642–9646.

Birktoft, T. J. & Blow, D. M. (1972). Structure of crystalline α-chymotrypsin. J. Mol. Biol. 68, 187–240.

Blake, C. (1983). Exons—present from the beginning? Nature (London) 306, 535–537.

Blouquit, Y., Calvin, M. C., Rosa, R., Prome, D., Prome, J.-C., Pratbernou, F., Cohen-Solal, M., & Rosa, J. (1988). Sequence of the human erythrocyte phosphoglycerate mutase by microsequencer and mass spectrometry. J. Biol. Chem. 263, 16906–16910.

Cech, T. R. (1986). RNA as an enzyme. Sci. Am. 255(5), 76–84.

Cech, T. R. & Bass, B. L. (1986). Biological catalysis by RNA. Annu. Rev. Biochem. 55, 599–629.

Clewell, D. B. & Gawron-Burke, C. (1986). Conjugative transposons and the dissemination of antibiotic resistance in streptococci. Annu. Rev. Microbiol. 40, 635–659.

Coates, M. L. (1975). Hemoglobin function in the vertebrates: An evolutionary model. J. Mol. Evol. 6, 285–307.

Danson, M. J. (1988). Archaebacteria: The comparative enzymology of their central metabolic pathways. Adv. Microbiol. Physiol. 29, 165–231.

Darnell, J. E. & Doolittle, R. F. (1986). Speculations on the early course of evolution. Proc. Natl. Acad. Sci. USA 83, 1271–1275.

deDuve, C. (1991). Blueprint for a Cell. Neil Patterson Publishers, New York.

Delbaere, L. (1975). Tertiary structural differences between microbial serine proteases and pancreatic serine enzymes. Nature (London) 257, 758–763.

Evans, P. R., Farrants, G. W., & Hudson, P. I. (1981). Phosphofructokinase: Structure and control. Phil. Trans. R. Soc. Lond. Ser. B 293, 53–62.

Fabry, S., Heppner, P., Dietmaier, W., & Hensel, R. (1990). Cloning and sequencing the gene encoding 3-phosphoglycerate kinase from mesophilic *Methanobacterium bryantii* and thermophilic *Methanothermus fervidus*. Gene 91, 19–25.

Fothergill-Gilmore, L. A. & Michels, P. A. M. (1992). The evolution of glycolysis. Prog. Biophys. Mol. Biol. 59, 105–235.

Fothergill-Gilmore, L. A. & Watson, H. C. (1989). The phosphoglycerate mutases. Adv. Enzymol. Relat. Areas Mol. Biol. 62, 227–313.

Gilbert, W. F. (1978). Why genes in pieces? Nature 271, 501.

Gilbert, W. (1986). The RNA world. Nature (London) 319, 618.

Go, M. (1981). Correlation of DNA exonic regions with protein structural units in haemoglobin. Nature (London) 291, 90–92.

Hammes, G. G. (1981) In: Fluorescence Methods, pp. 257–287. Academic Press, London.

Hensel, R., Zwickl, P., Fabry, S., Lang, J., & Palm, P. (1989). Sequence comparison of glyceraldehyde-3-phosphate dehydrogenases from the three urkingdoms: Evolutionary implication. Can. J. Microbiol. 35, 81–85.

Higgins, D. G. & Sharp, P. M. (1988). CLUSTAL: A package for performing multiple sequence alignments on a microcomputer. Gene 73, 237–244.

Joulin, V., Peduzzi, J., Romeo. P.-H., Rosa, R., Valentin, C., Dubart, A., Lapeyre, B., Blouquit, Y., Garel, M.-C., Goossens, M., Rosa, J., & Cohen-Solal, M. D. (1986). Molecular cloning and sequencing of the human erythrocyte 2,3-bisphosphoglycerate mutase cDNA: Revised amino acid sequence. EMBO J. 5, 2275–2283.

Kruger, K., Grabowski, P. J., Zaug, A. J., Sands, J., Gottschling, D. E., & Cech, T. R. (1982). Self-splicing RNA: Autoexcision and autocyclization of the ribosomal RNA intervening sequence of *Tetrahymena*. Cell 31, 147–157.

Le Boulch, R., Joulin, V., Garel, M.-C., Rosa, J., & Cohen-Solal, M. (1988). Molecular cloning and nucleotide sequence of murine 2,3-bisphosphoglycerate mutase cDNA. Biochem. Biophys. Res. Commun. 156, 874–881.

Lively, M. O., El-Maghrabi, M. R., Pilkis, J., D'Angelo, G., Colosia, A. D., Ciavola, J.-A., Fraser, B. A., & Pilkis, S. J. (1988). Complete amino acid sequence of rat liver 6-phosphofructo-2-kinase/fructose-2,6-bisphosphatase. J. Biol. Chem. 263, 839–819.

Marchionni, M. & Gilbert, W. (1986). The triosephosphate isomerase gene from maize: Introns antedate the plant-animal divergence. Cell 46, 133–141.

Margulis, L. & Guerrero, R. (1991). Kingdoms in turmoil. New Scientist 129, 46–50.

Matsumura, M., Katakura, Y., Imanaka, T., & Aiba, S. (1984). Enzymatic and nucleotide sequence studies of a kanamycin-inactivating enzyme encoded by a plasmid from thermophilic bacilli in comparison with that encoded by plasmid pUB 110. J. Bacteriol. 160, 413–420.

McMullen, B. A. & Fujikaza, K. (1985). Amino acid sequence of the heavy chain of human α-factor XIIa (activated Hageman factor). J. Biol. Chem. 260, 5328–5341.

Mezquita, J. & Carreras, J. (1981). Phylogeny and ontogeny of the phosphoglycerate mutases. I. Electrophoretic phenotypes of the glycerate-2,3-P_2 dependent phosphoglycerate mutase in vertebrates. Comp. Biochem. Physiol. 70B, 237–245.

Moller-Zinkham, D., Barner, G., & Thauer, R. K. (1989). Function of methanofuran, tetrahydromethanopterin and coenzyme F420 in *Archaeoglobus fulgidus* AQ314. Arch. Microbiol. 152, 362–368.

Muirhead, H., Clayden, D. A., Barford, D., Lorimer, C. G., Fothergill-Gilmore, L. A., Schiltz. E., & Schmitt, W. (1986). The structure of cat muscle pyruvate kinase. EMBO J. 5, 475–481.

Noguchi, T., Yamada, K., Inoue, H., Matsuda, T., & Tanaka, T. (1987). The L- and R-type isozymes of rat pyruvate kinase are produced from a single gene by use of different promoters. J. Biol. Chem. 262, 14366–14371.

Noller, H. F., Hoffarth, V., & Zimniak, L. (1992). Unusual resistance of peptidyl transferase to protein extraction procedures. Science 256, 1416–1419.

Pathy, L. (1985). Evolution of the proteases of blood coagulation and fibrinolysis by assembly from modules. Cell 41, 657–663.

Perler, F. B., Comb, D. G., Jack, W. E., Moran, L. S., Qiang, B. Q., Kucera, R. B., Benner, J., Slatko, B. E., Nwankwo, D. O., & Hempstead, S. K. (1992). Intervening sequences in an archaeal DNA polymerase gene. Proc. Natl. Acad. Sci. USA 89, 5577–5581.

Perutz, M. F. (1978). Electrostatic effects in proteins. Science 201, 1187–1191.

Poorman, R. A., Randolph, A., Kemp, R. G., & Heinrikson, R. L. (1984). Evolution of phosphofructokinase—gene duplication and creation of new effector sites. Nature (London) 309, 467–469.

Potter, S. & Fothergill-Gilmore, L. A. (1992). Purification and properties of pyruvate kinase from *Thermoplasma acidophilum*. FEMS Microbiol. Lett. 94, 235–239.

Quigley, F., Manin, W. F., & Cerff, R. (1988). Intron conservation across the prokaryote-eukaryote boundary: Structure of the nuclear gene for chloroplast glyceraldehyde-3-phosphate dehydrogenase from maize. Proc. Natl. Acad. Sci. USA 85, 2672–2676.

Reeck, G. R., De Haen, C., Teller, D. C., Doolittle, R. F., Fitch, W. M., Dickerson, R. E., Chambon, P., McLachlan, A. D., Margoliash, E., Jukes, T. H., & Zuckerkandl, E. (1987). "Homology" in proteins and nucleic acids: A terminology muddle and a way out of it. Cell 50, 667.

Reinhold-Hurek, B. & Shub, D. A. (1992). Self-splicing introns in tRNA genes of widely divergent bacteria. Nature 357, 173–176.

Sakoda, S., Shanske, S., DiMauro, S., & Schon, E. A. (1988). Isolation of a cDNA encoding the B isozyme of human phosphoglycerate mutase (PGAM) and characterization of the PGAM gene family. J. Biol. Chem. 263, 16899–16905.

Schwartz, R. M. & Dayhoff, M. O. (1978). Origins of prokaryotes, eukaryotes, mitochondria, and chloroplasts. Science 199, 395–403.

Sharp, P. A. (1987). Splicing of messenger RNA precursors. Science 235, 766–771.

Shih, M-C., Lazar, G., & Goodman. H. M. (1986). Evidence in favor of the symbiotic origin of chloroplasts: Primary structure and evolution of tobacco glyceraldehyde-3-phosphate dehydrogenases. Cell 47, 73–80.

Takenaka, M., Noguchi, T., Inoue, H., Yamada, K., Matsuda, T., & Tanaka, T. (1989). Rat pyruvate kinase M gene. J. Biol. Chem. 264, 2362–2367.

Tilghman, S. M., Tiemeier, D. C., Seidman, J. G., Peterlin, B. M., Sullivan, M., Maijel, J. V., & Leder, P. (1978). Intervening sequence of DNA identified in the structural portion of a mouse β-globin gene. Proc. Natl. Acad. Sci. USA 75, 725–729.

Tjivikua, T., Balleste, P., & Rebek, J. (1990). A self-replicating system. J. Amer. Chem. Soc. 112, 1249–1250.

Tsujino, S., Sakoda, S., Mizuno, R., Kobayashi, T., Susuki, T., Kishimoto, S., Shanske, S., DiMauro, S., & Schon, E. A. (1989). Structure of the gene encoding the muscle-specific subunit of human phosphoglycerate mutase. J. Biol. Chem. 264, 15334–15337.

White, M. F. & Fothergill-Gilmore, L. A. (1989). Sequence of the gene encoding phosphoglycerate mutase from *Saccharomyces cerevisiae*. FEBS Lett. 229, 383–387.

White, M. F. & Fothergill-Gilmore, L. A. (1989). Mutase versus synthase: The phosphoglycerate mutase family studied by protein engineering. Biochem. Soc. Trans. 18, 257.

White, M. F., Fothergill-Gilmore, L. A., Kelly, S. M., & Price, N. C. (1993). Dissociation of the tetrameric phosphoglycerate mutase from yeast by a mutation in the subunit contact region. Biochem. J. 295, 743–748.

Whittaker, R. H. (1959). On the broad classification of organisms. Q. Rev. Biol. 34, 210–226.

Winn, S. I., Watson. H. C., Harkins. R. N., & Fothergill, L. A. (1981). Structure and activity of phosphoglycerate mutase. Phil. Trans. R. Soc. London Ser. B 293, 121–130.

Woese, C. R. & Fox, G. E. (1977). Phylogenetic structure of the prokaryotic domain: The primary kingdoms. Proc. Natl. Acad. Sci. USA 74, 5088–5090.

Woese, C. R., Kandler, O., & Wheelis, M. L. (1990). Towards a natural system of organisms: Proposal for the domains Archaea, Bacteria, and Eucarya. Proc. Natl. Acad. Sci. USA 87, 4576–4579.

Yanagawa, S., Hitomi, K., Sasaki, R., & Chiba. H. (1986). Isolation and characterization cDNA encoding rabbit reticulocyte 2,3-bisphosphoglycerate synthase. Gene 44, 185–191.

RECOMMENDED READINGS

Cech, T. R. (1986). RNA as an enzyme. Sci. Am. 255(5), 76–84.

Cold Spring Harbor Symposia on Quantitative Biology (1987). Evolution of catalytic function. Vol. 52. Many articles are relevant.

Doolittle, R. F. (1986). Of Urfs and Orfs: A primer on How to Analyze Derived Amino Acid Sequences. University Science Books. Mill Valley, CA.

Chapter 2

The Evolution of the Mitochondrial Genome

CECILIA SACCONE and ELISABETTA SBISÀ

Principles of Medical Biology, Volume 1B
Evolutionary Biology, pages 39–72

ISBN: 1-55938-802-1

INTRODUCTION

The first experimental evidence of the existence of a mitochondrial genetic system dates back to the early 1960s when protein synthesis in isolated organelles was first described. Mitochondrial (mt) DNA was then discovered in 1965; since then studies on mitochondrial biogenesis have flourished in many countries. These studies led to the notion that mitochondrial biogenesis is due to the cooperation of two genetic systems: the main cellular nucleus–cytoplasm genetic system which codes for the greatest part of mitochondrial components; and the mitochondrial genetic system which yields a modest but essential contribution.

Over the last decade the development of new genetic and molecular biology technologies has greatly widened our knowledge of the organization and expression of this cytoplasmic genome. Human mtDNA was the first eukaryotic genome to be completely sequenced in 1980. Although at that time the main features of the mitochondrial genome were already known, thanks to the numerous biochemical and genetic studies carried out in various laboratories, its sequencing added another dimension to our knowledge, thus opening new and otherwise intractable approaches in both theoretical and experimental studies. The discovery made it possible to carry out studies of the molecular bases of several mitochondrial defects including human mitochondrial diseases. The most spectacular advance was in the field of molecular evolution. mtDNA has become the most popular molecule for evolutionary studies and has had a strong impact also in other science areas such as paleontology, taxonomy, and many branches of applied biology.

In this chapter, the main properties of the mitochondrial genome and its evolutionary behavior are summarized. Since, however, the mitochondrial genome is not the same in various organisms, and since the amount of information available thus far is considerable, only certain aspects of the subject will be dealt with in this chapter. The features common to all systems are first described and the better characterized specific aspects relevant to some groups of organisms are also reported. The major emphasis, however, is on the mitochondrial genomes of yeast and mammals. This is because yeast, in particular *S. cerevisiae*, is a facultative aerobe and thus suitable for genetic studies. Mammalia, like other metazoa, possess a small and simple genome whose sequencing has led to new views in several fields of applied biology, including human health.

ORIGIN AND FUNCTION OF MITOCHONDRIA

The Endosymbiont Theory

The fact that nearly all eukaryotic cells harbor two physically and functionally distinct genomes—those of the nucleus and the mitochondrion (photosynthetic eukaryotes harbor also a third one of plastid)—has been usually explained in two ways. According to the endosymbiont or exogenous hypothesis, nuclear genome and organellar genome initially inhabited different types of cells which then formed symbioses. Subsequently, alterations of the genomes and the exchange of genetic material between genomes took place. The organellar genome, though reduced in size, transferred most of the information to the nuclear genome. Conversely, the autogenous hypothesis postulates that nuclear and organellar genomes, originally within a single type of cell, were then enwrapped by membranes, becoming physically compartmentalized and functionally specialized.

Although both hypotheses suppose that nuclear and organellar genomes are derived from a common ancestor, the endosymbiont theory assumes that each genome has had a long period of independent evolution as a separate genomic lineage, whereas the autogenous hypothesis assumes that each genome diverged within a single cellular lineage from a genome or genomes already present in that lineage. Direct evidence in favor of either hypothesis is not yet available. However, the reconstruction, even if partial, of the evolutionary history of each genome may offer valid arguments in support of one of the two theories.

Although the available data are compatible with the endosymbiontic theory, it seems unlikely that this hypothesis will ever become truly universal. The relevant evidence in favor of the endosymbiotic hypothesis is based on two types of data: on the one hand, morphological and biochemical studies, and on the other, data derived from protein and nucleic acid sequence analyses, and from comparative studies of nuclear and organellar genomic organization and molecular biology. All these data have shown a great similarity between both plastid and mitochondrial genomes, and eubacterial genomes. In some cases, it has also been possible to draw evolutionary trees in support of a given phylogenetic relationship.

Plastid genomes are generally simpler than mitochondrial genomes which show in terms of size, structure, and genetic organization in various organisms considerable variation. Thus, the evidence of the exogenous origin of plastids, in particular that obtained from cyanobacteria or oxygenic photosynthetic prokaryotes, is much more compelling. For mitochondria, the great variability of structures can be explained by assuming that mitochondrial genomes undergo rapid evolutionary changes or, alternatively, that different endosymbiotic lineages contributed to the mitochondria of different eukaryotic taxa. The close evolutionary affinity between mitochondria, *Paracoccus*, and certain *Rhodospirillaceae* shown by several investigators appears rather convincing (John and Whatley, 1975).

The prevailing view, therefore, supports an exogenous origin for both plastid and mitochondrial genomes. All other hypotheses, aimed at finding the most probable prokaryotic ancestor of the eukaryotic cell and the mechanisms for the origin of compartments equipped with a specific genome within the eukaryotic cell, require an even greater number of assumptions and rather weak arguments (Gray and Doolittle, 1982).

Once the endosymbiotic theory has been assumed as the most plausible, there are still a number of questions that are very difficult to answer. The real evolutionary events that took place in the primitive eukaryotic cell after the establishment of stable endosymbiotic events cannot, and very likely never will, be reconstructed. The existing situation, characterized by the fact that the majority of endosymbiont genes were transferred to the nucleus, whereas a few but constant number of genes remained associated with the organellar genome, indicates that such compartmentalization provides an evolutionary advantage for the eukaryotes. However, no compelling reason has been given why mtDNA should have persisted throughout evolution. The fact remains that not all nuclear genes for mitochondrial proteins derive from those of the exogenous prokaryote colonizing the primitive eukaryotic cell. In the course of evolution, the mitochondria differentiated and specialized, according to the metabolic requirements of the organism (or tissue). Thus, new genes could have been acquired by the nuclei.

A few years ago, Saccone's group demonstrated that the divergence time between the cytoplasmic and the mitochondrial form of glutamine synthetase (GS) in *Drosophila melanogaster* does not exceed 600 million years; that is, the divergence time between vertebrates and invertebrates. Since a mitochondrial form of GS exists in the uricotelic organisms which, like insects, use the mtGS to convert ammonia to glutamine, the suggestion has been made that the nuclear gene for the mtGS was a rather recent acquisition, probably by retroposition (Pesole et al., 1991).

Admittedly, the relationships between the nuclear and mitochondrial genome remain almost unknown. But before long, the techniques of modern biology will hopefully be able to clarify this important and fascinating aspect of mitochondrial biogenesis, and hence, to throw light on the evolutionary origin of the organellar genome.

Mitochondrial Functions

The primary role of mitochondria in virtually all respiring eukaryotic cells is electron transport and oxidative phosphorylation; namely, processes that provide cells with ATP, the principal source of cellular energy.

Mitochondria are formed by two membranes, an outer membrane which is highly permeable, and an inner membrane where electron transport and oxidative phosphorylation take place. The inner space is called the matrix: it contains the enzymes involved in the citric acid cycle, enzymes catalyzing fatty acid oxidation, and certain amino acid reactions and other enzymes which are specific for a given organism or tissue. The matrix space also contains the mitochondrial DNA and the genetic apparatus for replication, transcription, and translation of the mitochondrial genome. The energy of metabolic fuels is made available in mitochondria in the form of ATP high-energy bonds. For this, both chemical and electrical gradients (ΔpH and $\Delta\Psi$), generated by electron flow in the respiratory chain, are used in the ATP-catalyzed reaction. Since mitochondria represent closed compartments in the cell, the cytosolic-mitochondrial relationships are made possible by the existence of specialized transport systems (translocators, carriers) which allow the transport of metabolites from the cytosol to the mitochondria and vice versa. The carrier family, common to all cells, includes only a few translocators whose existence is necessary for cell survival. This is exemplified by the ATP/ADP translocator, the Pi carrier and the carriers for the translocation of citric acid intermediates.

Mitochondrial permeability mechanisms include compounds that are not involved in energy metabolism. This is true of vitamins and coenzymes which have been shown to enter mitochondria on the basis of different mechanisms. Ascorbate, pyridoxin, and biotin, for example, have been shown to passively diffuse into mitochondria without accumulating. However, pyridoxal 5-phosphate and OH-cobalamin enter mitochondria via diffusion, and can accumulate by binding to specific, high-affinity binding proteins. In addition, a carrier-mediated translocation mechanism has been demonstrated for the oxidized folate and coenzyme A, as well as for thiamine derivatives (Lanoue and Schoolwerth, 1984; Kramer and Palmieri, 1992).

The mechanism of transport of other substrates required for mitochondrial reactions, such as the nucleotides used in mitochondrial replication and transcription processes, remain poorly understood. Hitherto, particular attention has been paid to the problem of translocation through the mitochondrial membranes of proteins which are synthesized in the cytosol, and are able to reach the mitochondrial compartments (see below).

THE PECULIAR FEATURE OF THE MITOCHONDRIAL GENOME

Size and Structure

The mitochondrial genome shows great variability in terms of structure, gene content, organization, and mode of expression in different organisms. The extraordinary diversity probably reflects the different evolutionary pathways that had generated the segregation of genetic information in the two different cellular compartments of nuclei and mitochondria.

The circular double-stranded structure appears to be the rule for mtDNA, but it exhibits an extraordinary variability in length particularly in lower eukaryotes, and in plants, as shown in Table 1. The size ranges from 19 kbp (kilo basepairs) in fungi to 2500 kbp in plants. A linear double helix molecule, 46 kbp long, is present instead in Ciliata; for example, *Tetrahymena pyryformis* and *Paramecium aurelia*. In the algae, *Chlamydomonas reinhardtii*, besides the linear (major species) mitochondrial genome (16 kbp long), circular (minor species) mtDNA might also be present (Ma et al., 1992). In the kinetoplasts of the Tripanosomatidae, *Crithidia fasciculata*, *Leishmania tarentolae*, and *Trypanosoma brucei*, the mitochondrial genome is composed of an intricate network of catenated circles. The network contains two

Table 1. Genome Size of mtDNAs

	Organism	Size (kbp)	μm	Shape
Fungi	*Schizosaccharomyces pombe*	19	6	Circle
	Saccharomyces cerevisiae	78	21–25	Circle
Protozoa	*Tetrahymena pyriformis*	46	15	Linear
	Leishmania tarentolae	21	9.9	Circle (maxi)
Plantae	*Brassica campestris*	200	—	Circle
	Citrullus vulgaris	2500	—	Circle
Metazoa	*Caenorhabditis elegans*	14	4.8	Circle
	Xenopus laevis	17	5.8	Circle

types of circles: (1) thousands of minicircles (1–3 kbp, depending on the species), which constitute 90 to 95% of the DNA and lack long amino acid coding frames; and (2) only 50 to 100 maxicircles (20–40 kbp) which, in other organisms, are the equivalent of mtDNA (Benne, 1985).

In higher plants the mitochondrial genome ranges from 200 kbp to 2500 kbp in size, and shows great variability (e.g., sevenfold). This has also been demonstrated within the same family. Nevertheless, this peculiar feature is neither related to the number of repeated sequences, nor to detectable translation products. The entire genetic complexity is contained in a circular chromosome (master) that may be resolved into subgenomic circles by recombination through directly repeated sequences. This mechanism is also involved in the rapid and extensive rearrangements that characterize the evolution of plant mtDNA. In *Brassica campestris*, one of the better characterized plant mitochondrial genomes, the master circle, 218 kbp long, recombines through a direct repeated sequence, 2 kbp long, to generate two subgenomic circles of 135 and 83 kbp (Figure 1). This tripartite structure is common among other plants, but more complex structures have been reported. The mitochondrial genome of wheat (430 kbp) contains at least 10 repeats, and in maize the master circle (570 kbp) contains six major repeats, five direct and one inverted,

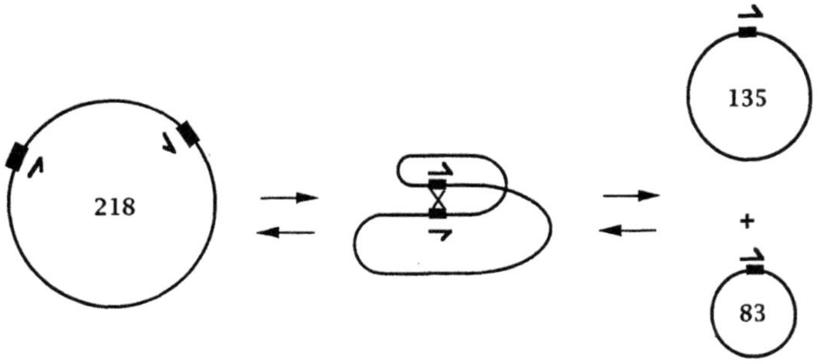

Figure 1. Organization of the *B. campestris* mitochondrial genome into three circular molecules and their interconversion via recombination. Solid boxes indicate repeats.

Table 2. Genome Size of Metazoa mtDNAs

Metazoa	*Organism*	*Size (bp)*
Nematoda	*Caenorhabditis elegans*	13794
Insecta	*Drosophila yacuba*	16019
Echinodermata	*Paracentrotus lividus*	15697
Amphibia	*Xenopus laevis*	17553
Aves	*Gallus domesticus*	16775
Mammalia	*Homo sapiens*	16569

which give rise to multiple circular size classes. In *Brassica hirta*, no recombination repeats have been found, suggesting that the multipartite genome structure, though common, is not a constant feature of higher plant mitochondria. Moreover, frequent acquisitions of genes from nuclei and chloroplasts take place, resulting in the formation of mosaic genomes (Newton, 1988; Gray et al., 1989; Levings and Brown, 1989; Palmer, 1990 for reviews).

In all metazoa the mitochondrial genome size has a constant length (about 16 kbp) and is extremely reduced compared to that of other eukaryotes (Table 2). The smallest size pertains to the nematode *Caenorhabditis elegans* with 13,794 bp, while the largest is present in the amphibious *Xenopus laevis* with 17,553 bp. More recently, numerous cases where the length of mtDNA exceeds the normal and rather constant average size have been described in the literature. Such length variations can span an almost twofold size range (14–28 kbp) or even more (39 kbp as in sea scallop). However, it seems that in such cases duplication of some genomic regions has occurred. Length variations are more frequent in invertebrates and in poikilothermic vertebrates. Such length variations appear to be generated very rapidly and can be distributed both within (heteroplasmy) and between individuals. In general, length differences are confined to the control region, but cases in which length variations appear to be dispersed and/or to include structural genes have also been described (Saccone and Sbisà, 1989; Wolsthenholme, 1992 for reviews).

Genetic Content

With only a few exceptions mitochondrial genomes have a rather uniform gene content, despite great variation in the arrangements of the genes in the genomes. All the mitochondrial genomes studied thus far code for two ribosomal (r) RNA species; only the plant ribosomes contain mtDNA coded 5S rRNA. The ribosomal proteins are nuclear-coded (except in fungi and plants where one or few ribosomal proteins are coded by the mtDNA).

The number of transfer (t) RNA species is variable among different organisms. Animal and fungal mtDNAs code for a complete set of genes, whereas in the green algae *Chlamydomonas*, and in the protozoa *Tetrahymena*, a significant number of tRNAs are coded in nuclear DNA and imported into the mitochondria from the

Table 3. Mitochondrial DNA Coded Proteins

	NADH-DH	*Cytb*	*COX*	*ATPase*	*Ribosomal*	*Maturase*
Neurospora crassa	7	1	3	3	1	—
Saccharomyces cerevisiae	—	1	3	3	1	5
Leishmania tarentolae	3	1	3	—	—	—
Chlamydomonas reinhardtii	5	1	1	—	—	—
Marchantia polymorpha	7	1	3	3	16	—
Ascaris suum	7	1	3	1	—	—
Homo sapiens	7	1	3	2	—	—

cytoplasm. Evidence from several plants indicates that only some tRNAs are encoded by the mitochondrial genome, thus leaving open the possibility of a tRNA import also in these organisms. In particular, some tRNAs are nuclear coded and some are chloroplast-like species that originated in the evolution of the chloroplast genome (Dietrich et al., 1992 for review). In trypanosomatid protozoa, no tRNA genes have been detected so far (Benne, 1985; Simpson, 1987).

In metazoa a set of 22 mt tRNA genes are sufficient to decode all the mitochondrial genetic code since, owing to a more extended U:N wobble, they recognize a degenerate triplet code through a slightly different mechanism; that of the universal genetic code. A single tRNA can apparently read all codons of a four-codon family. This is possible because the U present in the first wobble position of the anticodon can bind to all four third-codon positions, or because two nucleotide pairs are sufficient to stabilize the anticodon–codon interaction.

The messenger (m) RNA species code for subunits of the oxidative phosphorylation complexes and, rarely, for additional proteins involved in RNA processing (maturase) or associated with the small ribosomal subunits. Table 3 lists the number of the subunits for each enzyme complex.

Coding information with respect to cytochrome b (Cyt b) and cytochrome oxidase (CO) subunits I, II, and III is present in all genomes except in *Chlamydomonas*. The metazoa and *Marchantia polymorpha* mtDNAs code for seven NADH-dehydrogenase (ND) subunits; homologous genes are present in the filamentous fungi *Neurospora crassa*, whereas the trypanosomatids (*Leishmania, Trypanosoma, Crithidia*), the unicellular green alga (*Chlamydomonas*), and higher plants appear to specify for only a subset of the mtND subunits. In higher plants, it is not clear how many subunits are encoded by the mitochondria, since three ND activities are present. In yeast, having a functionally different ND complex, all genes are present in the nucleus.

The mitochondrial genomes also code for two subunits of the ATPase; these are called ATPase 8 and 6. The ATPase 8 gene has not been found in the nematode genome and both ATPase subunits have not been found in trypanosomatids. Yeast and plants also code for the subunit 9; the *Neurospora* and *Aspergillus* contain two distinct copies of this gene, one in the nucleus and another in the mitochondria.

Some plants code for the α subunit of the F1-ATPase complex. An extreme example of deviation of the genetic content is the *Chlamydomonas* which lacks the COII, the COIII, and all the ATPase subunits and codes for a reverse transcriptase-like protein.

In some organisms, namely trypanosomatids and plants, some amino acid sequences of mitochondrial proteins cannot be correctly deduced from the corresponding gene sequences. The analysis of gene expression in mitochondria reveals the existence of an extremely interesting process, termed "RNA editing," that alters transcript sequences by inserting and/or deleting uridines (in trypanosomatids), or by converting C residues to U residues (in plants) at precise sites within coding regions, thereby giving rise to correct amino acid sequences. This mechanism therefore uncovers reading frames which are cryptic in the genome, and renders the mRNAs translatable. The major problem relates to the specificity of the mechanism in the editing process. There are essentially two possibilities: one is that determinants of specificity for editing are contained in primary sequence motifs, and/or higher order structures of pre-edited mRNA itself; alternatively, the information for guiding the process may reside at other sites within the genetic system. Quite recently, small antisense transcripts, called guide RNAs (gRNAs), that direct editing by base pairing with the preedited RNA, have been found in trypanosomatid mitochondria. They are encoded in maxicircles and minicircles of the *Leishmania tarentolae* and contain sequences that represent precise complementary versions of the mature mRNAs within the edited regions (Blum et al., 1991).

Another type of RNA editing (C insertion) has been demonstrated to occur in slime mold (*Physarum polycephalum*) mitochondria. The editing has been found also in nonmitochondrial genetic systems such as the C→U conversion in the apolipoprotein B mRNA in mammalian cells, and the G insertion in paramyxo viruses (Simpson and Schaw, 1989; Gray et al., 1992 for reviews).

Gene Organization

Yeasts and Plants

Despite the relatively constant gene content, gene order and organization varies strikingly from phylum to phylum. For example, the mitochondrial genome organization of the *S. cerevisiae* is very loose (the two/third contains AT-rich noncoding sequences), and several genes, in particular, the apocytochrome b, cytochrome oxidase subunit I, and the 21S rRNA, are discontinuous (Figure 2). The number of introns and of GC- and AT-rich mini inserts within the genes are strain-dependent. Some introns of these split genes code for proteins involved in RNA processing or intron transposition. Gene distribution is asymmetric because all genes, except one tRNA, are coded for by the same strand, and 16 out of the 25 tRNAs are clustered. Seven replication origins (ori sequences) are scattered along the yeast genome (4 on one strand and 3 on the other). The peculiar feature of these regions is that they

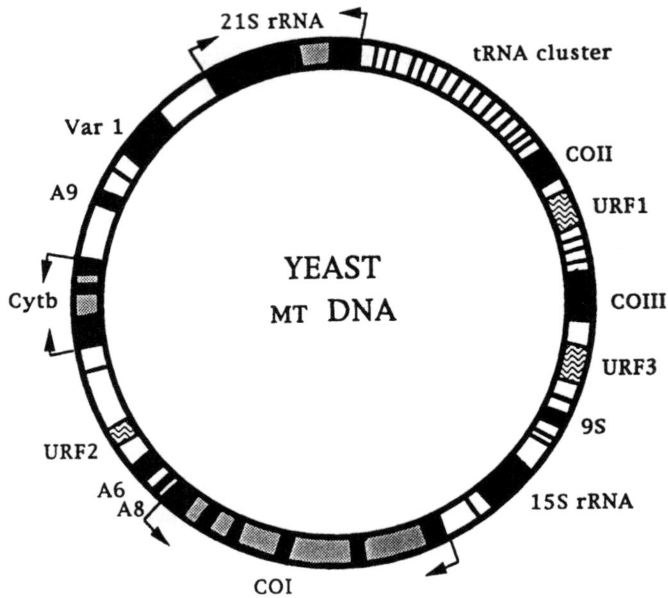

Figure 2. Organization of yeast mtDNA. Black bars = mRNA and tRNA genes; grey shaded areas = introns; waved areas = unassigned reading frames (URF). Cytb = cytochrome b; CO I, II, III = cytochrome oxidase subunits; A6, A8, A9 = ATPase subunits; 15S, 21S = rRNAs; 9S = RNA component of mtRNAse P; var 1 = protein of the small ribosomal subunit.

are homologous and are formed by three GC-rich clusters separated by four AT-rich stretches. These regions can be folded in a hairpin structure, due to the presence of inverted repeats, and contain little open reading frames and a sequence *r*, from which events of transcription and replication start (de Zamaroczy and Bernardi, 1985; Grivell, 1987).

In plants where mitochondrial genomes are much larger than their fungal or animal counterparts, gene organization is highly dispersed. Extensive noncoding sequences separating the coding regions and introns have been detected in several genes. The mtDNAs show a tri- or multipartite structural organization and, due to frequent recombinations, they are continuously rearranged (Figure 1). A consequence of this peculiar evolutionary pattern is that the genome organization varies greatly in linear gene order. Thus, highly conserved coding sequences are often flanked by completely different sequences in the mtDNAs of different, and even closely related, plant species. In particular, tRNA genes are isolated and flanked by long noncoding stretches. Moreover, chloroplast DNA sequences and plasmid-like sequences are present in the mtDNA (Gray et al., 1992).

Metazoa

The gene structure and organization of metazoa markedly differ from those of yeast and plant mtDNAs. The most distinctive feature of the mitochondrial genome is its extremely compact gene organization. Apart from the replication origin regions, the genome is saturated with discrete gene products (Figure 3). Genes coding for transfer, ribosomal, and messenger RNAs lack intronic sequences and untranslated regions; they are often contiguous but sometimes slightly overlap or are separated by only few nucleotides. The gene order and organization varies from phylum to phylum. Gene rearrangement consists of a different distribution of genes between the two strands (polarity inversion), gene transpositions, and gene loss (in nematoda) (Table 4).

Comparing the mitochondrial gene organization in nematoda, insecta, echinodermata and vertebrata, the most remarkable feature is the lack of correlation of tRNA distribution in the four phyla (Figure 4). In vertebrates the tRNA genes are scattered along the molecule and make up a sort of punctuation functioning as signals for the RNA processing (see below). In echinodermata, 15 out of the 22 tRNA genes are in a cluster, and in nematoda all genes are coded for by the same strand.

The gene rearrangement appears to be a peculiar feature of the invertebrata, but recent reports of duplication in and around the control region of the mitochondrial genome in reptiles and fish, and of rearranged gene orders in amphibians, birds, and marsupials, have suggested that the vertebrate mitochondrial genome may be more fluid than it was thought. Numerous observations indicate that tRNAs must be considered as mobile elements, and the finding that tRNAs are present at the end of duplications and deletions supports the hypothesis that tRNA genes are involved in gene rearrangement.

Events of duplication and remolding of tRNA genes during the evolutionary rearrangement of mitochondrial genomes have been suggested to occur in the sea urchin, *Paracentrotus lividus*, on the basis of comparative analyses (Cantatore et al., 1987). In particular, a high homology between the Leu^{CUN} and the Leu^{UUR} tRNA genes was observed, together with an altered location of the Leu^{CUN} gene in the mtDNA of the sea urchin compared to vertebrates. In addition, a sequence (72

Table 4. Characteristics of Animal Mitochondrial Genomes

Common Features	*Peculiar Features*
Size*	Gene order
Shape	Gene distribution between strands
Gene Content*	Variability in the length and sequence of the control region
Compact gene arrangement	tRNA distribution
Single copy of gene/genome*	Codon usage

Note: *With few exceptions.

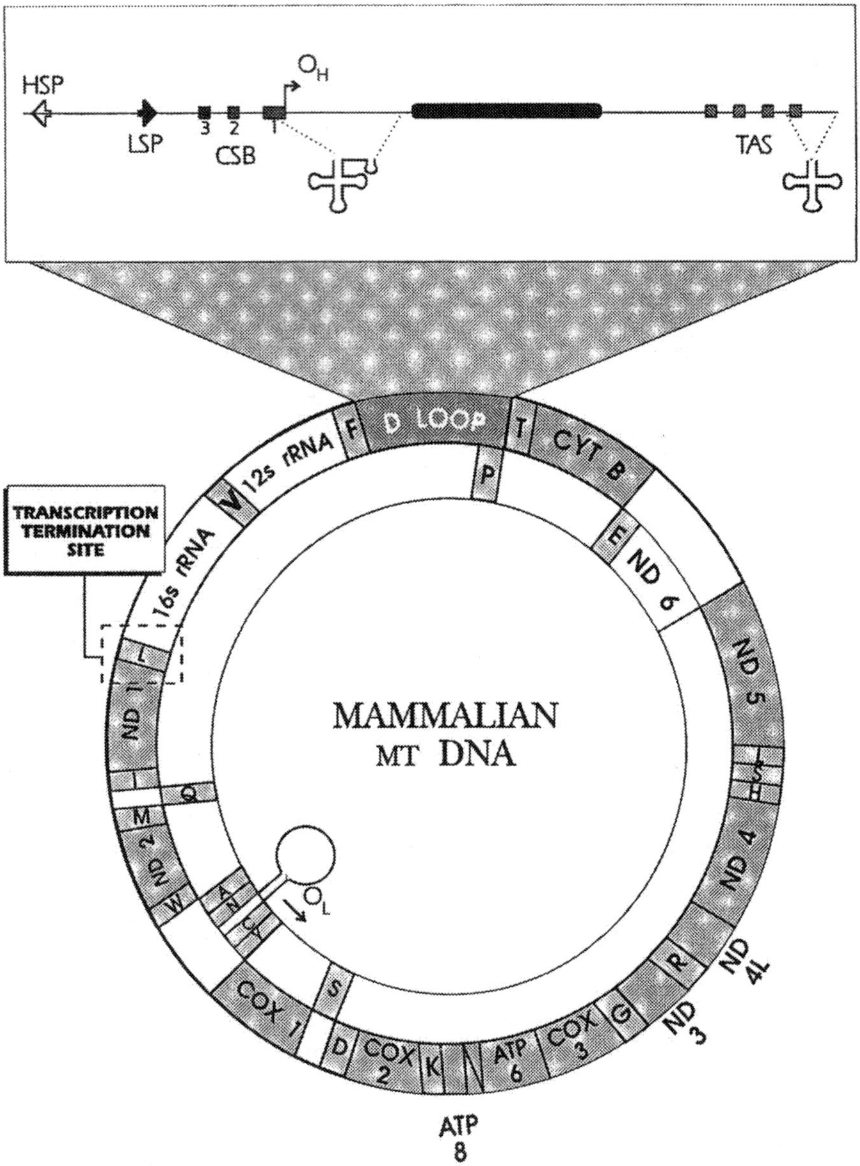

Figure 3. Organization of mammalian mtDNA. ATP6, ATP8 = ATPase subunits; CO I, II, III = cytochrome oxidase subunits; Cytb = cytochrome b; 1, 2, 3, 4, 4L, 5, 6 = NADH dehydrogenase subunits; 12S, 16S rRNAs = ribosomal RNAs; A = ala; C = cys; D = asp; E = glu; F = phe; G = gly; H = his; I =ile; K = lys; L = leu; M = met; N = asn; P = pro; Q = gln; R = arg; S = ser; T = thr; V = val; W = trp; Y = tyr transfer RNA genes. Gene distribution between the two strands and the main regulatory regions are indicated. O_L = L strand replication origin. In the expanded D-loop region, O_H = H strand replication origin; HSP = H strand promoter; LSP = L strand promoter; CSB = conserved sequence box; TAS termination associated sequences; the secondary cloverleaf-like structures and the conserved central domain (black bar) are indicated.

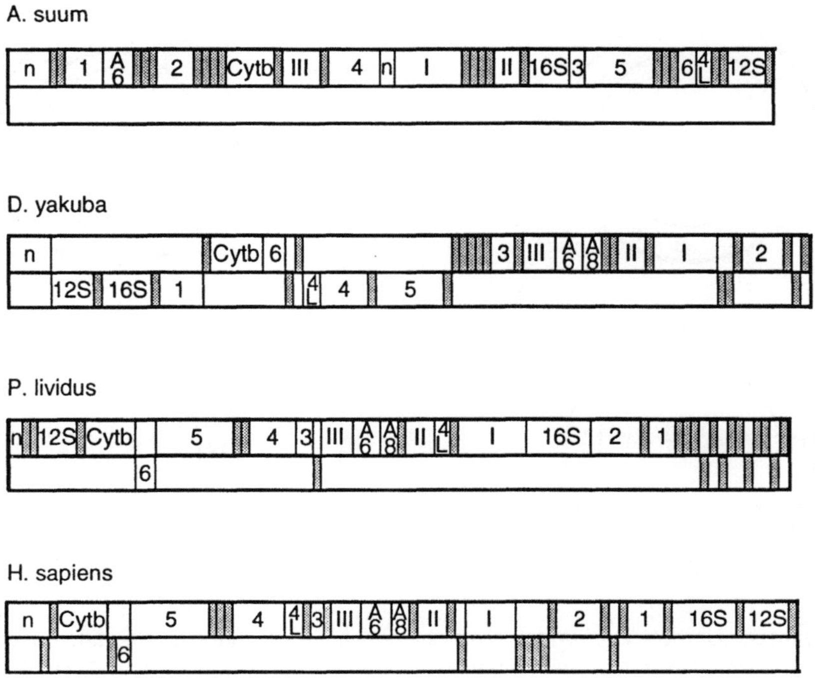

Figure 4. Comparative gene organization in several metazoa mtDNAs. n = main non-coding regulatory region; Cytb = cytochrome b; 1,2,3, 4, 4L, 5, 6= NADH dehydrogenase subunits; I, II, III = cytochrome oxidase subunits; A6, A8 = ATPase subunits; 12S, 16S = rRNAs; grey shaded boxes = tRNAs. The gene distribution between the two strands is indicated.

bp long) containing a trace of the old Leu^{CUN} gene at its original location was observed in the sea urchin, but this codes for an extra amino acid sequence at the amino termini of the ND5 gene. Such data have been explained by assuming that during evolution, a tRNA gene lost its function and became part of a protein coding gene. This loss was accompanied by the gain of a new tRNA through duplication and divergence from a tRNA gene specific for a different family of codons, namely the Leu^{UUR} tRNA gene. On the basis of these assumptions, the possible phylogenetic interrelationship between metazoa has been suggested. The unrooted tree depicted in Figure 5 can give one possible explanation for the differences in gene order among metazoan phyla.

The Noncoding Regions in Metazoa

A peculiar feature of metazoa mitochondrial genomes is the presence of a main noncoding region which contains the regulatory elements for the replication and

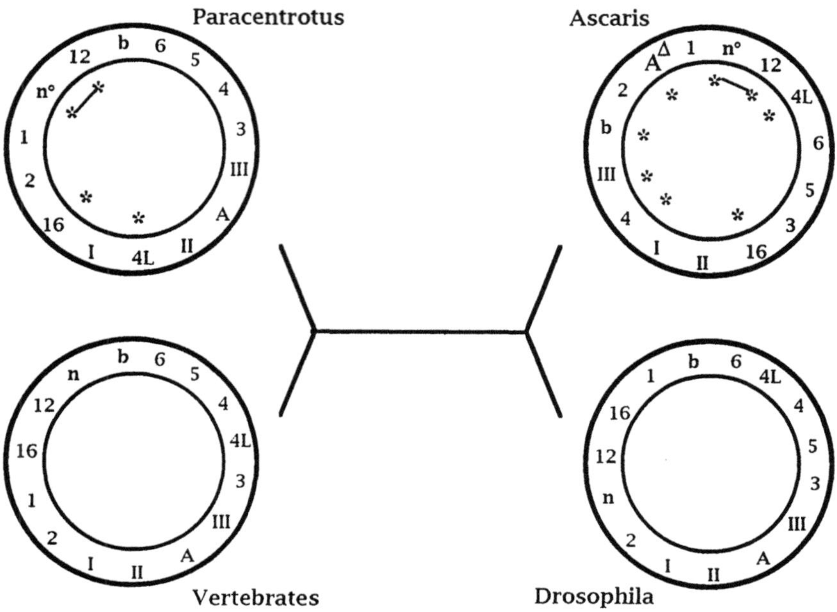

Figure 5. Unrooted tree which gives one possible explanation for the different gene order of mtDNA in four animal phyla. n = non coding region; 1, 2, 3, 4, 4L, 5, 6 = NADH dehydrogenase subunits; 12, 16 = small and large rRNAs; I II, III = cytochrome oxidase subunits; b = cytochrome b; A = ATPase subunits. The tRNA genes are not considered. * = genes that moved; ° = events on the 12- and n-related regions may be different in *Ascaris* and *Paracentrotus*. Δ = loss of ATPase subunit 8.

expression of the mtDNA. Regardless of the high degree of conservation of the coding genes, this region shows great variability of length and base composition. It ranges from 121 to 132 bp in sea urchins to 879 (mouse) to 1838 bp (rabbit) in mammals, to 2135 bp in amphibia and 1029 to 5100 in *Drosophila.* In invertebrates, the structure and evolution of the regulatory region has not yet been characterized.

In *Ascaris* and *Drosophila* the noncoding region is called the AT-rich region because of the extremely high A+T content. In *Ascaris*, the main AT region is 899 bp long, and a smaller noncoding sequence of 117 bp, which can be folded into a stem and loop structure, is also present. In *Drosophila*, the AT region is extremely polymorphic; it varies in sequence and length both in different species (1029 bp in *D. virilis* and 5100 bp in *D. melanogaster*) and in individuals of the same species, or in different mtDNA molecules of a single fly. The putative promoters and the replication origin are contained in two conserved regions one of which can form a hairpin structure.

In echinodermata the main noncoding region (132 bp) is located in the tRNA cluster. It appears as a condensate version of the vertebrate replication origin and the nascent strand coincides with a very stable stem–loop structure.

In vertebrata the main noncoding region is called the D-loop-containing region, because, starting from the strand replication origin (O_H), the heavy strand displaces (D) the parental one, and creates a triple-strand structure. This region, ranging from 898 to 2135 bp, also contains the promoters for both the heavy (HSP) and light (LSP) strands. The two strands are called heavy (H) and light (L) according to their isopycnic sedimentation in a cesium-chloride gradient.

The D-loop, despite its functional importance, is the most rapidly evolving part of mtDNA (see section on the evolution of the D-loop region). However, detailed comparative analyses of this region reveal several peculiar features that are well conserved in the evolution of vertebrates (Saccone et al., 1991a). In particular, the 5′ and 3′ ends contain thermodynamically stable secondary cloverleaf-like structures and short conserved sequences, namely CSBs (conserved sequence boxes) and TASs (termination-associated sequences) which are associated with the start and stop sites of the nascent H strand (Figure 3). How function in this region is preserved in spite of such high diversity of primary structure remains to be clarified.

The other noncoding region contains the origin of the light strand replication (O_L); it is only 30 bp long and is flanked by five tRNA genes. This region can be folded in a stable stem and loop structure which is very conserved in all vertebrata except in birds. In fact, a sequence equivalent to the O_L has not been found in the same position in both chicken and quail mitochondrial genomes.

In rat and human, the two origins of replication show an intrinsic DNA curvature, which correlates with periodic distribution of dinucleotides in the sequence, and is involved in protein interactions (Pepe et al., 1989).

Mitochondrial Genetic Code

The mitochondrial genetic code is slightly modified with respect to the universal one, and it also varies from phylum to phylum. The distribution of codons with unusual amino acid specificities is summarized in Table 5. Rather than a stop codon, UGA specifies tryptophan in all mitochondrial genetic codes except in plants. AGA and AGG code for arginine in fungi and plants, as in the nuclear genetic code; however, in all other invertebrates they specify serine (with the exception of *Drosophila* mtDNA, where AGG is not present). In vertebrates, AGA and AGG codons do not occur internally in any mitochondrial protein gene, but in some cases they are found at the 3′ ends. This finding and the lack of tRNAs recognizing these codons support the hypothesis that they are termination codons. In plant mitochondria, CGG specifies tryptophan instead of arginine, and UGA is a terminator codon, as in the nuclear code.

Another peculiar feature of mitochondria is that the protein genes employ unorthodox translation initiation codons. In addition to AUG, all four AUN codons have been found at the 5′ termini of mRNAs (Gadaleta et al., 1989) among mammalia and echinoderms. In few cases, GUG appears to be used as an initiation codon in the mtDNAs of rat, mouse, chicken, sea urchin, and sunflower. In

Table 5. Deviation from the Universal Code in Mitochondria

		AUA	*CUN*	*AGR*	*CGG*	*UGA*	*AAA*
Fungi	*N. crassa*	Ile	Leu	Arg	Arg	Trp	Lys
	S. cerevisiae	Met	Thr	Arg	Arg	Trp	Lys
Nematoda		Met	Leu	Ser	Arg	Trp	Lys
Insecta		Met	Leu	Ser	Arg	Trp	Lys
Echinodermata		Ile	Leu	Ser	Arg	Trp	Asn
Plantae		Ile	Leu	Arg	Trp	Ter	Lys
Chordata		Met	Leu	Ter	Arg	Trp	Lys

Drosophila yakuba, GUG and AUAA have been proposed as the translation initiation codons. In nematoda, none of the mRNAs starts with the canonical AUG, but AUA, AUU, UUG and GUU are used as initiation codons.

The mitochondrial termination codons are UAA and UAG; in vertebrates, they also are AGA and AGG. In metazoa, some gene transcripts end with the incomplete termination codons U or UA, and the complete termination codons are created by post-transcriptional polyadenylation.

REPLICATION AND EXPRESSION OF THE MITOCHONDRIAL GENOME

Replication

The replication of the mitochondrial genome of yeast starts from multiple origins contained in the ori regions. The process is bidirectional and RNA-primed. RNA polymerase and a primase synthesize the primers starting from the *r* sequences, and the initiation of DNA synthesis of one strand is activated by transcription and/or replication on the other one. The possible involvement of mitochondrial factors in the processing of transcripts starting on the *r* strand has been suggested (Grivell, 1987).

The replication of mtDNA has been extensively studied in vertebrates. Vertebrate mitochondrial genomes contain two distinct origins of replication at a distance of about 10 kbp. The origin of the H strand synthesis (O_H) is located within the D-loop-containing region; the L strand origin (O_L) within a cluster of five tRNA genes. The replication starts with the elongation of the nascent H strand and the expansion of the D-loop. When the displacement exposes the O_L as a single-strand template, the synthesis of the L strand starts in the opposite direction. Both H- and L-strand replications require the presence of RNA primers which are produced by mtRNA polymerase at the level of O_H, and by a mitochondrial primase at the level of O_L. In particular, in the D-loop region, as suggested, the replication priming for O_H and the transcription of the L strand initiate from precisely the same site—the LSP. Although the initiation of replication and transcription appears to be modulated in a coordinated fashion, the precise mechanism remains unclear. An interesting finding is that the transition from RNA to DNA synthesis occurs at the level of short CSBs, which were proposed to be regulatory signals for the start of the nascent strand in a region which can be folded in a thermodynamically stable cloverleaf-like structure. It has been proposed that the RNase MRP (mitochondrial RNA processing) is responsible for endonucleolytic cleavage of RNA primers during mtDNA replication. *In vitro*, the RNase MRP is capable of processing mt D-loop RNA substrates at multiple discrete sites (Karwan et al., 1991). An intriguing problem is the localization of MRP in the cellular compartments, since the amount of the MRP RNA component is very low in organelles (Kiss and Filipowicz, 1992; Topper et al., 1992).

The mtDNA primase can recognize the stem and loop structure of the O_L and initiate RNA priming within the T-rich loop of the template; the RNA/DNA transition occurs near the stem base (Clayton, 1991 for review).

Interestingly, the transcriptional events in both origins of replication are symmetric in the rat where stable complementary transcripts have been characterized (Sbisà et al., 1990; Sbisà et al., 1992).

The DNA polymerase γ (nuclear coded) is the only DNA polymerase activity isolated so far from mammalian mitochondria. It has an associated $3' \rightarrow 5'$ exonuclease activity which might confer a proofreading capacity (Indsorf and Bogenhagen, 1989).

Transcription and Regulation

The peculiar features of the gene structure and organization previously described are reflected in the mode of expression of the genomes. The pathways of mitochondrial transcription and post-transcriptional processing have been extensively investigated in animals, fungi, and recently in plants.

Several transcription units (at least 13) are present in the loose gene organization of yeast mtDNA. In yeast, the transcripts are synthesized as polycistronic precursors starting within a highly conserved nonanucleotide motif that constitutes the promoter element. An RNase P, containing an essential 9S RNA component, coded for by the mtDNA, cuts the polycistronic RNAs at the level of a conserved dodecamer motif present in nearly all protein coding genes. The processing of the split genes appears to be very complex as both mitochondrial products, called maturases, coded by their own introns, and nuclear-coded components are involved. Yeast mRNAs are not polyadenylated and usually have flanking untranslated regions (Grivell, 1989 for review).

The large size of plant mtDNAs, compared to their coding functions, suggests their mode of expression might more closely resemble that of yeast than of mammalian mtDNA. It has been demonstrated that nearly all coding and noncoding sequences are transcribed, although transcripts of many noncoding segments occur only in low abundance, thus suggesting a rapid turnover.

As previously described, plant mitochondrial genome sustains extensive reshuffling of genes by recombination through directly repeated sequences. In this context, how transcriptional regulatory elements, located upstream and downstream of the coding regions, are maintained is unknown. In plants, multiple transcription units are present; most genes seem to be expressed as monocistronic transcripts. But dicistronic messages resulting from cotranscription of adjacent mitochondrial genes have also been found in wheat, rice, maize, sunflower, broad bean, tobacco, and *Oenothera*. However, transcriptionally linked genes in one plant (e.g., COIII and orf 25 in rice) may be unlinked in a related plant species (e.g., wheat). Moreover, different genes in the same plant can have both simple and complex transcript patterns. The same holds for a given gene in different plants. It has been also observed that several separate initiation sites for a single gene can be present. Multiple transcripts of the same gene differ in length of 5′ and/or 3′ noncoding regions and can reflect multiple transcription initiation, or termination, or processing events at different sites within noncoding regions of the primary transcripts (e.g., maize). Different sites of transcription initiation within the same plant species show a low degree of primary sequence conservation. For some mitochondrial genes the transcriptional patterns appear more complex due to different posttranscriptional RNA processing mechanisms, including a series of splicing events for removing introns, and processes for the editing of cryptic reading frames (Gray et al., 1992 for review).

The transcription mechanism in the sea urchin *Paracentrotus lividus* takes place via multiple and probably overlapping transcription units. The wide variation in the steady-state levels of the mature mRNAs indicates that the mitochondrial expression is also regulated at the posttranscriptional level (Cantatore et al., 1990).

The vertebrate mtDNA is extremely compact and transcribed, starting from two distinct promoters (one for each strand) in a polycistronic way (Cantatore and Saccone, 1987; Clayton, 1991 for review). The main peculiar feature of the mitochondrial transcription is its complete symmetry, a phenomenon which has been discovered in HeLa cells even before the sequence and the genetic content of the genome was determined. The two strands of mtDNA are transcribed in opposite direction, starting from the two promoters at the 5′ end of the main regulatory region of the mitochondrial genome.

It is not yet known how starting from polycistronic transcripts of both strands some RNAs are completely degraded while others have different levels of stability. The different concentrations of the mtRNA species in cells is probably the result of numerous events which, at different levels (transcriptional, post-transcriptional and translational), regulate the expression of the mitochondrial genome. Analyses of the kinetics of RNA labeling have suggested that L transcripts are synthesized more efficiently than H transcripts. Nevertheless, L transcripts are far less stable and rapidly degraded, with the exception of those corresponding to the structural genes (tRNAs and ND6) and to the D-loop region. Studies on the stability of the different RNA species have demonstrated that the half-life of tRNAs, mRNAs, and rRNAs are different, even if they belong to a common precursor. Controls, both at the initiation (more efficient promoter) and termination sites (presence of termination factors), can regulate the differential expression of RNA species.

Gene expression is also largely regulated post-transcriptionally. Differential stability presumably explains the different steady-state amounts of individual mRNAs derived from the same transcript. Translational regulations should occur, since the different mRNAs appear to be translated with different efficiency. The other possibility to be taken into account is that the mitochondrial products of the respiratory complexes are controlled by the availability of the cytoplasmic subunits of the same complex (Attardi et al., 1989).

The development of transcriptionally faithful *in vitro* systems has greatly streamlined the characterization of promoters and the purification of the components of the transcription mechanism in vertebrates and in fungi. Transcription in both systems is directed by at least two protein factors: an RNA polymerase, and a transcription factor (mtTF1) which is essential to support a correct and efficient transcription initiation. The mtTF1 binds to the promoters and has the capacity to unwind and bend DNA. It is interesting that the relative binding affinity is higher for the LSP, which is the most active promoter. A protein similar to mtTF1 has been isolated from yeast mitochondria (Fisher et al., 1991).

It is noteworthy that a transcription terminator factor, binding downstream at the 16S rRNA end, has been purified. This protein recognizes a sequence of 13 bases within the $tRNA^{Leu}$; this activity could explain the higher abundance of rRNA species compared to mRNAs (all located downstream at the termination site) in mitochondria. Interest in this peculiar "regulatory region" is increased by the fact that a mutation in the target sequence has been linked to the mitochondrial MELAS myopathy. The mutation is an A→G transition (position 3243) that correlates with a reduced affinity of the termination factor for the MELAS template (Hess et al., 1991).

The polycistronic mode of mitochondrial transcription implies that the activity of one or more endoribonucleases in the processing of the primers is involved in the replication and in the maturation of the structural genes. According to the tRNA excision processing model, an RNAse P-like activity recognizes the secondary structure of the tRNAs which punctuate the genome. The removal of tRNA from the primary transcripts generates the mature forms of the RNA species which are subsequently polyadenylated. Although this processing model is reasonable, its validity has not yet been tested directly.

Ribonuclease P is an ubiquitous ribonucleoprotein which has been isolated and extensively studied in bacteria, nuclei, and yeast mitochondria (Darr et al., 1992 for review). In human, mouse, and rat an RNAse P activity copurifies with the closely related MRP ribonuclease activity (see section on replication). These enzymes contain both protein and RNA components essential for the processing activity. Interestingly, the RNA components of these ribonucleases can be folded into conserved similar structures, suggesting a possible common mechanistic conservation (Altman et al., 1993; Karwan, 1993; Schmitt et al., 1993).

In vivo mapping performed by Sbisà et al. (1992) demonstrated the presence of novel processing intermediates (e.g., RNA precursor containing mRNA+tRNA species). This suggests that the mechanism of RNA processing in mitochondria proceeds stepwise, thereby producing several precursors of the mature forms (Nardelli et al., 1994). In other metazoa, however, the sites and the mechanisms of replication and transcription have not yet been elucidated. A very interesting issue in the regulation and expression of the mitochondrial genome is how the nuclear coded enzymes and the mitochondrial substrates evolved in concert.

NUCLEUS-MITOCHONDRIA INTERACTIONS

Both replication and expression of the mitochondrial genome are controlled by nuclear genes. The nuclear genomes specify all the enzymes of the mitochondrial matrix; that is, the proteins involved in the replication, transcription, RNA processing, translation, and protein (and RNA) import. Furthermore, the nucleus encodes the information for the subunits of the enzyme complexes which, together with the mtDNA coded subunits, constitute the oxidative phosphorylation apparatus. In this context, the biogenesis of mitochondria clearly involves intricate control circuits which are not yet understood. In some organisms, particularly yeast, the study of mitochondrial genetics is readily feasible. Indeed, a considerable amount of information on the interactions between the nucleus and the mitochondrion is available. For higher eukaryotes, this kind of information is far more scanty.

Nuclear Genes Affecting Mitochondrial Biogenesis

Since 1953, when Ephrussi described nonrespiring mutants in yeast, now referred to as nuclear *petite (pet)*, nearly a thousand mutants have been identified. Similarly, less numerous mutants have been described for *Neurospora crassa*. The mutants have been divided into four major groups (Attardi and Schatz, 1988 for review): (1) mutations in genes encoding enzymes of intermediary metabolism (e.g., cytochrome c, ADP/ATP translocase, heme biosynthesis enzymes, subunits of F1-ATPase, etc.); (2) mutants with altered components of the mitochondrial genetic system (e.g., RNA polymerase, ribosomal proteins, etc.); (3) mutants with altered components of the mitochondrial import machinery; and (4) pet mutants carrying lesions in proteins controlling the expression of specific mitochondrial genes. Studies of mutants of the latter group indicate that mitochondrial enzyme complexes may be more subtly regulated than previously realized, since nuclear genes may control the translation and/or the processing of specific mt mRNAs.

An interesting observation is that one of the factors which, in *Neurospora crassa*, controls the correct splicing of individual mtRNA precursors (the CIT-18 gene product) has been identified as one of the two distinct forms of mt tyrosyl-tRNA synthetase. In *S. cerevisiae* a similar mechanism has been described involving a different synthetase, the leucyl-tRNA synthetase or its derivative, which mediates the splicing of mitochondrial pre-mRNA. Similar regulatory mechanisms probably operate in animal cells where the enzymes involved in mtRNA processing are also coded by nuclear genes, but no experimental evidence is yet available.

Another fine regulatory mechanism mediated by nuclear genes is the existence of many mitochondrial proteins that are involved in energy metabolism as two or several differentially regulated isoenzymes. This was first shown in yeast and then also discovered in other lower eukaryotes, like *Dictyostelium discoideum*, and then in plants and mammals, where such a phenomenon seems to be linked to tissue-specific expression of mitochondria. Thus, it could explain tissue-specific defects in human mitochondrial diseases (Kadenbach et al., 1987).

Mitochondrial Signals Affecting Nuclear Genes

The data regarding the influence of the mitochondrial genetic system or mitochondrial metabolites on nuclear genes are much less abundant. However, recent research in yeast has suggested that some nuclear genes may be influenced by the absence or damage of mtDNA. For example, in *S. cerevisiae* the molecular mechanism by which heme and catabolites regulate the expression of nuclear genes for mitochondrial proteins has been described. At the 5′ end of nuclear genes, enhancer-like sequences, called "upstream activator sites" (UAS), mediate the activation of transcription by heme or its inhibition by catabolite repression. The activation of UAS requires, in turn, transacting factors coded in other loci (Grivell, 1989 for review). Interaction between such factors and enhancer-like elements, might be one of the basic mechanisms by which nuclear and mitochondrial genes are coordinately expressed.

In mouse the well-documented case of a cell surface antigen (MTF) represents the only evidence of a mtDNA-generated signal transmitted outside the organelle. The MTF is maternally transmitted and blocked by rhodamine 6G, a dye interfering selectively with mtDNA. The specificity of this surface antigen is determined jointly by a class I histocompatibility antigen, nuclear coded, and the product of the ND1 subunit, a mitochondrial coded gene (Fischer Lindahl et al., 1991).

Protein Sorting

In metabolically active cells, mitochondria are the most abundant organelles constituting 10 to 20% of the total cellular proteins. This means that several hundred different proteins coded for by nuclear DNA, and synthesized in the cytosol, should reach their specific mitochondrial compartment. Mitochondria are themselves made up of different compartments: an outer and an inner membrane; an intermembrane space; and a matrix. Thus, the problem to be solved is how the imported polypeptides are targeted to a specific site.

Experimental evidence, including *in vitro* synthesis of mitochondrial proteins in the transcription–translation system directed by polyA-RNA, indicates that the majority of mitochondrial proteins to be imported possess length variable presequences of 15 to 70 amino acids which are usually present at the N-terminus of the polypeptide chains. Though such presequences differ considerably in their primary structure, one common feature is the high content of positively charged and hydroxylated amino acids having the capability of forming amphipatic structures in the form of either α-helices or β-sheets. It has been suggested that amphiphilic helices may promote binding of the precursor to the mitochondrial surface for the recognition of the appropriate receptor. However, the presence of a cleavable presequence is not an essential requirement for the import process since several mitochondrial proteins, particularly those directed to the outer membrane, do not possess a cleavable presequence. In addition, the presence or absence of a presequence on a specific protein may vary from species to species. Thus, it seems likely that many proteins contain regions which can potentially act as mitochondrial presequences inside their tertiary structures.

In general, it appears that all proteins, in order to be imported into the mitochondria, are preprogrammed for entry into the interior of the organelle independent of the signals which determine their final intramitochondrial location. Some general rules have been established from several studies performed in various organisms which can be summarized as follows (Glover and Lindsay, 1992 for review):

1. The outer membrane precursor proteins do not possess, in general, a cleavable presequence. Their translocation is receptor-mediated and occurs independent of the mitochondrial membrane potential. Like all other mitochondrial proteins, they are programmed to enter the matrix space, but their import is stopped at the level of the outer membrane by suitable, hydrophobic "stop-transfer" sequences. After binding to the outer membrane the proteins become protease resistant and assemble into the mature oligomeric state in the lipid bilayer.
2. Both intermembrane space and inner membrane proteins are translocated in a receptor- and membrane potential-dependent way at the level of particular "contact sites" between the two membranes. The intermembrane space proteins reach the matrix where their N-terminal matrix targeting signal is removed by a matrix processing proteinase. The proteins are then reexported to the intermembrane space, by means of a secondary targeting signal present in their sequence. The mature products are generated by specific proteases present in the intermembrane space.

The inner membrane proteins may not be exposed to the matrix compartment to any appreciable extent, as they are directly embedded into the inner membrane. When the presequence is present it is removed by the matrix processing protease.

3. The proteins of the matrix also recognize appropriate receptors and are imported through the contact sites by an energy-dependent mechanism. Since they do not contain "stop-transfer" sequences, these proteins are directly processed in the matrix space where they acquire the final oligomeric structure.

The nuclear-coded mitochondrial proteins, being post-transcriptionally translocated, must be not only transported specifically into the organelles, but also maintained in the cytosol in a "translocation-competent" state which ensures their transport. There are indeed several cytosolic factors involved in translocation and import of the precursors.

Genetic and biochemical studies have demonstrated that a highly conserved group of cytoplasmic proteins, the so called "heat-shock" proteins, are responsible for the translocation competence of mitochondrial precursors. These proteins should act as molecular chaperons to convey the precursor proteins in a suitable conformation to the surface of the organelles. It seems that other cytoplasmic proteins, some of which containing RNA, are also required for an efficient translocation process.

It has long been known that many enzymes are present in both mitochondria and cytoplasm. In general, these isozymes are coded by different nuclear genes. However, studies in yeast have shown that many isozyme products are encoded by single nuclear genes. In these cases, there are generally two major transcription and translation processes, and a differential initiation of either transcription or translation can generate two forms of protein products that differ by an N-terminal extension (Narjarian et al., 1987).

The mechanism of transport of RNA into the mitochondria is still unclear. For tRNA species the experimental evidence available so far suggests that they enter the organelles as part of a ribonucleoprotein complex. Other nuclear RNA species might reach the mitochondrion in association with proteins which should move through the contact sites as the other proteins to be translocated.

In conclusion, the mechanism of protein import appears rather complex involving specific signals present in the sequence itself, a number of cytosolic and mitochondrial factors, as well as energy producing processes.

EVOLUTION OF THE MITOCHONDRIAL GENOME IN MAMMALS

In recent years the sequencing and comparison of numerous mitochondrial genomes from animals, in particular from mammals, have revealed a number of interesting and useful properties of this genome which have had a strong impact in various fields of applied research. Because of its maternal inheritance and the lack of recombination, animal mtDNA has become the most popular molecule in the field of molecular evolution.

In this section, the peculiar evolutionary features of the mammalian mitochondrial genomes together with some examples of their applications are discussed.

Base Composition

The mammalian mitochondrial genome has a GC content ranging from 36% (mouse) to 44% (human) (Figure 6). The most remarkable feature in the primary structure is the asymmetry in base composition, in particular the uneven distribution of G and C between the two strands. The mtRNAs species of the sense strand (H strand) are poor in G. This indicates that such a compositional feature places constraints on the expression of the mitochondrial genome.

If the different RNA species are considered separately, this bias in base composition is present in all functional sequences coded by the sense strand. However, in the case of rRNA genes, the percentage of the four bases among mammals is more uniform than in messenger and transfer RNA genes, owing to their high degree of conservation (Saccone and Sbisà, 1989).

Codon usage for the mitochondrial-coded proteins clearly shows that G-ending codons are avoided in H strand transcripts and confirms the asymmetric base composition of mtDNA. Figure 7 indicates the degree of asymmetry of the third codon positions in the quartets (family codons) of mt mRNA genes in humans. The G-C asymmetry is always much higher than the A-T asymmetry, with the exception of ND4L.

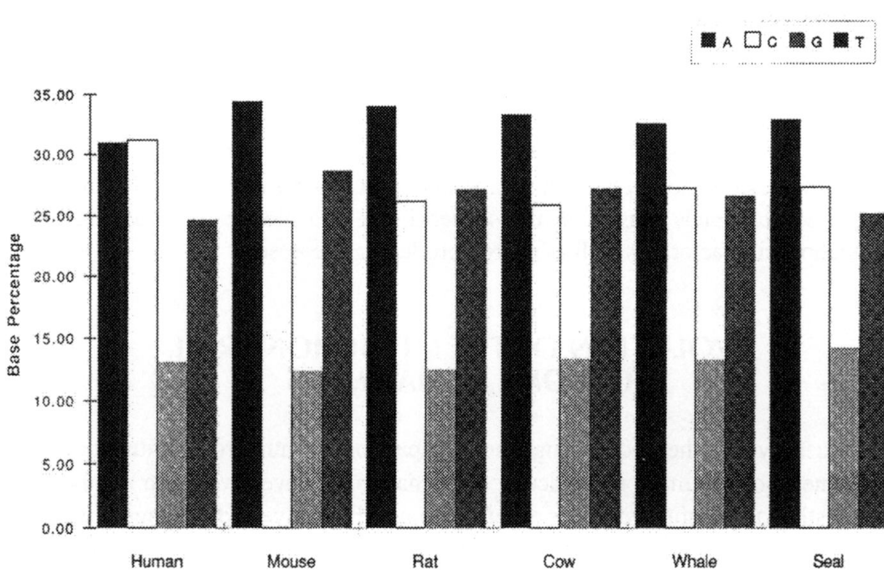

Figure 6. Average base composition in mammalian mtDNA.

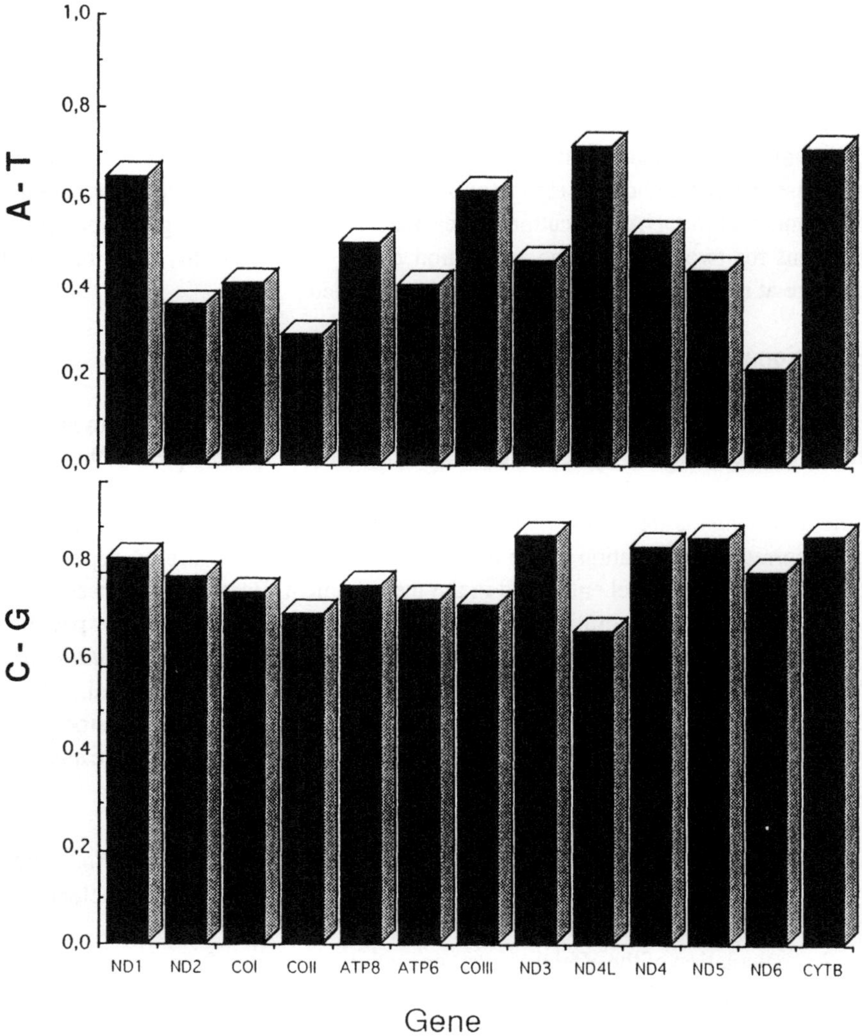

Figure 7. Asymmetry index of the third codon positions in the quartet family codons of mRNA genes in humans. The asymmetry index has been calculated as | A–T | / (A+T) (A/T asymmetry) and | C–G | / (C+G) (C/G asymmetry) where A, C, G, and T are the occurrence of the four bases in the third codon positions of quartet family codons.

Such a strong compositional bias, in particular the trend to avoid G in the third codon position, might also explain several deviations of the mitochondrial code with respect to the universal one. In animal mitochondria, the initiation codon often used is AUA instead of AUG, which is a G-ending codon. One of the three nonsense codons, UGA, becomes an additional tryptophan codon. The canonical UGG tryptophan triplet would involve the use of two contiguous G, a situation which is very rare in the sense strand of mtDNA. In other words, the compositional bias must have been strong enough to influence the genetic code itself.

Different models can be suggested to explain the peculiar compositional bias of the mammalian genome. One involves metabolic discrimination between nucleotide bases and/or replication errors followed by biased repair. Another is where the mechanism of mtDNA replication leaves one DNA strand as a single, unprotected filament for two-thirds of the replication cycle. The possibility, too, of direct damage at the level of the RNAs cannot be excluded.

Evolutionary Rate

It has been widely recognized that mtDNA evolved more rapidly than nuclear DNA. This particularly concerns nucleotide substitutions which represent the major if not the unique mechanism of evolution in mammals; insertions, deletions, or other mechanisms are minor and very rare, except in the D-loop region (see below).

A correct determination of the nucleotide substitution rate strictly depends on the mathematical model employed in the measurement. Based on the observation that the nucleotide substitution pattern of mtDNA in mammals was rather peculiar (because of the asymmetry in base composition), a new model—the *Stationary Markov Clock* (SMC)—was put forward by Saccone's group (Saccone et al., 1990). This does not follow the *a priori* assumptions that are at the basis of other stochastic models available in the literature. The SMC model used in the quantitative analysis of mtDNA allows:

1. determination of the four-nucleotide substitution rate matrix, R_{ij}, which describes the propensity of nucleotide j to be substituted by nucleotide i independently of time, t;
2. calculation of the effective average evolutionary rate of sequence sites (i.e., the rate of silent substitutions in mRNA genes); and
3. construction of phylogenetic trees.

By using this method, the genetic distances between primates (Holmes et al., 1989), several orders of mammals (Pesole et al., 1991; Saccone et al., 1991a), and, more recently, the taxonomic relationship between human populations and thus the coalescence time of mtDNA of our species, have also been determined (Pesole et al., 1992) (see below).

The average evolutionary rate of the mitochondrial genome in mammals shows that the rate of silent substitution is definitely higher in mitochondrial genes than in nuclear genes, whereas the replacement rate varies from gene to gene, and in several cases can be lower in mitochondria than in nuclei. This specifically applies to the subunits of the cytochrome-oxidase complex, where the evolution of some nuclear coded subunits is faster than that of mitochondrial coded subunits (Saccone et al., 1991b).

As for the regulatory regions, comparisons between mitochondrial and nuclear elements are meaningless due to the peculiar mechanisms of replication and expression occurring in mitochondria, as well as the diverse structure of the two genomes.

Evolution of the D-Loop Region

As mentioned already, mammalian mtDNA possesses a main noncoding region containing the regulatory elements of replication and expression of the genome. In contrast with the high degree of conservation of the rest of the genome, the D-loop region shows great variability both in length and base composition among mammals. The exception is in the central part involving about 250 nucleotides.

On the basis of both degree of conservation and base content, the D-loop has been divided into three domains. In the central (C) domain, corresponding to one of the best conserved tracts of mtDNA, the percentage of G increases to 19%, a value slightly higher than that found in the genome region coding for rRNA (~18%), tRNA (~15%), and mRNA (~12%). (GG)n repeats (n ranging from 3 to 6) were almost exclusively found here. The left (L) and right (R) domains containing the main regulatory elements, such as L and H promoters and the CSB blocks, have evolved rapidly in a species-specific manner which generates heterogeneity both in length and in base composition. They are prone to insertion and deletion of elements, and to the generation of short repeats by replication slippage. However, the presence of blocks of conserved sequences, like CSBs and TASs, and similar cloverleaf structures in these regions indicate a basic similarity in the regulatory mechanisms of the mitochondrial genome in all mammalian species (Brown et al., 1986).

Figure 8 shows a schematic alignment of the D-loop regions in some mammals. Primates are characterized by an insertion of 136 bases in the L domain, where the majority of polymorphic sites are localized (E. Sbisà, unpublished observation). Due to their peculiar evolution and different length, the L and R domains cannot be used to measure the genetic distance between species. Conversely, the central domain behaves as a good molecular clock and allows one to calculate divergence times between mammals, and to construct phylogenetic trees (Figure 9) (Saccone et al., 1991a).

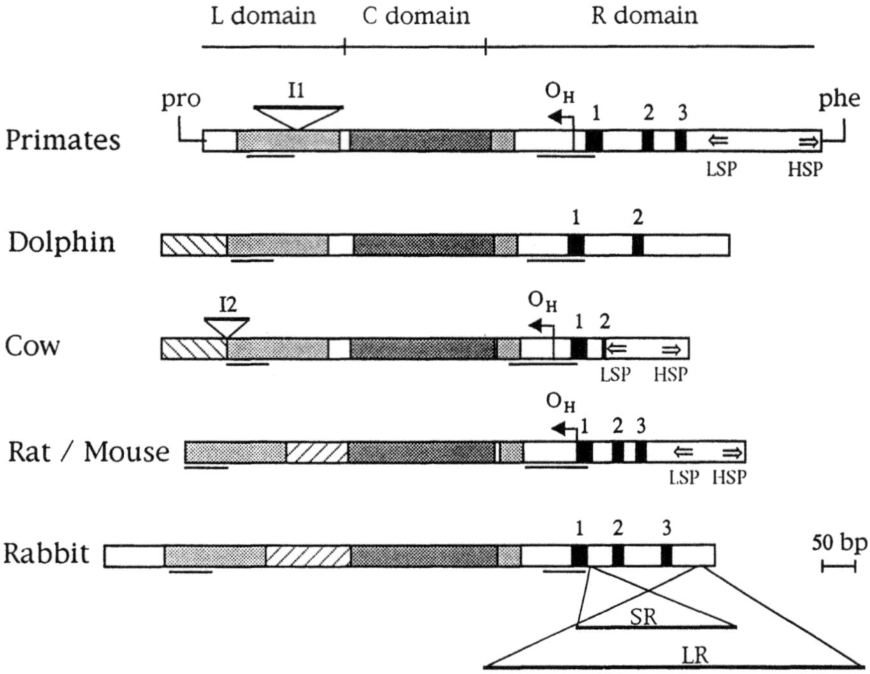

Figure 8. Schematic representation of the alignment of the D-loop containing region in mammals. The areas shaded with the same pattern correspond to sequence similarities. The underlined regions indicate the cloverleaf-like secondary structure. I1, I2 = insertion elements; SR, LR = short and long repeats; O_H = H strand replication origin; HSP = H strand promoter; LSP = L strand promoter; 1, 2, 3= CSBs.

Evolution of Human Mitochondrial Genome

The study of mtDNA evolution in humans has gained popularity in several fields of applied biology. In anthropology, for example, the mtDNA has been used to date the origin of modern man and to provide further insight into the evolutionary histories of populations.

The D-loop-containing region, particularly its left and right domains, being the most variable part of the mitochondrial genome, have been sequenced in different human populations. The results have been used to estimate the time of the ancestral mtDNA (coalescence time) from which all extant human races derive. Wilson and his group working on this part of the mitochondrial genome put forward the monogenetic model of the origin of modern man, according to which *Homo sapiens* is a novel species that spread from a single African woman, the *Black Eve*, about 150,000 years ago (Vigilant et al., 1991).

Employing the same sequences of the D-loop region and applying a corrected version of the *Stationary Markov Clock*, Saccone and her group took into account

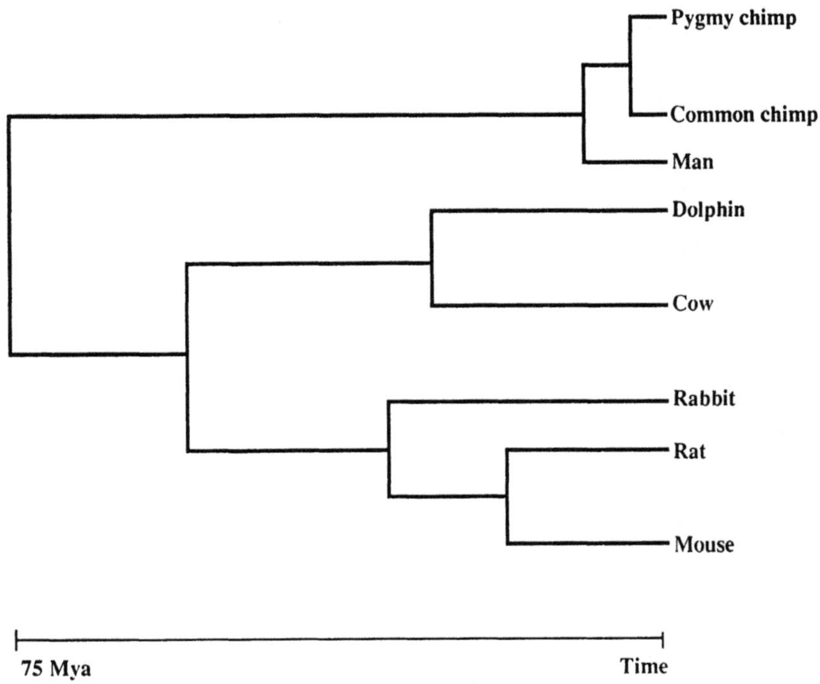

Figure 9. Phylogenetic tree of the D-loop region central domain.

invariable sequence sites, and were able to confirm the African origin of modern man. However, the time of the *Black Eve* was pushed back to at least 400,000 ± 100,000 years ago. According to these results, the mitochondrial data are more consistent with an expansion beyond Africa at about the period of *Homo erectus*, with continuity to the present time of regional populations, rather than with the replacement of archaic human populations outside of Africa by a second and more recent wave of anatomically modern humans (Pesole et al., 1992).

Another interesting aspect of mtDNA evolution in humans is the study of mitochondrial polymorphism and, in particular, the pattern of base substitutions between sequences. By analyzing the base substitutions of human D-loop regions the average R_{ij} rate matrix was constructed. This matrix, as previously stated, provides information about the substitution rate of each base into the three others. As shown in Table 6, the matrix is asymmetric. The total transition to the transversion ratio is 11.4 ± 2.2, but the transitions are not equivalent. T-C is higher than the A-G transition rate, the latter having G to A significantly higher than A to G. Transversions involving G are practically nil.

The nonequivalence between T-C and A-G transition types (T-C higher than A-G) was first observed by Brown and Simpson (1982) in the evolutionary analysis

Table 6. The Average Nucleotide Substitutions[a] Rate Matrix Calculated by Comparing Human D-Loop Sequences

	A ↵	*C* ↵	*G* ↵	*T* ↵
A	—	9.6 ± 1.7	104.2 ± 9.3	0.8 ± 0.9
C	10.9 ± 1.9	—	2.3 ± 1.8	127.2 ± 7.3
G	43.9 ± 4.1	0.8 ± 0.6	—	0.0 ± 0.0
T	0.6 ± 0.6	83.9 ± 4.8	0.0 ± 0.0	—

Note: [a]Sub · year · 10^{-9}.

of the rat COII gene. The same phenomenon, although in the opposite direction (A-G higher than T-C) was found in the evolution of the sea urchin mitochondrial genes (De Giorgi et al., 1991). Such behavior has been explained on the basis of the asymmetric mechanism of mtDNA replication which, as previously reported, implies that one of the old strands remains for three-fourths of the cycle in single-stranded structure and thus may be more prone to mutation than the other. A similar question has been raised to explain the different base compositions of the compartments of nuclear genome in warm-blooded vertebrates (Holmquist, 1990).

The study of mitochondrial genome evolution has acquired great importance in human pathology. Several mutations of human mtDNA have been correlated with a number of mitochondrial diseases known as encephalomyopathies. Table 7 lists the types of base substitutions detected so far. In most cases an A-G transition is involved in both mRNA and tRNA genes, suggesting that this transition type is less

Table 7. Mutation Types in Human Mitochondrial Diseases

Mutation	*Gene*	*Position*	*Disease*
A → G	ND2	4917	LHON*
G → A	ND5	13708	LHON*
G → A (Arg → His)	ND4	11778	LHON
G → A (Ala → Thr)	ND1	3460	LHON
T → C (Leu → Pro)	ND1	4160	LHON
T → G (Leu → Arg)	ATP6	8993	NARP
A → G	Lys	8344	MERFF
A → G	Thr	15923	FIRED
A → G	Thr	15924	FIRED
A → G	Leu	3243	MELAS
G → A	Leu	3260	cardiomyopathy

Note: Abbreviations: LHON, Leber's hereditary optic neuropathy; NARP, neurogenic muscle weakness ataxia and retinitis pigmentosa; MERFF, myoclonus epilepsy with ragged red fibers; FIRED, fatal infantile respiratory enzyme deficiency; MELAS, mitochondrial encephalopathy with lactic acidosis and stroke-like episodes;
*Indicates mutations occurring at the same time with that at position 11778.

neutral than the T-C type. Beside nucleotide substitutions, large deletions and, in some cases, duplications have also been detected in other mitochondrial myopathies (Wallace, 1992 for review).

An accumulation of multiple deletions has been observed in aging; that is, a biological phenomenon where the role of mitochondria seems crucial (Muller-Hocker, 1992 for review). The deletions usually pertain to the 10 kb region lying between the H and L strand replication origins. This is the region involved in D-loop expansion during the replication of mtDNA.

Studies of this type have aroused much interest and the need has been expressed by the community of researchers for a structured collection of these data into a specialized database.

SUMMARY

The primary role of mitochondria, the organelles present in virtually all respiring eukaryotic cells, is electron transport and oxidative phosphorylation; two processes that provide the cells with ATP, the principal source of cellular energy. Mitochondria are formed by two membranes, an outer membrane having mainly permeability properties and an inner membrane where electron transport and oxidative phosphorylation take place. The inner space, called the matrix, contains metabolic enzymes, the mtDNA, and the genetic apparatus for its replication and expression.

Mitochondrial biogenesis is the result of cooperation between two genetic systems: the main cellular nucleus–cytoplasm genetic system, which codes for the major portion of mitochondrial components; and the mitochondrial genetic system, which yields a modest but essential contribution. In eukaryotic cells, the presence of two (three in photosynthetic eukaryotes) physically and functionally distinct genomes can be explained by the endosymbiont theory which postulates that nuclear and organellar genomes inhabited different types of cells, which subsequently formed symbioses.

The mitochondrial genome shows great variability in terms of structure, gene content, organization, and mode of expression in different organisms. The size of the mitochondrial genome ranges from about 16 kbp in metazoa to 2500 kbp in plants. The double stranded structure is circular, but in some organisms (e.g., Ciliata) the genome is linear. In spite of such variability, the gene content is rather uniform. All mitochondrial genomes studied thus far code for: two rRNA species (in plants also 5S rRNA); a more or less complete set of tRNA species; and a few mRNA species (13 in metazoa) for subunits of the mitochondrial respiratory complexes. Differences in the organization of the genes within the genome in different organisms is associated with a different mode of expression.

Replication and expression of the mitochondrial genome is controlled by nuclear genes which specify all the proteins involved in these processes. The functions of these genes have been elucidated mainly by means of genetic experiments in facultative aerobic organisms (e.g., yeast). Recent evidence indicates that mitochondrial signals may affect nuclear genes. A special cytoplasmic apparatus is involved in the import of nuclear-coded proteins into mitochondria and their targeting to the different mitochondrial compartments.

The mitochondrial genomes of animals, particularly in mammals, have been widely studied. Their sequencing has revealed a number of interesting and useful properties which have already had a powerful impact on biotechnology. Because of its reduced size, the maternal inheritance and the lack of recombination, animal mtDNA has become the most popular molecule in the field of molecular evolution. The study of mtDNA evolution in humans has provided information about the date of the origin of modern man and the evolutionary histories of populations. Several mutations of human mtDNA are involved in a number of mitochondrial diseases.

REFERENCES

Altman, S., Kirsebom, L., & Talbot S. (1993). Recent studies of ribonuclease P. FASEB J. 7, 7–14.

Attardi, G. & Schatz, G. (1988). Biogenesis of mitochondria. Annu. Rev. Cell Biol. 4, 289–333.

Attardi, G., Chomyn, A., King, M. P., Kruse, B., Loguercio Polosa, P., & Murdter, N. N. (1989). Regulation of mitochondrial gene expression in mammalian cells. Biochem. Soc. Trans. 18, 509–513.

Benne, R. (1985). Mitochondrial genes in trypanosomes. Trends Genet. 1, 117–121.

Blum, B., Sturm, N. R., Simpson, A. M., & Simpson, L. (1991). Chimeric gRNA-mRNA molecules with oligo(U) tails covalently linked at sites of RNA editing suggest that U addition occurs by transesterification. Cell 65, 543–550.

Brown, G. G. & Simpson, M. V. (1982). Novel features of animal mtDNA evolution as shown by sequences of two rat cytochrome oxidase subunit II genes. Proc. Natl. Acad. Sci. USA 79, 3246–3250.

Brown, G. G., Gadaleta, G., Pepe, G., Saccone, C., & Sbisà, E. (1986). Structural conservation and variation in the D-loop containing region of vertebrate mitochondrial DNA. J. Mol. Biol. 92, 503–511.

Cantatore, P. & Saccone, C. (1987). Organization, structure and evolution of mammalian mitochondrial genes. Int. Rev. Cytol. 108, 149–206.

Cantatore, P., Gadaleta, M. N., Roberti, M., Saccone, C., & Wilson, A. C. (1987). Duplication and remolding of tRNA genes during the evolutionary rearrangement of mitochondrial genomes. Nature 329, 853–855.

Cantatore, P., Roberti, M., Loguercio Polosa, P., Mustich, A., & Gadaleta, M. N. (1990). Mapping and characterization of *Paracentrotus lividus* mitochondrial transcripts: Multiple and overlapping transcription units. Curr. Genet. 17, 235–245.

Clayton, D. A. (1991). Replication and transcription of vertebrate mitochondrial DNA. Ann. Rev. Cell Biol. 7, 453–478.

Darr, S. C., Brown, J. W., & Pace, N. R. (1992). The varieties of ribonuclease P. TIBS 17, 178–182.

De Giorgi, C., De Luca, F., & Saccone, C. (1991). Mitochondrial DNA in the sea urchin *Arbacia lixula*: nucleotide sequence differences between two polymorphic molecules indicate asymmetry of mutations. Gene 103, 249–252.

de Zamaroczy, M. & Bernardi, G. (1985). Sequence organization of the mitochondrial genome of yeast. A review. Gene 37, 1–17.

Dietrich, A., Weil, J. H., & Marechal-Drouard, L. (1992). Nuclear-encoded transfer RNAs in plant mitochondria. Annu. Rev. Cell Biol. 8, 115–131.

Fischer, R. P., Lisowsky, T., Breen, G. A. M., & Clayton, D. A. (1991). A rapid, efficient method for purifying DNA-binding proteins: denaturation-renaturation chromatography of human and yeast mitochondrial extracts. J. Biol. Chem. 266, 9153–9160.

Fischer Lindahl, K., Hermel, E., Loveland, B. E., & Wang, C.-R. (1991). Maternally transmitted antigen of mice: a model transplantation antigen. Ann. Rev. Immunol. 9, 351–372.

Gadaleta, G., Pepe, G., De Candia, G., Quagliariello, C., Sbisà, E., & Saccone, C. (1989).The complete nucleotide sequence of the *Rattus norvegicus* mitochondrial genome: Cryptic signals revealed by comparative analysis between vertebrates. J. Mol. Evol. 28, 497–516.

Glover, L. A. & Lindsay, J. G. (1992). Targeting proteins to mitochondria: A current overview. Biochem. J. 284, 609–620.

Gray, M. W. & Doolittle, W. F. (1982). Has the Endosymbiont hypothesis been proven? Microbiol. Rev. 46, 1–42.

Gray, M. W., Cedergren, R., Abel, Y., & Sankoff, D. (1989). On the evolutionary origin of the plant mitochondrion and its genome. Proc. Natl. Acad. Sci. USA 86, 2267–2271.

Gray, M. W., Hanic-Joyce, P. J., & Coveno P. S. (1992). Transcription, processing and editing in plant mitochondria. Ann. Rev. Plant Physiol. Plant Mol. Biol. 43, 145–157.

Grivell, L. A. (1987). In: Genetic Maps 1987 (O'Brien, S. J., Ed.), Vol. 4, pp. 290–297. Cold Spring Harbor Laboratory, Cold Spring, New York.

Grivell, L. A. (1989). Nucleo-mitochondrial interactions in yeast mitochondrial biogenesis. Eur. J. Biochem. 182, 477–493.

Hess, J. F., Parisi, M. A., Benett, J. L., & Clayton, D. A. (1991). Impairment of mitochondrial transcription termination by a point mutation associated with the MELAS subgroup of mitochondrial encephalomyopathies. Nature 351, 236–239.

Holmes, E. C., Pesole, G., & Saccone, C. (1989). Stochastic models of molecular evolution and the estimation of phylogeny and rates of nucleotide substitutions in the hominoid primates. J. Hum. Evol. 18, 775–794.

Holmquist, G. P. (1990). In: Advances in Mutagenesis Research (Obe, G., Ed.), Vol. 2, pp. 95–126. Springer, Berlin-Heidelberg.

Indsorf, N. F. & Bogenhagen, D. F. (1989). DNA polymerase γ from Xenopus laevis. The identification of a high molecular weight catalytic subunit by a novel DNA polymerase photolabeling procedure. J. Biol. Chem. 264, 21491–21497.

John, P. & Whatley, F. R. (1975). Paracoccus denitrificans and the evolutionary origin of the mitochondrion. Nature 254, 495–498.

Kadenbach, B., Kuhn-Neutwig, I., & Buge, U. (1987). Evolution of a regulatory enzyme: Cytochrome-c oxidase (complex IV). Curr. Topics Bioenerg. 15, 113–161.

Karwan, R., Bennett. J., & Clayton D. A. (1991). Nuclear RNase MRP processes RNA at multiple discrete sites: Interaction with an upstream G box is required for subsequent downstream cleavages. Genes Dev. 5, 1264–1276.

Karwan, R. (1993). RNase MRP/RNase P: A structure-function relation conserved in evolution? FEBS Lett. 319, 1–4.

Kiss, T. & Filipowicz, W. (1992). Evidence against a mitochondrial location of the 7-2/MRP RNA in mammalian cells. Cell 70, 11–16.

Kramer, R. & Palmieri, F. (1992). In: Molecular Mechanisms in Bioenergetics (Ernster, L., Ed.), pp. 359–384. Elsevier Science Publishers, B.V.

Lanoue, K. F. & Schoolwerth, A. C. (1984). In: Bioenergetics (Ernster, L., Ed.), pp. 221–268. Elsevier Science Publishers, B.V.

Levings III, C.S. & Brown, G. G. (1989). Molecular biology of plant mitochondria. Cell 56, 171–179.

Ma, D. P., King, Y.-T., Kim, Y., Luckett Jr., W. S., Boyle, J. A., & Chang, Y.-F. (1992). Amplification and characterization of an inverted repeat from the Chlamidomonas reinhardtii mitochondrial genome. Gene 119, 253–257.

Muller-Hocker, J. (1992). Mitochondria and aging. Brain Pathol. 2, 149–158.

Nardelli, M., Tommasi, S., D'Erchia, A. M., Tanzariello, F., Tullo, A., Primavera, A. T., De Lena, M., Sbisà, E., & Saccone, C. (1994). Detection of novel transcripts in the human mitochondrial DNA region coding for ATPase8-ATPase6 subunits. FEBS Lett. in press.

Narjarian, D., Dihanich, M. E., Martin, N. C., & Hopper, A. K. (1987). DNA sequence and transcript mapping of MOD5: features of the 5′ region which suggest two translation starts. Mol. Cell. Biol. 7, 185–191.

Newton, J. K. (1988). Plant mitochondrial genomes: organization, expression and variation. Ann. Rev. Plant Physiol. Plant Mol. Biol. 39, 503–532.

Palmer, J. D. (1990). Contrasting modes and temps of genome evolution in land plant organelles. Trends Genet. 6, 115–120.

Pepe, G., Gadaleta, G., Palazzo, G., & Saccone, C. (1989). Sequence-dependent DNA curvature: Conformational signal present in the main regulatory region of the rat mitochondrial genome. Nucl. Acids Res. 17, 8803–8819.

Pesole, G., Bozzetti, M. P., Lanave, C., Preparata, G., & Saccone, C. (1991). Glutamine synthetase gene evolution: A good molecular clock. Proc. Natl. Acad. Sci. USA 88, 522–526.

Pesole, G., Sbisà, E., Preparata, G., & Saccone, C. (1992). The evolution of the mitochondrial D-loop region and the origin of modern man. Mol. Biol. Evol. 9, 587–598.

Saccone. C. & Sbisà, E. (1989). In: Organelles in Eukaryotic Cells (Tager, M., Azzi, A., Papa, S., & Guerrieri, F., Eds.), pp. 127–142. Plenum Publishing, New York.

Saccone, C., Lanave, C., Pesole, G., & Preparata, G. (1990). In: Methods in Enzymology (Doolittle, R. F., Ed.), Vol. 183, pp. 570–583. Academic Press, New York.

Saccone, C., Pesole, G., & Sbisà, E. (1991a). The main regulatory region of mammalian mitochondrial DNA: Structure-function model and evolutionary pattern. J. Mol. Evol. 33, 83–91.

Saccone, C., Pesole, G., & Kadenbach, B. (1991b). Evolutionary analysis of the nucleus-encoded subunits of mammalian cytochrome c oxidase. Eur. J. Biochem. 195, 151–156.

Sbisà, E., Nardelli, M., Tanzariello, F., Tullo, A., & Saccone, C. (1990). The complete and symmetric transcription of the main noncoding region of rat mitochondrial genome: *In vivo* mapping of heavy and light transcripts. Curr. Genet. 17, 247–253.

Sbisà, E., Tullo, A., Nardelli, M., Tanzariello, F., & Saccone, C. (1992). Transcription mapping of the Ori L region reveals novel precursors of mature RNA species and antisense RNAs in rat mitochondrial genome. FEBS Lett. 296, 311–316.

Schmitt, M. E., Bennett, J. L., Dairaghi, D. J., & Clayton, D. A. (1993). Secondary structure of RNase MRP RNA as predicted by phylogenetic comparison. FASEB J. 7, 208–213.

Simpson, L. (1987). The mitochondrial genome of Kinetoplastid protozoa: Genomic organization, transcription, replication, and evolution. Ann. Rev. Microbiol. 41, 363–382.

Simpson, L. & Shaw, J. (1989). RNA editing and the mitochondrial cryptogenes of kinetoplastid protozoa. Cell 57, 355–366.

Topper, J. N., Bennett, J. L., & Clayton, D. A. (1992). A role for RNAse MRP in mitochondrial RNA processing. Cell 70, 16–20.

Vigilant, L., Stoneking, M., Harpending, H., Hawkes, K., & Wilson, A. C. (1991). African populations and the evolution of human mitochondrial DNA. Science 253, 1503–1507.

Wallace, D. C. (1992). Mitochondrial genetics: A paradigm for aging and degenerative diseases? Science 256, 628–632.

Wolstenholme, D. R. (1992). Animal mitochondrial DNA: Structure and evolution. Int. Rev. Cytol. 141, 173–216.

Chapter 3

Thermal Polyamino Acids in the Origins of Life and Biomedical Applications

SIDNEY W. FOX and PETER R. BAHN

INTRODUCTION

The historical treatment of origin-of-life theory has been reviewed by Oparin (1957), Florkin (1975), Fox (1988) and others. It is a long and involved story because of the widespread interest of humans in all walks of endeavor who desire to know their ultimate origins, and because of the theological and scientific dogmas that have attempted to satisfy this need.

Principles of Medical Biology, Volume 1B
Evolutionary Biology, pages 73–85

ISBN: 1-55938-802-1

To mention only some outstanding features here, we need especially to recall first that virtually all of the major religions of the world have documented histories of a mythological nature as answers to the universal questions asked. The problem was transferred to the arena of scientific debate in the spontaneous generation era. Most prominent in the study of spontaneous generation was the famous debate in which Louis Pasteur treated the then current thinking in a context of experimental test.

According to Vallery-Radot (1922), Pasteur showed the fallacy of production of simple life from various infusions by demonstrating that unsterilized preparations were simply good culture media for microbes from the air. Pasteur is alleged to have said: "Never will the doctrine of spontaneous generation recover from the mortal blow struck by this simple experiment." This statement has survived in various interpretations as indicating that Pasteur disproved spontaneous generation. He, however, did not disprove spontaneous generation in a sterile environment, sterility being a necessary condition of the primitive Earth at a time when no organisms were present to contaminate it.

The famous debate left a positive residue for the future. Pasteur is also reported to have said:

> There is the question of so-called spontaneous generation. *Can matter organize itself?* (italics added). In other words, are there beings that can come into the world without parents, without ancestors?

Today we have the evidence that matter of the right sort does organize itself into cells (Fox et al., 1959), and that such cells have in some degree or kind all of the vital properties of modern living cells (Fox, 1971,1992; Mader, 1990).

IN THE BEGINNING OF THE BEGINNING

The approach to the problem of life's origin has required, first of all, identification of the stage from which molecular evolution began, such that cellular evolution could continue therefrom. By that same line of heretical reasoning, recognition of the solution obtained has been inhibited by general, but not complete, lack of recognition of that beginning. In looking back, it indeed seems logical to many that initial informed molecules were nucleic acids, especially DNA ("naked gene") in a role much as we see it today.

The possibility that the initial informed molecules were proteins has been largely, but not entirely, ignored by the seemingly logical view that the matrix was random. Experiments showed that initial (thermal) protein possessed, in at least root form, virtually all of the biological functions that we find in today's organisms (Fox, 1971; Mader, 1990; Fox, 1992). It was further observed that the matrix was highly nonrandom (Fox, 1960b; Dose and Rauchfuss, 1972; Tyagi and Ponnamperuma, 1990; Melius and Srisomsap, 1991). Other experiments showed that amino acids themselves are relevant to their ordering, whereas nucleotides and polynucleotides are not (Tyagi and Ponnamperuma, 1990); the case for thermal proteins as initial forms of biological information was thus further strengthened. The underlying mechanism has also been shown to be one of self-ordering of amino acids, a process inherited into today's mechanism of protein synthesis (Ivanov and Foertsch, 1986) (see Figure 1). The production of ordered proteins from nucleic acids without prior proteins has never been demonstrated, whereas the production of polynucleotides (Jungck and Fox, 1973) and polypeptides (Nakashima and Fox, 1980) from appropriate thermal proteins has been accomplished in the laboratory.

These observations of macromolecular synthesis, both resulting from the same lysine-rich proteinoid, provide the basis for understanding the origin of the contemporary nucleic acid genetic coding system (Fox, 1981). Fundamental to all the progress and a gradual understanding thereof is the special recognition that evolution itself is a forward synthetic process (Young, 1984) and also that evolution sometimes proceeds through growth and decay.

For activities such as enzymatic behavior (Rohlfing and Fox, 1969) or hormonal potency (Fox and Wang, 1968) in thermally constructed proteins, the degrees of potency are much less than in the modern molecular versions. For other behavior, such as the cytological tendency for reproduction (Fox, 1971; Lehninger, 1975) or

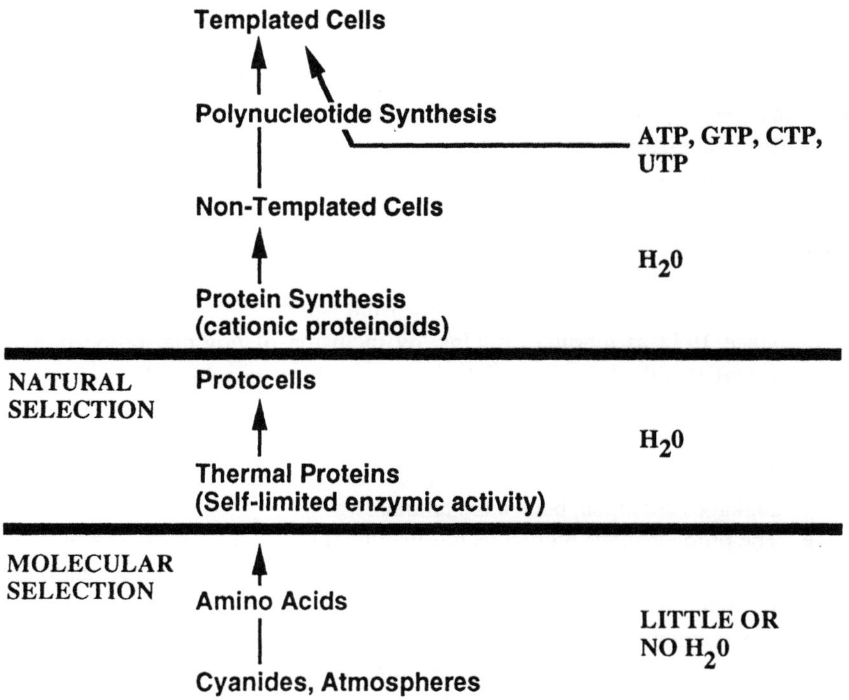

Figure 1. Information flow in the origin of life.

degree of electrical response to varied stimuli (Przybylski and Fox, 1986), the versions identified in the laboratory as prebiotic are fully comparable in power to those in modern cells.

One dichotomy in thinking between two groups of scientists in this problem area has been as profound in the more recent scientific era as the difference in perspective between the religionist and the natural scientist, or as between the spontaneous generators and their opponents. These two groups have consisted of those who adhere to a proteins-first context and those who reason, without adequate justification and without verification, that DNA arose first or RNA arose first.

THE EXPERIMENTAL ERA

After the spontaneous generation era, the next experiments worthy of note are those of the Mexican cosmologist, Alfonso L. Herrera. Herrera found, on reaction of plausibly postulated organic intermediates, the production of two amino acids: a polymer of undeclared constitution and a pigment; and many examples of cellular structure. Of the latter, he sent photomicrographs to many investigators around the world (Oparin, 1957). He used as intermediates compounds found 30 years later to be constituents of interstellar matter.

Herrera's experiments were done in the open environment, much as would be expected for spontaneous occurrences on the primitive Earth. In this respect, his experiments were more plausible than those of Miller, who employed an atmosphere in closed flasks that he and others later repudiated. Herrera's experiments were published in summary form in *Science* (Herrera, 1942); the much more thoroughly publicized experiments of Miller were reported 11 years later (Miller, 1953).

The *principal findings*, excluding neuroscientific ones (Fox, 1992), that have occurred up to this point are:

- Amino acids are cocondensed into what *Chemical Abstracts* has indexed since 1972 as thermal proteins, by inclusion of minimal proportions of aspartic acid, glutamic acid, lysine, or combinations thereof.
- All amino acids can be simultaneously cocondensed.
- The products are highly nonrandom.
- The products are predominantly peptides.
- Linkages are alpha, beta, and gamma.
- The products have numerous biofunctional activities.
- Starting with L or D amino acids makes little difference in the qualitative biofunctionality of some of the polymers.
- The polymeric products can be fractionated to discrete components.
- Each of the components yields an amino acid analysis that is stoichiometric.
- The compositions of the reaction mixtures differ from those of the polymers or the components, sometimes markedly so.

- Each of the components contains its own amino acid sequence.
- The reactions are highly reproducible by any of the criteria used to assess that.

The prospect for a modern chemical understanding of how life began has been one of sufficient unattainability that the problem was long avoided. Some of the reasoning has been conditioned by the biblical dogma that life came from a formless void. While the scientific thinking essentially provided roots for the idea that life must have begun from macromolecules, such molecules were often thought of as DNA and RNA which in turn were considered solely informational [e.g. a "naked gene" (DNA)]. The biblical and scientific dogmas thus reenforced each other (Fox, 1981) by virtue of the fact that each type has designated the matrix as random (Miller and Orgel, 1974; Morris et al., 1974; Fox, 1981b).

THE CONTROVERSIAL ERA

On the one hand, we have a situation in which experiments stemming from amino acids through terrestrial protoprotein to cellular life have been regarded as providing since the mid-1980s a theory of life's origins (Florkin, 1975; Kleinsmith and Kish, 1988; Salthe, 1989; Berra, 1990). On the other hand, in disregard of those experiments, others have visualized only nucleic acids as the first informational biomolecules, even though they have been unable to provide experimental answers to conform to the difficult, if not impossible, problems posed by such thinking.

The proteins-first approach to the origin of life is the only one that has yielded protocells from theoretically defensible precursors. This conceptualization follows from observation (Fox et al., 1959) of self-organization of thermal polyamino acids, designated by Chemical Abstracts since 1972 as *thermal proteins* (TPs).

The principles of the origins theory include the following:

1. Amino acids can be thermally pyrocondensed to polymers containing any number of types of amino acid if at least a small proportion of aspartic acid or glutamic acid is in the reaction mixture. This was discovered experimentally.

2. The products obtained are highly nonrandom in both composition and sequence. This principle was discovered experimentally. No DNA/RNA was involved. The self-ordering of amino acids is the fundamental process. Evidence from studies of sequence indicate that it has been inherited into modern protein synthesis (Ivanov and Foertsch, 1986; Tyagi and Ponnamperuma, 1990).

3. Over a wide range of composition, the thermal copolyamino acids, when wetted, assemble themselves into microspheres that serve as models for the first cells on Earth, called *protocells*.

4. TPs control the synthesis and patternization of polynucleotides (Jungck and Fox, 1973). The *Central Dogma* is the reverse of this synthetic evolutionary translation of TPs into polynucleotides. As in the production of protocells, the

direction of early evolution is synthetic; the analytic approach to such problems has so far taught nothing.

5. The synthesis of both polynucleotides and polypeptides by lysine-rich TPs explains the origin of the modern genetic coding mechanism.

6. In the TPs and their microspheres are found the roots of virtually all modern biological functions (Fox, 1971, 1980; Mader, 1990).

After it was learned how amino acids could be thermally copolymerized, similar experiments were tried with mononucleotides (Schwartz and Fox, 1964). Heat was used because heat is ubiquitous in the Earth; not just near volcanoes, as has been asserted by some (e.g., Miller and Orgel, 1974). Also, heat is not destructive like lightning which is an especially unlikely source of energy for reactions in closed systems within the Earth; indeed one cannot see how lightning could penetrate significantly below the surface of the planet.

In testing various nucleotides by heating, with the helpful addition of polyphosphoric acid, only cytidylic acid underwent significant polymerization (Schwartz and Fox, 1964). Moreover, the origin of cytidylic acid or other nucleotides was not

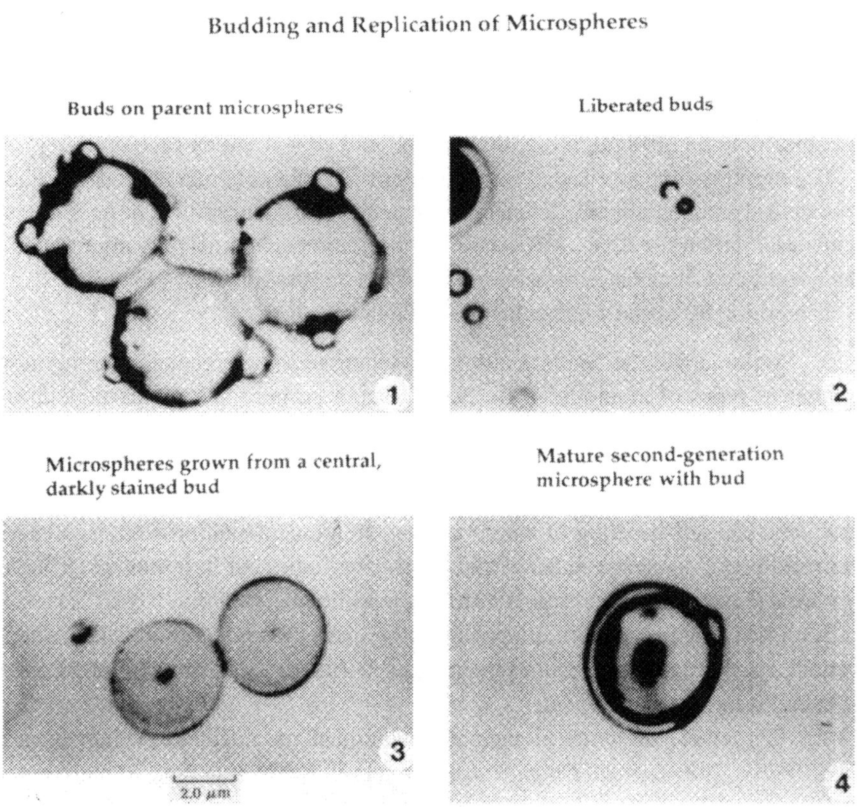

Figure 2. Budding and replication of microspheres.

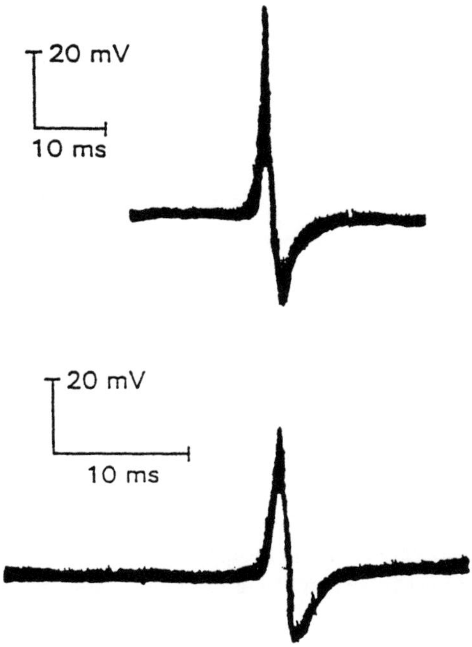

Microsphere Spiking

Figure 3. Action potential resembling that of neuron. (*Upper*) Spiking in crayfish stretch receptor neuron. (*Lower*) Spiking in microsphere of 2:2:1-proteinoid. (By Dr. A. Przybylski).

then and is not now visualized, nor how it could yield cells, metabolism, or genetic coding, while all of these are found in a string of products from TPs (Florkin, 1975).

Aside from the virtual impossibility of visualizing, let alone demonstrating, how nucleic acid could initiate bioevolution, the problem of randomness enters into the selection of the original molecular matrix.

The opposition to the flowsheet presented (Figure 1) is found to be focused on the question of randomness. By unquestioning acceptance of randomness, many eminent scientists have reenforced the supposition of randomness for thermal polycondensation of amino acids in the absence of instructions from DNA/RNA. These scientists are listed in Table 1.

The view against nonrandomness (order) has been most intensively articulated by Miller and Orgel (1974) as follows:

> The probability of obtaining a specified polypeptide containing 100 amino acids in a sample of random polypeptides is 20^{-100}. The chance of obtaining a peptide with glutamic acid at the end, on the other hand, is 20^{-1}. Thus the degree of nonrandomness in thermal polypeptides so far demonstrated is minute compared with the nonrandomness of proteins. It is deceptive to suggest that thermal polypeptides are similar to proteins in their nonrandomness.

Table 1. Proponents of Randomness

Cairns-Smith (1982)[1]	Eigen and Schuster (1977)[1]
Crick (1981)[2]	Oparin (1957)[2]
Crick et al. (1970)[1]	Monod (1971)[2]
De Duve (1991)	Morris et al. (1974)
Dyson (1985)[1]	Prigogine and Stengers (1984)
Elgen (1971)[2]	Weiner and Maizels (1987)[1]

Notes: [1]From Tyagi and Ponnamperuma (1990).
[2]From Fox and Windsor (1984).

Nevertheless, no evidence for randomness in polypeptides has been offered, undoubtedly because there is none. Molecules are not properly likened unto uniformly sized playing cards that follow the laws of probability. Rather, each molecule has its individual shape, distribution of electronic charges, hydrophobic zones, and other properties; their interactions and reactions are specific and are the basis for specificity and variety in the biological world. The efforts of the above cited scientists including Miller and Orgel are thus misplaced from chemistry to probability theory, placing irrelevant hypothetical considerations before rigorous chemical results.

BIOMEDICAL ADVANCES

The biomedical and other industrial applications of TPs are possible for the same fundamental and sometimes contested reason that a laboratory synthesis of a protoorganism was possible (Fox 1988; Grobman, 1988; Salthe, 1989; Berra, 1990), thus verifying theory. The reason is the finding of rigorously demonstrated nonrandomness in thermal proteins. Why some chemists hold randomnist views is puzzling since the assumption of randomness denies the existence of stereochemistry, thus equating a group of molecules like the family of amino acids to playing cards, which of course are all of the same shape. An emphasis on randomness is also equivalent in an origin-of-life context to a belief in instant creation, inasmuch as molecular evolution (Fox, 1953) was found to be the basis for biological evolution (Fox, 1960a).

While the above argument for randomness may yet have staying power due to its longtime deep roots, the biomedical applications already are making themselves felt:

Memory Enhancement

Selected hydrophobe-rich TPs, especially some containing tryptophan, affect memory retention in mice subjected to training in footshock avoidance at particular entrances to a maze (Fox and Flood, 1992). Memory scores of 20% to 80% were recorded for mice into which the hydrophobic polymers were injected one week before the mice were tested.

In the controls, a simple TP, poly(asp,glu) devoid of hydrophobic amino acid residues, registered no enhancement of memory.

The polymers were tested by intracerebroventricular injection. Future research is being focused on the design of a soluble analog that can pass the blood–brain barrier and thus be used orally for testing in humans.

Synthetic NGF-Like Growth Factors

The work of Hefti et al. (1991) demonstrates that TPs of the same class as those studied earlier for memory enhancement have, when added topically to cultures of rat forebrain neurons, promoted prolonged survival. It would be easy to assume that the work reported on in 1992 followed from that of the study published in 1991 (Hefti et al.). The fact is that the memory enhancement and the growth of laboratory protocells were observed first, each of these having been stimulated by observations in the model systems (Hsu et al., 1971). Both types of study were undertaken in the light of the Hebb hypothesis that extensions of cellular memory bases, formed in some way, would yield a higher level of memory storage and retention.

This research is now aimed at producing a memory-basing protocellular mass that can be delivered to humans in an oral form.

Antiaging

In the same study of Hefti et al. (1991), neuronal cells in culture were found to age several times more slowly when appropriate TPs were added to the cultures. Investigations to determine the generality of this property are in progress.

Blood Substitutes

The use of ethanolamine-derivatized thermal polyaspartic acid has been recorded as promising blood plasma expanders (Neri et al., 1973). The materials have, however, displayed limited, shelf life (Neri, personal communication). This explains the observation that materials of such promising characteristics have not been marketed. Research on extending the shelf stability of these TP preparations is under way.

The possibility of producing artificial blood cells arises from the discovery of Dose and Zaki (1971) that heme can be cocondensed with amino acids to yield hemoTPs. These authors indicated oxygen-carrying capacity by demonstration of catalase activity. Such studies have been extended by Hsu and Fox (1976).

MSH Analogs

Since the principal class of hormones is that of the proteins, it is not surprising that TPs were examined early for hormonal activity. Such activity was found in analogs of MSH prepared by simple thermal polycondensation of amino acids (Fox and Wang, 1968).

Enzymes

The type of biological activity first sought in the TPs was that of enzymes. Representatives of the major classes of enzyme as listed by the International Union of Biochemistry were found (Rohlfing and Fox, 1969; Fox and Dose, 1977).

The activities are much weaker than those found in the counterpart contemporary enzymes; this relationship also obtains to the hormonal activity, and may be explained on a basis of need for bioevolution to rely solely on L-isomers from the thermal polymers (Bonner, 1992). At the same time, other properties, such as electrical response to stimuli and ease of reproduction, are in their respective strengths comparable to those of modern forms. Of special relevance to the origins problems is the power of lysine-rich TPs to synthesize polypeptides (Nakashima and Fox, 1980) and polynucleotides (Jungck and Fox, 1973).

The informational relationship thus suggested is (ancient thermal) proteins → nucleic acids → (modern) nucleic acids → proteins. This reversal of direction from ancient evolution to modern evolution is seen throughout these studies, and helps to explain their singularity (Young, 1984).

Microencapsulation

The first of the applications of TPs to be commercialized is that of microencapsulation for drug delivery. As is well known, a liposome industry exists for this purpose. Liposomes are made from lipids; the resultant vesicles lack stability, tend to be larger than TP microspheres, and are not amenable to the range of physical properties available with TPs. Accordingly, interest has increased in the use of TPs for formation of microspheres that have all of the desired properties.

Several patents exist in the field and others are in process. At least one new company has built on this base from a patent (Steiner and Rosen, 1990) that credits the basic pioneering work of Fox and associates (Fox and Brooke, 1979; Steiner and Rosen, 1990; Weiss, 1992). Rosen, an investigator interested in diabetes research, has cooperated in producing microencapsulated insulin. From this beginning, the end of drug delivery by needle is visualized (Weiss, 1992) since proteinoids yield orally useable small capsules. In addition to the Steiner and Rosen patent, which has been licensed commercially, Fox and Veltri (1990) hold a patent for microencapsulated adriamycin, and Fox and associates are at work on highly improved modes of drug delivery through TPs. Much advance work has been recorded by Milstein and associates at Enisphere Technologies (Santiago et al., 1993).

SUMMARY

The principal findings of this approach to molecular evolution are as follows: Amino acids are cocondensed into thermal proteins by inclusion of aspartic acid, glutamic acid, lysine, or combinations thereof. All amino acids can be simultaneously cocondensed. The products are highly nonrandom. The products have numerous biofunctional activities. The polymeric products can be fractionated to discrete components. Experiments stemming from amino acids through terrestrial protoprotein to cellular life have been regarded since the mid-1980s as providing a theory of life's origins. The proteins-first approach to the origin of life is the only one that has yielded protocells from theoretically defensible precursors.

The biomedical applications are already making themselves felt: Selected hydrophobe-rich thermal proteins, especially some containing tryptophan, affect memory retention. Thermal proteins of the same class have, when added topically to cultures of rat forebrain neurons, promoted survival. Neuronal cells in culture were found to age several times more slowly when appropriate thermal proteins were added to the cultures. The use of some thermal proteins as blood expanders looks promising. Thermal proteins were examined for hormonal activity and such activity was found in analogs of MSH prepared by thermal polycondensation of amino acids. The type of biological activity first found in thermal proteins was that of enzymes. Thermal proteins are being commercialized for microencapsulation in drug delivery. Several patents exist in this field and others are in process.

REFERENCES

Berra, T. M. (1990). Evolution and the Myth of Creationism. Stanford University Press, Stanford CA.

Bonner, W. A. (1991). The origin and amplification of biomolecular chirality. Origins of Life 116, 59–111.

Cairns-Smith, A. G. (1982). The Genetic Takeover and Mineral Origins of Life. Cambridge University Press, Cambridge.

Crick, F. (1970). Central dogma of molecular biology. Nature 227, 561–563.

Crick, F. (1981). Life Itself. Simon and Schuster, New York.

Crick, F. H. C., Brenner, S., King, A., & Pieczienick, P. (1976). A speculation on the origin of protein synthesis. Origins of Life 7, 389–397.

De Duve, C. (1991). Blueprint for a Cell: The Nature and Origin of Life. Carolina Biological Supply, Burlington, NC.

Dose, K. & Rauchfuss, H. (1972). On the electrophoretic behavior of polymers of amino acids. In: Molecular Evolution: Prebiological and Biological (Rohlfing, D. L. & Oparin, A. I., Eds.), pp. 199–217. Plenum Press, New York.

Dose, K. & Zaki, L. (1971). Peroxidative and catalase activity of hemoproteinoids. Z. Naturforsch. B 26, 144–148.

Dyson, F. (1985). Origins of Life. Cambridge University Press, Cambridge.

Eigen, M. (1971). Selforganization of matter and the evolution of biological macromolecules. Naturwissenschaften 58, 463–523.

Eigen, M. & Schuster, P. (1972). The Hypercycle. Springer-Verlag, Berlin.

Florkin, M. (1975). Ideas and experiments in the field of prebiological chemical evolution. In. Comprehensive Biochemistry 29B (Florkin, M. & Stotz, E. H., Eds.), pp. 231–260. Elsevier, Amsterdam.

Fox, S. W. (1953). A correlation of observations suggesting a familial mode of molecular evolution as a concomitant of biological evolution. American Naturalist 87, 253–256.

Fox, S. W. (1960a). How did life begin? Science 123, 200–208.

Fox, S. W. (1960b). Self-organizing phenomena and the first life. Yearbook Soc. Gen'l Systems Res. 5, 57–60.

Fox, S. W. (1971). Chemical origins of cells-2: Chem. Eng. News 49 (50), 46–53.

Fox, S. W. (1980). The origins of behavior in macromolecules and protocells. Comp. Biochem. Physiol. 67B, 423–436.

Fox, S. W. (1981a). A model for protocellular coordination of nucleic acid and protein syntheses. In: Science and Scientists (Kageyama, M., Nakamura, K., Oshima, T., Uchida, T., Eds.), pp. 39–45. Japan Scientific Societies Press, Tokyo.

Fox, S. W. (1981b). Creationism, the random hypothesis and experiments. Science 213, 290.

Fox, S. W. (1988). The Emergence of Life. Darwinian Evolution from the Inside. Basic Books, New York.

Fox, S. W. (1992). Thermal proteins in the first life and in the "mind-body" problem. In: Evolution of Information Processing Systems (Haefner, K., Ed.), pp. 203–228. Springer-Verlag, Berlin.

Fox, S. W. & Brooke, S. (1979). Microcapsules in protobiology and industry. In: Microencapsulation (Kondo, T., Ed.), pp. 257–290. Techno, Tokyo.

Fox, S. W. & Dose, K. (1977). Molecular Evolution and the Origin of Life. Marcel Dekker, New York.

Fox, S. W. & Flood, J. F. (1992). Effects of thermal peptides on retention of footshock avoidance training in mice. Peptides 13, 1079–1081.

Fox, S. W. & Veltri, R. W. (1990). Microencapsulated Antitumor Agent. U.S. Patent No. 4,963,364.

Fox, S. W. & Wang, C-T. (1968). Melanocyte-stimulating hormone activity in thermal polymers of alpha-amino acids. Science 160, 547–548.

Fox, S. W. & Windsor, C. R. (1984). Reproducibility of amino acid compositions in repeated copolymerization of amino acids. Int'l J. Quantum Chem. Quantum Biol. Symp. 11, 103–108.

Fox, S. W., Harada, K. & Kendrick, J. (1959). Production of spherules from synthetic proteinoid and hot water. Science 129, 1221–1223.

Grobman, A. (1988). Cover of Fox, S. W. The Emergence of Life. Basic Books, New York.

Hefti, F., Junard, E. O., Knusel, B., Strauss, W. L., Strang, P. F., Przybylski, A., Vaughan, G. & Fox, S. W. (1991). Promotion of neuronal survival in vitro by thermal proteins and poly(dicarboxylic) amino acids. Brain Res. 541, 273–283.

Herrera, A. L. (1942). A new theory of the origin of life. Science 96, 14.

Hsu, L. L. & Fox, S. W. (1976). Interactions between diverse proteinoids and microspheres in simulation of primordial evolution. BioSystems 8, 89–101.

Hsu, L. L., Brooke, S. & Fox, S. W. (1971). Conjugation of proteinoid microspheres: A model of primordial communication. Current Trends in Mod. Biol. 4, 12–25.

Ivanov, O. C. & Foertsch, B. (1986). Universal regularities in protein primary structure: Preference in bonding and periodicity. Origins of Life 17, 35–49.

Jungck, J. R. & Fox, S. W. (1973). Synthesis of oligonucleotides by proteinoid microspheres acting on ATP. Naturwissenschaften 60, 425–427.

Kleinsmith, L. J. & Kish, V. M. (1988). Principles of Cell Biology. Harper and Row, New York.

Lehninger, A. L. (1975). Biochemistry. Worth and Co., New York.

Mader, S. (1990). Biology, 3rd ed. Wm. C. Brown, Dubuque, IA.

Melius, P. & Srisomsap (1991). Nonrandom thermal polymerization of amino acids. J. Appl. Polym. Sci. 42, 1167–1168.

Miller, S. L. (1953). A production of amino acids under possible primitive earth conditions. Science 117, 528–531.

Miller, S. L. & Orgel, L. E. (1974). The Origins of Life on the Earth. Prentice-Hall, Englewood Cliffs, NJ.

Monod, J. (1971). Chance and Necessity (transl. by Wainhouse, A.). A.A. Knopf, New York.

Morris, H. M. (1974). Scientific Creationism. Creation-Life Publishers, San Diego.

Nakashima, T. & Fox, S. W. (1980). Synthesis of peptides from amino acids and ATP with lysine-rich proteinoid. J. Molec. Evolution 15, 161–168.

Neri, P., Antoni, G., Benevenuti, F., Cocola, F., & Gazzei, G. (1973). Synthesis of alpha, beta-poly (2-hydroxyethyl)-DL-aspartamide, a new plasma expander. J. Med. Chem. 16, 893–897.

Oparin, A. I. (1957). The Origin of Life on the Earth. Oliver and Boyd, Edinburgh.

Orgel, L. E. (1973). The Origins of Life. John Wiley & Sons, New York.

Prigogine, I. & Stengers, I. (1984). Order Out of Chaos. Bantam Books, New York.

Przybylski, A. & Fox, S. W. (1986). Electrical phenomena in proteinoid cells. In: (Gutmann, F. & Keyzer, H., Eds.), pp. 377–396. Plenum Press, New York.

Rohlfing, D. L. & Fox, S. W. (1969). Catalytic activities of thermal polyanhydro-alpha-amino acids. Advances in Catalysis 20, 373–418.

Salthe, S. (1989). Review of the Emergence of Life. Quart. Rev. Biol. 64, 343–344.

Santiago, N., Milstein, S., Rivera, T., Garcia, E., Zaidi, T., Hong, H., & Bucher, D. (1993). Oral immunization of rats with proteinoid microspheres encapsulating influenza virus antigens. Pharm. Research 10, 1243–1247.

Schwartz, A. W. & Fox, S. W. (1964). Thermal synthesis of internucleotide phosphodiester linkages. Biochim. Biophys. Acta 87, 694–696.

Steiner, S. & Rosen, R. (1990). Delivery Systems for Pharmacological Agents Encapsulated with Proteinoids. U.S. Patent No. 4, 925, 673.

Tyagi, S. & Ponnamperuma, C. (1990). Nonrandomness in prebiotic peptide synthesis. J. Molec. Evolution 30, 391–399.

Vallery-Radot, P. (1922). Fermentations et Generations Dites Spontanees, Tome II. Masson et Cie, Paris.

Weiner, A. M. & Maizels, N. (1987). t-RNA structures tag the 3′ ends of genomic RNA molecules for replication: Implications for the origin of protein synthesis. Proc. Natl. Acad. Sci. USA 84, 7383–7387.

Weiss, R. (1992 July). End of the Needle. Discover, pp. 64–67.

Young, R. S. (1984). Prebiological evolution: The constructionist model for the origin of life. In: Molecular Evolution and Protobiology (Matsuno, K., Dose, K., Harada, K., & Rohlfing, D. L., Eds.), pp. 45–48. Plenum Press, New York.

RECOMMENDED READINGS

Fox, S. W. (1988). The Emergence of Life. Darwinian Evolution from the Inside. Basic Books, New York.

Fox, S. W. & Dose, K. (1977). Molecular Evolution and the Origin of Life. Marcel Dekker, New York.

Ho, M. W. & Saunders, P. T. (eds.) (1984). Beyond Neo-Darwinism. An Introduction to the New Evolution Paradigm. Academic Press, Harcourt Brace & Jovanovich, Publishers, New York.

Miller, S. L. & Orgel, L. E. (1974). The Origin of Life on the Earth. Prentice-Hall, Englewood Cliffs, NJ.

Oparin, A. I. (1957). The Origin of Life on the Earth. Oliver and Boyd, Edinburgh.

Orgel, L. E. (1973). The Origins of Life. John Wiley & Sons, New York.

Shapiro, R. (1986). Origins, A Skeptic's Guide to the Creation of Life on Earth. Summit Books, New York.

Chapter 4

Mammalian Sex Chromosomes:

Modern Variants Provide Clues to Reconstruct the Evolution of Form and Function

JACLYN M. WATSON and
JENNIFER A. MARSHALL GRAVES

Principles of Medical Biology, Volume 1B
Evolutionary Biology, pages 87–114

ISBN: 1-55938-802-1

INTRODUCTION

Vertebrate Sex Chromosomes

Mammalian sex chromosomes systematically differentiate the sexes. The X and Y chromosomes are largely unpaired at male meiosis and are readily identified by comparing the mitotic chromosome complements of male and female individuals. The X chromosome is large and bears thousands of genes. In females, one X chromosome is genetically silenced by a unique chromosome-wide repression of gene expression, termed X chromosome inactivation (XCI), which equalizes the dosage of X-linked gene products between the sexes. The Y chromosome is small and bears few active genes other than the testis-determining factor (TDF), which determines male sexual development.

Higher vertebrates exhibit a wide variety of sex chromosome mechanisms including the mammalian XX female/XY male system. There are species of birds, reptiles, and fish which have female/male XX/XY, XX/XO, ZW/ZZ, and OW/OO systems in which either male or female is the heterogametic sex, as well as species with indistinguishable sex chromosomes, or sex determination resulting from environment stimuli, such as temperature (Bull, 1983).

Theories of Sex Chromosome Evolution in Vertebrates

The sex chromosomes of an ancestral vertebrate were almost certainly genetically equivalent (Ohno, 1967). Therefore, the highly dimorphic sex chromosomes characteristic of modern mammals and many other vertebrate groups must have evolved from an homologous chromosome pair. The origin of the differences between the X and Y chromosomes of mammals is usually attributed to a cluster of key sex-determining genes located on the Y, but not the X. The integrity of the Y-linked cluster is protected from recombination with the X in the XY male by a chromosome rearrangement that suppresses recombination and isolates the Y chromosome. In birds, which have a ZZ male/ZW female system, the origins of ZW differentiation are attributed to the protection of a cluster of sex-determining genes on the W, but not the Z, from recombination with the Z in the female. As a result of genetic isolation, mutations leading to the alteration or absence of gene product could accumulate in the region of the sex determining gene(s). Then region(s) containing dysfunctional genes could become genetically inactive and accumulate "junk" DNA, or be lost by deletion.

The loss of Y-linked alleles is thought to have selected for the evolution of an X chromosome inactivation system that compensates for the difference in the dosage of X-linked genes between the sexes in mammals. However, in birds and reptiles, the loss of W-linked alleles has not been accompanied by a Z inactivation system, and the dosage of Z-linked genes is uncompensated between males and females (Chandra, 1991; Chandra and Nanjundiah, 1992).

A series of intermediate steps in the differentiation of sex chromosomes is apparent in the sex chromosomes in various snake families (Becak et al., 1964; Ohno, 1967). In one family the Z and W appear to be identical. However, in many members of another closely related family the W and Z are differentiated by a structural rearrangement which alters the position of the centromere. In another, more highly evolved family, the differences between the sex chromosomes are extreme, the W being greatly reduced in size.

Although the process of differentiation of dimorphic sex chromosomes in birds, reptiles, and mammals is superficially similar, the apparently concurrent evolution of a mechanism of dosage compensation for X-linked genes in mammals, but not Z-linked genes in birds, suggests that the two processes did differ significantly. This was a point of controversy with some authors supporting the basic similarity between the two processes (e.g., Ohno, 1967; Jones and Singh, 1985) and others proposing a sequence of events applicable only to mammals (e.g., Lyon, 1974a,b; Lucchesi, 1978). More recently, gene mapping has made it obvious that there is no homology between the mammalian X and the reptilian and avian Z chromosomes. The few Z-linked genes known in birds are autosomal in mammals and X-linked mammalian genes are autosomal in birds. Thus, it is unlikely that the X/Y and Z/W sex chromosome systems had a common origin, and more likely that different vertebrate groups have independently evolved sex chromosomes, probably from different autosomal pairs in a common ancestor (Graves and Schmidt, 1992).

Mammalian Sex Chromosomes

Eutherian ("placental") mammals have an XX female/XY male chromosomal sex-determining system, in which the X and Y are morphologically and genetically very different, sharing only a tiny "pseudoautosomal" region (PAR) which pairs at meiosis. The X chromosome constitutes about 5% of the haploid genome (some thousands of genes), and is completely conserved; until very recently there was not a single exception to "Ohno's Law" that genes found on the X in one eutherian species will be X linked in all other species (Ohno, 1967; O'Brien and Graves 1991; Disteche et al., 1992).

X chromosome inactivation (XCI) is a chromosome-wide inactivation mechanism, although the PAR and some X-Y shared genes escape inactivation on the human (but not the mouse) X (Ballabio and Willard, 1992). The involvement of the X in this unique, chromosome-wide inactivation may account for the conservation of the X by protecting the X chromosome from rearrangements (Ohno, 1967). In females, one or other X chromosome becomes genetically inactive, late-replicating, and cytologically heterochromatic in each somatic cell early in embryogenesis, and XCI is very stable and somatically heritable. Inactivation is suggested to emanate from an X-controlling center (XCE), and a candidate gene, *XIST*, has recently been proposed to fulfil this function (Brown and Willard, 1990; Brockdorff et al., 1991).

The Y chromosome is much smaller and heterochromatic, containing only a few identifiable genes. One of these determines testis formation, the first step in the pathway of male sexual development. A candidate sex-determining gene, *SRY*, has recently been cloned from the human Y chromosome (Sinclair et al., 1990).

Marsupial mammals also possess highly differentiated X and Y chromsomes. The sex chromosomes do not pair at meiosis and apparently share no region of homology (Sharp, 1982). The marsupial X chromosome is somewhat smaller than the eutherian X, constituting about 3% of a similar-sized haploid genome, and bears a number of genes which are also X-linked in the Eutheria (Dawson and Graves, 1986; Dobrovic and Graves, 1986). The regulation of expression of X-linked genes in marsupials is also controlled by XCI. However, the system of XCI is fundamentally different from that in eutherians, being paternal, incomplete, and tissue-specific; such a system may be ancestral to both groups of therian mammals (Cooper, 1971; discussed by Graves, 1983). The marsupial Y chromosome is very tiny and also bears very few identifiable genes, including the marsupial homologue of *SRY* (Foster et al., 1992).

In monotremes, comparisons of the mitotic chromosomes of male and female individuals suggest that the sexes are differentiated basically by XX female/XY male (Platypus), or a simple derivative of this in the echidna species (Murtagh, 1977; Wrigley and Graves, 1988a). The monotreme X and Y chromosomes share a large region of homology, pairing extensively at the first metaphase division of male meiosis (Murtagh, 1977; Watson, 1990). The platypus X and echidna X_1 chromosomes are cytologically and genetically equivalent (Watson et al., 1992). They are slightly larger (approximately 6% of the haploid genome) than the basic eutherian X, and much larger than the marsupial X chromosome.

Cytological studies (DNA replication asynchrony) suggest that XCI may regulate the expression of X-linked monotreme genes (Murtagh, 1977; Wrigley and Graves, 1988b). On the basis of its presence in males and absence in females, the Y chromosome is presumed to be male or at least testis-determining. However, it has not been possible to directly demonstrate the role of the Y in sex determination.

Significance of Studies of the Mammalian Sex Chromosomes

Sex chromosomes provide unique examples of chromosome evolution and function, in sex determination and regulation of activity. The gene which determines sex in humans and other mammals has been actively sought for decades because of its theoretical and practical importance. An understanding of the initiation of this major developmental pathway in mammals has clinical applications in diagnosing and possibly treating abnormalities of human sexual development, as well as the potential to facilitate our handling of animal reproduction. Furthermore, the variety of chromosomal mechanisms of sex determination among vertebrates suggests that sex chromosomes may have evolved independently many times, and may have provided an isolating mechanism in speciation.

X chromosome inactivation involves the stable, somatically heritable, genetic silencing of an entire X chromosome in females, early in embryogenesis. It is a major example of a coordinated, multistep genetic regulatory system, and has both clinical and theoretical implications. A thorough understanding of the mechanism of X chromosome inactivation will assist in the diagnosis and perhaps the treatment of some X-linked diseases, and could provide insights into how the expression of genes on different chromosomes are coordinately expressed or repressed in different tissues or at different developmental stages.

Finally, the analysis of modern variants of the mammalian XX female / XY male system of sex determination has extended our knowledge of chromosome evolution by providing insights into the process of X/Y differentiation. Thus, an understanding of the functioning of the different mechanisms may explain the divergence of vertebrate groups in terms of reproductive isolation by differences in the determination of sex.

The Comparative Approach to Mammalian Genetics

The comparative approach to mammalian genetic analysis takes advantage of the differences in genetic form and function which have evolved between groups of mammals in order to identify common features which must have preexisted in a species ancestral to the groups.

There are three groups of modern mammals. The *Eutheria* (so-called placental mammals) and the *Metatheria* (marsupials) diverged approximately 130 to 150 million years ago (Hope et al., 1990). The *Prototheria* (which are represented only by the order *Monotremata*) diverged from the eutherian–metatherian line of descent even earlier in the 200-million-year-old history of the mammals (Kemp, 1982).

The marsupials are represented by approximately 240 species belonging to three American and 13 Australian families (Kirsh and Calaby, 1977). Individual species exhibit a wide range of body size and form, and, in Australia, occupy most ecological niches. The young are born in a very immature state after a short period of gestation. Development is completed outside the mother's body, attached to a teat. Frequently, but not invariably, a pouch protects the developing offspring.

Monotremes are represented by only three species, the platypus (*Ornithorhynchus anatinus*) and two echidna species, *Tachyglossus aculeatus* and *Zaglossus bruijnei*. They are highly specialized, found only in the Australasian region and are generally considered to belong to the non-therian mammalian subclass, *Prototheria* (Hopson, 1970; for a dissenting view Kemp, 1983). The monotremes lay eggs and the immature hatchlings suckle milk from the mother's teatless mammary area.

Comparisons between eutherian groups permit conclusions to be drawn about the genomic arrangements and mechanisms of gene regulation in an ancestral eutherian mammal. Similarly, the genetic characteristics of ancestral marsupial and monotreme mammals can be deduced. Then, comparisons between more distantly related groups may be made in order to provide information on more ancient evolutionary events. Eutherian–marsupial comparisons will enable the characterization of the "therian" ancestor, which can then be compared with the prototherian monotremes to gain an understanding of genomic organization and function in a common mammalian ancestor 200 million years ago. By tracing the changes which have taken place in the different lineages, it may be possible to correlate changes in gene arrangements with functional changes, and to gain an understanding of the role of newly acquired mechanisms of gene regulation, both as they affect the function of the genome, and as they relate to further genomic rearrangements. Thus, comparative studies can identify more primitive modes of genetic control.

The analysis of the simpler mechanism may then facilitate our understanding of how the more complex system actually operates, and how, and perhaps why, additional modifications have evolved.

Methods of Comparative Gene Mapping

Classical methods of genetic mapping were limited to humans, and species which bred rapidly and prolifically in captivity, and to genetic markers whose expression was readily detectable and which could be assigned.

Genetic markers consisted of genes with polymorphic protein products (isozymes) which could be separated by gel electrophoresis and detected by linking the reactions catalyzed on the gel with visible change. More recently, the use of cloned gene sequences has vastly increased the number of genetic markers which can be assigned to chromosomes by classical methods. New cell and molecular techniques such as somatic cell hybridization, Southern hybridization, and *in situ* hybridization have made it possible to map gene sequences in almost any species.

Classical methods of gene mapping relied on the analysis of recombination rates between genes or unique DNA sequences (genetic markers) on the same chromosome to establish gene order and relative distance apart. Gene assignments were largely limited to the X chromosome; X-linkage being established by the differential inheritance of maternal and paternal alleles by male and female offspring.

Somatic cell genetic techniques revolutionized gene mapping by isolating individual chromosomes, or small numbers of chromosomes, of the species to be mapped. Interspecific cell hybrids were formed by fusing the somatic cells of two different species. Such hybrids preferentially segregate almost all of the chromosomes of one or the other species. Thus, genes could be assigned to chromosomes by correlating segregation of gene markers and chromosomes from different cell hybrids. Similarly, genes borne on the same chromosome could be identified by their cosegregation in cell hybrids ("synteny").

The application of recombinant DNA techniques has provided an almost inexhaustible source of new markers for classic genetic analysis as well as somatic cell genetic analysis. This is because variation in DNA base sequence is more frequent than amino acid variation in proteins, especially since most mammalian DNA is noncoding. Base sequence variants can be detected by Southern blot analysis if they affect sites recognized and cut by families of restriction enzymes. Recognition sequences may consist of anything from 4 bases (4 base cutters cut the DNA often and produce many smaller fragments) to 10 bases (rare cutters cut rarely and produce few larger fragments). DNA fragments of different sizes can be separated on agarose gels and transferred to nylon membranes (Southern blotting). DNA fragments which contain specific gene sequences or other sequences of interest can be detected by hybridizing the radioactivity-labeled, cloned sequence to the membrane. The probe molecules hybridize specifically to homologous or near-homologous sequences. Thus the DNA fragments containing these sequences can then be detected by autoradiography. Since the equivalent genes from different species are likely to differ in the position of restriction enzyme cut sites, analysis of fragment size on a Southern blot can distinguish between species origin of genes in a cell hybrid.

Other DNA sequence variants, particularly in the numbers of short repeated sequences, can be detected by polymerase chain reaction (PCR) in which short segments are replicated over and over ("amplified") from fixed ends. PCR has greatly facilitated a number of analyses, including the cloning of gene homologues in other species of gene sequences and site-specific DNA modifications (Pfeifer et al., 1990; Singer-Sam et al., 1990).

The availability of cloned DNA sequences has also provided new ways to localize genes or other DNA sequences by physical methods. *In situ* hybridization of labeled probe DNA to fixed denatured metaphase chromosomes, followed by autoradiography permits the direct visualization of the site of the gene (or other DNA sequence) on the chromosomes (Harper et al., 1981). Alternatively, metaphase chromosomes can be sorted by size using a fluorescence-activated chromosome sorter (FACS). Chromosomes are labeled with fluorescent dye which binds to the DNA, thus larger chromosomes take up more dye than smaller chromosomes. The FACS detects the fluorescent signal of each chromosome individually and channels them to different outlets according to the signal. Sorted chromosomes are identified microscopically and then analyzed by Southern blotting. DNA from sorted chromosomes can also be used as a probe under conditions of suppression hybridization, which eliminates repetitive DNA sequences. This new technique of chromosome painting promises rapid detection of large regions of homology between species (Weinberg et al., 1990).

Classic linkage analysis complements these methods of physical mapping. Data from family studies, somatic cell genetic analysis, and *in situ* hybridization have been integrated to construct detailed maps of several eutherian mammals, principally man, mouse and, increasingly, domestic mammals (reviewed by O'Brien and Graves, 1991).

Gene mapping in marsupials and monotremes has made use of the same techniques. It made slow progress initially, because of problems of breeding, and the difficulties of working with very unstable interspecies cell hybrids (Dawson and Graves, 1984; Watson and Graves, 1988). However, the development of molecular approaches to gene mapping in non-eutherian species (*in situ* hybridization, Southern blot analysis of hybrids, PCR cloning techniques) have permitted many gene assignments in the last few years (Graves et al., 1993).

Gene mapping in non-eutherian mammalian species is made simpler because the karyotype is highly conserved in both groups. Marsupial chromosomes have been well studied, and conserved 2n = 14 basic karyotype deduced (Rofe and Hayman, 1985). Similarly, the chromosomes of the three extant monotreme species have been shown to be highly conserved (Wrigley and Graves, 1988a).

Hybrid cell lines have been constructed which retain complete X chromosomes from a number of marsupial species, and revertants which have lost the X are available. Thus, genes can readily be assigned to or excluded from the X chromosome by the expression of marsupial form of an enzyme, or by Southern blot detection of a marsupial-specific DNA fragment. Their specific localization may be confirmed by *in situ* hybridization. Interspecific monotreme hybrids obtained contain either no recognizable monotreme chromosomes, or a part of chromosome 1 (Watson and Graves, 1988). Thus, the method most used for gene assignment in the monotremes has been *in situ* hybridization.

In situ hybridization using nonhomologous DNA probes is an exacting technique, requiring extreme care in the stringency conditions of hybridization and washing. The advantage of using nonhomologous probes is that it is possible to localize a relatively large number of genes to chromosomes quite quickly and, provided the gene sequence is moderately conserved (assessed by Southern analysis of "zoo blots"), low background and the presence of statistically significant "signal" provides a measure of confidence in the gene assignment. The statistical test used involves the calculation of the Zmax statistic which detects significantly overlabeled chromosomes, chromosome arms, or chromosome regions (Ewens et al., 1992).

Marsupial chromosomes are ideal for *in situ* hybridization, being large, few in number, and readily identified in unbanded chromosome preparations. This is not true of monotreme chromosomes. In unbanded mitotic preparations, only chromosomes 1–6, and the X chromosomes are readily identifiable. Banding techniques do not improve the situation, as many of the smaller chromosomes show virtually identical banding patterns. Thus, unbanded chromosomes are used for *in situ* hybridization and the distribution of silver grains is recorded on chromosomes 1–6 and the X; the other chromosomes are divided into two size classes, consisting of just over 50% of the genome, and the silver grains observed over these chromosomes are pooled.

The use of these methods for gene mapping in marsupials and monotremes have permitted us to localize a large number of genes which are X- and/or Y-linked in eutherian species. The surprising results of these still very incomplete gene mapping studies have enabled us to formulate hypotheses concerning the evolution of organization and function of the mammalian sex chromosomes.

THE X CHROMOSOME

Differences in X chromosome size, gene content, and inactivation between the three major groups of mammals have enabled us to deduce how the X and Y became different in evolution, and how the X of modern mammals functions.

Gene Content of the X Chromosome

Many genes mapping to the eutherian X chromosome (described in Table 1, XCR) have been found to be located also on the X chromosome in marsupials (Dawson and Graves, 1986; Dobrovic and Graves, 1986; Spencer et al., 1991a) and in monotremes (Watson, 1990; Watson et al., 1990; Spencer et al., 1991b; Watson

Table 1. Loci Located in the Conserved Suite of Mammalian X-Linked Genes

Gene Symbol[a]	*Marker Name*	*Location of Human Homologue*
AMG	Amelogenin (amelogenesis imperfecta 1)	Xp22.31-æp22.1
POLA	Polymerase (DNA directed) alpha	Xp22.1-p21.3
ZFX	Human X-linked homologue of ZFY	Xp22.3-p21.2
ZFY	Zinc finger protein	Ypter-p11.2
CYBB	Cytochrome beta-245, beta chain (chronic granulomatous disease)	Xp21
MAOA	Monoamine oxidase A	Xp21-p11
DMD	Dystrophy (Duchenne muscular dystrophy)	Xp21.2
OTC	Ornithine transcarbamylase	Xp21.1
SYN1	Synapsin 1	Xp11.2
TIMP	Tissue inhibitor of metalloproteinase	Xp11.3-p 11.23
	XAR (↑) XCR (↓) fusion point	
ARAF1	Murine sarcoma 3611 viral (v-raf) oncogene homologue1	Xp11.4- cen
GF1	Globin transcription factor 1	Xp11.23
UBE1	Ubiquitin activating enzyme E1	Xp11.23
ALAS	Aminolevulinate, delta-, synthase 2	Xp11.23
	Centromere	
AR	Androgen receptor (Testicular feminization syndrome)	Xq11.2-q12
PGK1	Phosphoglycerate kinase 1 (A)	Xq13-q21.1
GLA	Galactosidase alpha	Xq21.33
PLP	Proteolipid protein (Pelizaeus-Merzbachen Disease)	Xq21.4-q22
HPRT	Hypoxanthine guanine phosphoribosyl transferase (Lesch Nyhan Syndrome)	Xq26
F9	Blood coagulation factor IX (hemophilia B)	Xq26.3-q27.2
F8c	Blood coagulation factor VIIIc (hemophilia A)	Xq28
RCP	Red cone pigment (color blindness)	Xq28
G6PD	Glucose-6-phosphate dehydrogenase	Xq28
GDX	Unknown gene	Xq28
P3	Unknown gene	Xq28

Note: [a]Gene symbols follow nomenclature for the homologous human locus as proposed by the Human Gene Nomenclature Committee (Mc Alpine et al., 1989) and the Comparative Gene Mapping Committee (Lalley et al., 1989).

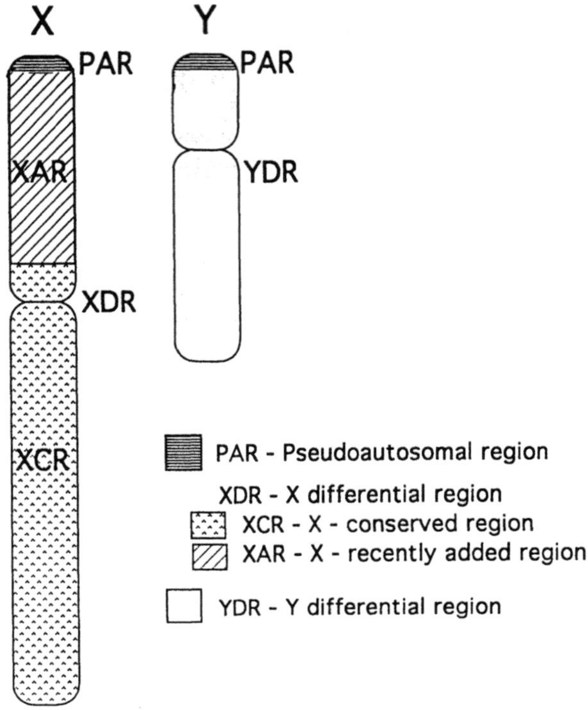

Figure 1. Human sex chromosomes.

et al., 1992). The observation that the X chromosomes of all three extant mammalian taxa share a common suite of genes suggests that this region has been conserved on the X chromosome for at least the 150 to 200 million years since these groups diverged. The location of this conserved region of the X chromosome (XCR) on the human X is indicated in Figure 1, and Figure 2a shows the location of the genes in human, tammar wallaby, and platypus. This conservation can no longer be ascribed to the inactivation of the X chromosome which protects it from rearrangement, because, in monotremes, some of the genes forming the conserved suite are located within the region of X–Y homology which presumably is not inactivated. Thus, in part at least, the conservation of the X chromosome may simply be a good example of the conservation of the mammalian genome as a whole (see comparative data listed by O'Brien and Graves, 1991).

The comparable sizes of the eutherian and monotreme X chromosomes initially suggested that the monotreme X chromosome may bear the full suite of eutherian X-linked genes, as one would expect if a large ancestral X had been conserved between eutherians and monotremes but disrupted in marsupials (Graves, 1987). However a number of genes (identified in Table 1, XAR) located on the X

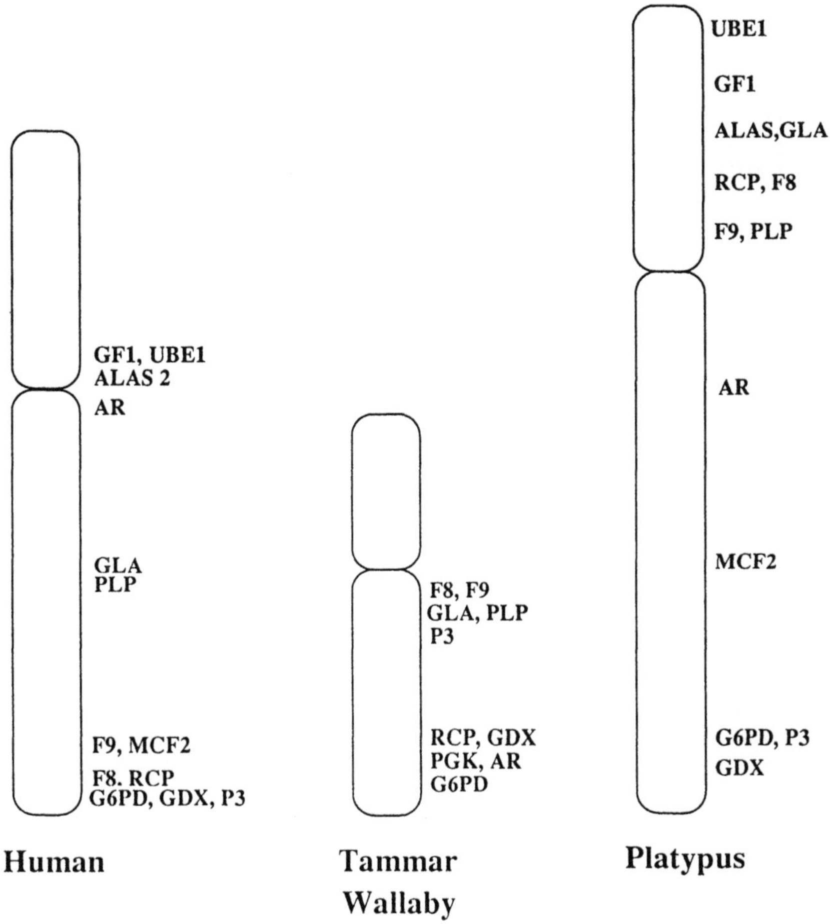

Figure 2a. Locations of gene loci conserved on the X chromosome in all mammalian lineages in humans, the Tammar Wallaby (a marsupial) and Platypus (a monotreme).

chromosome in all eutherian species tested and on the short arm of the human X chromosome, have been mapped to autosomes in both the marsupials and monotremes, where they form two distinct and very similar autosomal clusters (Figure 2b) (Sinclair et al., 1987; Sinclair et al., 1988; Spencer et al., 1991c; Watson et al., 1992a,b; reviewed by Graves and Watson, 1991). Since the marsupials and monotremes diverged independently from eutherians, this arrangement was probably present in the common mammalian ancestor. This means that the region was recruited to the X relatively recently in the eutherian lineage—about 150 million years ago (the time of the major eutherian radiations). The location of this recently added XAR region in humans is shown in Figure 1.

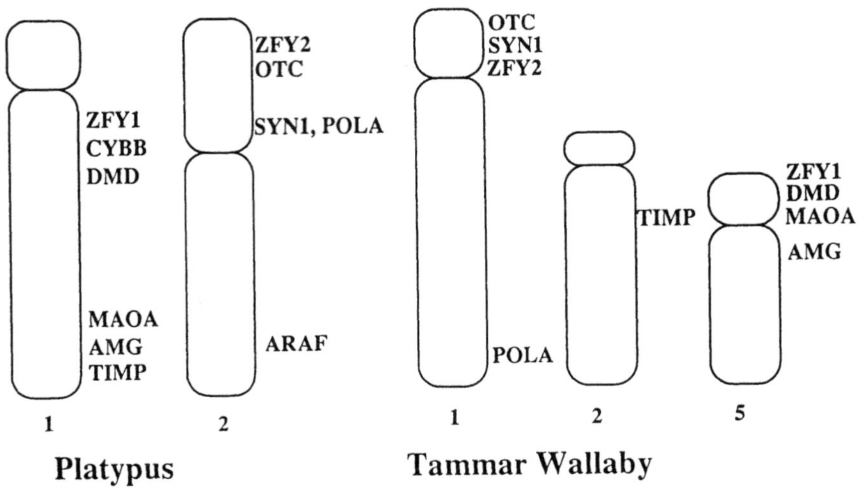

Figure 2b. Locations of genes in the XAR in Platypus and Tammar Wallaby.

The point of fusion of the original X conserved region (XCR) and recently added region (XAR) has been further refined by mapping genes near the centromere of the human X chromosome, in marsupials, and monotremes. Of these genes on the short arm of the human X, only three nearest the centromere (*ALAS*, *GF1*, and *UBE1*) form part of the conserved suite common to the X chromosome of all mammals. These results define an evolutionary fusion point at human Xp 11.33 between the ancestral region of the eutherian X chromosome and the newly acquired region (Figure 3).

There are some exceptional genes whose positions in the three mammalian groups cannot be explained by the hypothesis above. At least four genes (*HPRT*, *PGK*, *CYBB*, and *ARAF1* identified in Table 1) which are X-linked in both marsupial and eutherian mammals are autosomal in monotremes (VandeBerg and Cooper, 1978; Watson and Graves, 1989; Spencer et al., 1991c; Watson, unpublished data; Wilcox et al., in preparation).

CYBB is located near to *ZFY* and *DMD* in the platypus and apparently forms part of one of the clusters recently acquired by the eutherian X chromosome as the XAR (Figure 2). Therefore, it is probable that *CYBB* was originally autosomal and was translocated to the therian X chromosome after the prototherian divergence but before the eutherian—marsupial divergence. *ARAF* is on chromosome 2 which bears the other XAR region, but is distally located and does not appear to form part of the region. *HPRT* and *PGK* are located together on one of the unidentified Class A autosomes. *ARAF1*, *PGK*, and *HPRT* may have been part of the ancestral X chromosome, but were translocated to autosomes in the prototherian line of descent. Alternatively, they may have been originally autosomal, becoming part of the ancestral therian X chromosome by translocation(s) after the prototherian divergence.

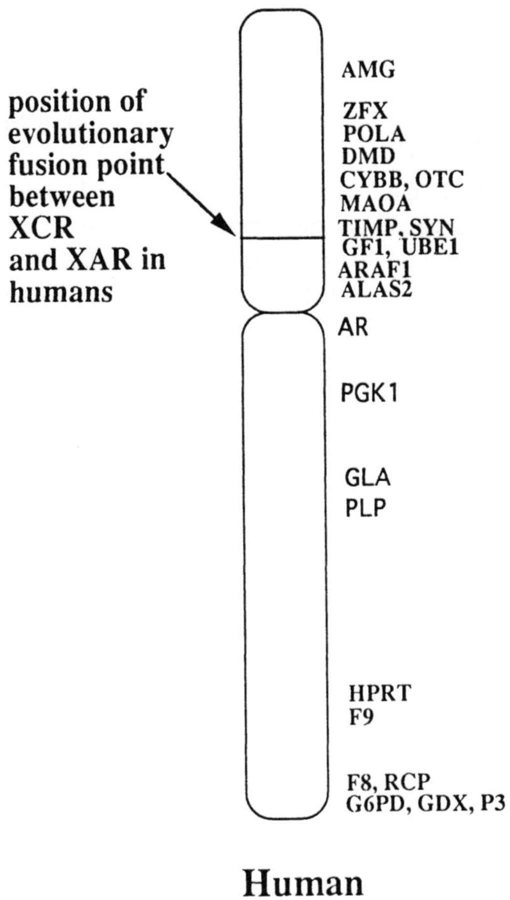

Figure 3. Locations of genes in the XCR and XAR in the human X chromosome.

A number of genes have been recently shown to be shared by the X and Y chromosomes in man, mouse, or both. These include genes in the pseudoautosomal region of the eutherian sex chromosomes, within which recombination occurs at meiosis, as well as a number of other gene loci distant from the pairing region. By comparing the location of these genes in other eutherian species, as well as marsupials and monotremes, we can deduce whether these genes were part of an ancestral pseudoautosomal region (PAR) or were part of the XAR. If these loci were part of an ancestral PAR, the genes should also be X- and Y-linked in marsupials and monotremes, whereas if they were part of the XAR they should map in the autosomal clusters.

The gene, *UBE1*, has related sequences on the mouse X and Y chromosomes but only an X-linked locus in humans. *UBE1* has been localized to the X and Y chromosomes in representative species of both macropodid and dasyurid marsupial species, and to the region of the X chromosome which is paired by the Y at meiosis, in the monotremes. Thus, *UBE1* was probably part of an X–Y-shared region in a common, ancestor to all three modern mammal groups 170 to 200 million years ago (Mitchell et al., 1992; Wilcox et al., in preparation).

Genes shared by the X and Y chromosomes of man and mouse include *ZFX/Y* and *AMG/AMGL*. In marsupials and monotremes, these genes are located in one of the two autosomal clusters of genes; so were part of the XAR translocated to the eutherian sex chromosomes after the marsupial divergence. The presence of these X–Y-shared genes indicates that the XAR was present on both the X and Y chromosomes. It was probably translocated to an ancient paired region of the proto X or Y, and subsequently transferred to the other by recombination.

The sequence divergence of X- and Y-linked gene homologues may precede the incorporation of the X-linked homologue into the system of XCI or conversely may be a consequence of X inactivation: This question may be analyzed by comparing, in different eutherian species, the locations of the alleles on the sex chromosomes, the functional homology of the alleles, and the activity status of the X-linked allele on the two X chromosomes in females. Comparisons of a number of genes shared by the sex chromosomes in man and mouse are outlined in Table 2. These data

Table 2. X–Y Shared Genes in Man, Mouse, and Marsupials

Gene Symbol	*Locus*	*Species*	*Gene Position*	*Y Chromosome Homologue*	*X Chromosome Inactivation*
STS	Steroid sulfatase	Human	Near PAR	Pseudogene	Partial
		Mouse	In PAR	Functionally equivalent	No (partial)
		Marsupial	Autosomal	—	—
ZFX/ZFY	Zinc finger protein	Human	Xp	Functionally equivalent	Not inactivated
		Mouse	X	Specialized tissue distribution	Inactive
		Marsupial	Autosomal	—	—
AMG	Amelogenin	Human	Near PAR	Not functionally equivalent	Possibly inactivated
		Mouse	X	None	Unknown
		Marsupial	Autosomal	—	—
UBE1	Ubiquitin activating enzyme	Human	Xp11.23	None	Not inactivated
		Mouse	X	Specialized tissue distribution	Inactivated
		Marsupial	X	Homologue	Unknown
RPS4	Ribosomal protein 4, small subunit	Human	Xq, near XIC	Expressed homologue	Not inactivated
		Mouse	X	None	Inactivated

suggest that inactivation of the X-linked allele follows the loss of functional homology of the Y-linked allele.

Comparisons of base sequence divergence of the X- and Y-linked alleles indicate that in humans the *AMG* alleles have been diverging for 25 million years (Watson et al., 1992d), while *ZFX/Y* and *RPS4X/Y* escaped recombination 100 million years ago (Fisher et al., 1991). These data suggest that, in human, *AMG* was part of the PAR until quite recently, while *ZFX/Y* and *RPS4X/Y* were genetically isolated before the major eutherian radiation 80 million years ago. In different eutherian lineages the fates of the X- and Y-linked alleles have been very different. In human, the X- and Y-linked alleles of *ZFX/Y* and *RPS4/Y* remain functionally homologous, whereas *AMG* and *AMGL* became differentiated. In the mouse there is no Y-linked homologue of either *AMG* or *RPS4*, and the X- and Y-linked homologues of the *ZFX/ZFY* are functionally distinct.

X Chromosome Inactivation: Dosage Compensation or Sex-Determining Function?

Males have only a single X chromosome, whereas females have two. However, only a single X chromosome in the somatic cells of female mammals is genetically active, thereby equalizing the dosage of X-linked gene products between the sexes. The inactivation of the paternally or the maternally derived X occurs at random in the early eutherian embryo. Inactivation is associated with late DNA replication, methylation of specific cytidine bases in the controlling regions of genes, and chromatin condensation (heterochromatinization), which makes the inactive X chromosome visible as dense sex chromatin mass, the Barr body, in the interphase nucleus.

Human X chromosome inactivation is accomplished by at least two independent mechanisms (DNA methylation and heterochromatinization) which can be dissected by reactivation studies with cell hybrids (Graves, 1983; Gartler et al., 1985). A third mechanism (late DNA replication) may be a necessary first step. The inactive X is late replicating in marsupials (Graves, 1967), and there is evidence for DNA replication asynchrony between the monotreme X chromosomes in some tissues (Murtagh, 1977; Wrigley and Graves, 1988b). However, while there is no convincing evidence that heterochromatinization is involved in inactivation in the platypus, or in adult tissues of the brushtailed opossum (McKay et al., 1987), sex chromatin bodies have been demonstrated in embryos of two distantly related marsupial species (Johnston and Robinson, 1987). DNA methylation is apparently not involved in XCI in the American opossum, since no differences in methylation in the 5′ region of G6PD were detected between the active and inactive alleles (Kaslow and Migeon, 1987). However, methylation differences were noted in the coding and 3′ regions of PGK in the red kangaroo and the hill kangaroo (Van Daal, 1986; Piper et al., 1993). These may be equivalent to internal methylation differences in human and mouse genes which do not correlate to activity. XCI is less stable in marsupial cells (Migeon et al., 1989) and may result from a simpler system which may represent a more flexible ancestral XCI.

The classic observation that in man and mouse XO individuals are females (with Turner's Syndrome) or XXY are male (Klinefelters) implies that the Y is male-determining, and that the number of X chromosomes plays no direct role in sex determination. However, it has been proposed that XCI inactivation may have evolved, not as a dosage compensation mechanism, but as a vital component of the sex-determining system (Chandra, 1985). All systems of dosage differences and dosage compensation known in reptiles and invertebrates are involved in sex determination (Chandra and Nandjundiah, 1992), and, in birds, gene dosage effects on sex have been demonstrated by the intersex phenotype of ZZW triploid chickens (Thorne and Sheldon, 1992).

The finding that a now-discredited candidate sex-determining gene, *ZFY*, had a related sequence on the X inspired speculation that a sex difference in dosage, effected by repression of gene activity on the inactive X in females, might control some aspects of sexual phenotype. The involvement of *ZFX* or *ZFY* in sex determination had to be abandoned when the marsupial and monotreme homologues of the zinc finger protein gene were found to be autosomal (Sinclair et al., 1988; Watson et al., 1992a). However, a recent report of several cases of human XY females with a normal Y, but a noninactivated duplication of the region of the short arm of the X-containing *ZFX* (Xq 21.3-p22.13), reawakened interest in a possible role for that region of the X chromosome in sex reversal. However, *ZFX* was excluded as the gene responsible by the observation of a male patient with an active duplicated *ZFX* region (Coles et al., 1992). This leaves a region at Xp 21.3 that must contain a gene(s) subject to inactivation and not present on the Y, which, in a double dose, either overrides the effect of TDF, or possibly acts via the ratio of active copies to autosomal sets to override the male-determining effect of the Y. This latter possibility is suggested by the female phenotype of a rare XXXXY triploid fetus (Maaswinkel Mooij et al., 1992). A similar example may be the sex reversal found in some lemming species, in which fertile XY females have a modified X with a dominant suppressor effect over the Y. The modified X contains a variant of the *Zfx/Zfy* gene (Lau et al., 1992) and perhaps other genes located near *ZFX* that are involved in human sex reversal.

Chandra originally suggested that sex determination and differentiation may be effected by the number of active copies of an X–Y-shared sex-determining gene, controlled by XCI. Thus, males would have two active copies and females only one (Chandra and Nandjundiah, 1992). That the dosage of X–Y-shared genes do have effects on embryo growth and sexual differentiation is seen by the effects on viability and growth rates in XO humans (Burgoyne, 1992). These growth and fertility differences are less severe in the mouse, perhaps because of the more complete inactivation of the X chromosome (Ashworth et al., 1991).

Although in marsupials, as in humans and mouse, the Y chromosome is testis-determining; the number of X chromosomes controls other aspects of sexual differentiation. In marsupials, determination of scrotum or mammary glands depends on the number of X chromosomes and is independent of the presence of the Y (Graves and Short, 1990; Sharman et al., 1990). Thus the dosage of an X-borne gene may be involved in the determination of scrotum or mammary gland. This gene could lie on the X only and be exempt from XCI in order to establish a 2:1 female to male dosage difference. Alternatively, this gene could be imprinted by its passage through the male gene line, so that the allele on the paternal X then determines the pouch. In either case, the marsupial system of XCI should not be considered an imperfect version of the eutherian system, but as a highly regulated system determining sexual dimorphisms (Cooper et al., 1993).

These gene dosage effects on sex determination and differentiation suggest that dosage differences were originally involved in mammalian sex determination, as it is in other vertebrates and insects.

X chromosome inactivation is apparently a chromosome-wide phenomenon controlled by spreading from an X inactivation center (XIC). The extent of spread from this center has been proposed to be controlled in marsupials, being more extensive in some tissues than in others (Graves and Dawson, 1988). There is much evidence that inactivity spreads from the mouse X into translocated autosomal regions and some suggestion of an age-dependent loss of control of spreading in mice (Wareham et al., 1987) . The XIC function has been mapped to the proximal region of the short arm of the human X (Brown et al., 1991a), and a candidate gene *XIST* has been isolated which is expressed from the inactive but not the active human X (Brown et al., 1991b; Brockdorff et al., 1991). The onset of expression of the mouse homologue *Xist* in embryogenesis correlates with the timing of X-inactivation (Kay et al., 1993). Perhaps the *XIST* locus, by controlling the spread of inactivation, and therefore, gene dosage of X-linked genes, may also have a role in sexual differentiation in marsupials.

Evolution of the Mammalian X Chromosome

All three extant groups of mammals share a highly conserved region of the X chromosome, amounting to approximately 3% of the haploid genome. The XCR region is therefore most likely to have been part of the X chromosome in a common ancestor, and to have been maintained on the X in all lineages, even though at least part of it is not protected by X–Y differentiation and X inactivation in monotremes. It is likely that at least part of this ancestral X chromosome was paired with the Y, since this arrangement is still observed in the monotremes today.

It is also possible that the ancestral X and Y chromosomes included other regions which were subsequently translocated to autosomes; for instance, large regions of the monotreme X, to which no gene markers have yet been assigned, may represent a part of the ancestral chromosome which was lost early in the therian line. Alternatively, it may represent an ancestral autosomal region translocated to the sex chromosomes in the monotreme lineage.

The common origin of mammals, reptiles, and birds, and the complete lack of homology between the mammalian X and Y and the bird Z and W sex chromosomes requires that the X and Y are derived from a homomorphic chromosome pair still represented by autosomes in modern birds. We suggest that the sex chromosomes of the mammalian ancestor contained the conserved region XCR represented on the X chromosome of the three extant groups. The X and Y of ancestral mammals were still genetically similar over a good deal of their length, as they are in modern monotremes. The observation that at least part of the conserved region is paired with the Y in monotremes suggests that the differentiation of the X and Y chromosomes occurred in a stepwise manner, rather than as the result of a single major rearrangement. A stepwise deletion and heterochromatinization of the Y chromosome appears to have progressed further in the therian mammals than in the monotremes.

Our observations that a large region of the eutherian X chromosome, containing several genes shared with the Y, is represented by two major autosomal clusters in both marsupials and monotremes, and implies that the addition of this region (XAR) to the X and Y in the Eutheria required at least two major rearrangements. Either two successive X-autosome translocations must have occurred, or the two autosomal regions were fused prior to their translocation to the sex chromosomes. In the absence of informative variation between eutherian sex chromosomes, it is impossible to distinguish between these hypotheses. However, it may yet be possible to find variants by mapping genes from the XAR in the little known edentates, the eutherian order which is thought to have diverged first. If the X and Y chromosomes were partially homologous at this early stage of eutherian radiation, as seems most likely, the addition of one or two autosomal blocks to a paired, non-inactivated pseudoautosomal region would not be particularly disruptive to sex chromosome function. Like the ancient PAR, the autosomal region(s) translocated into the pseudoautosomal region presumably became subject to progressive deletion from the Y chromosome.

The origin of the eutherian pseudoautosomal region must be questioned. Was it part of the ancestral sex chromosomes, or is it a relic of the autosomal regions added to the eutherian sex chromosomes? Possibly both regions are represented. The apparent absence of a pairing region between the marsupial sex chromosomes (Sharp, 1982), suggests that the process of X differentiation has gone to completion in the smaller marsupial sex chromosomes, but is incomplete in the larger composite eutherian sex chromosomes. This might suggest that the eutherian pseudoautosomal region is a relic of the autosomal region added to the sex chromosomes.

The relatively recent recruitment of a large region to the eutherian X chromosome raises questions about the evolution of X chromosome inactivation. Since the region was autosomal in a common mammalian ancestor, it could not have been differentiated with respect to the Y, nor would it have been part of the X inactivation system. After translocation gene loci in the enlarged pseudoautosomal region were progressively recruited into the system of X inactivation, as the corresponding Y-linked loci diverged, becoming nonfunctional or serving a different function. Alternatively, spreading of the X-inactivation system may initiate the divergence of the Y-linked allele (Graves and Schmidt, 1992).

There are some genes whose partial inactivation may reflect an intermediate stage in the evolution of XCI. For instance, the steroid sulphatase (*STS*)locus is only partially inactivated in humans (Migeon et al., 1982) being subject to some, but not all, of the multiple inactivation controls which operate in eutherian (Gartler et al., 1985; Riggs, 1990). In mouse, the *STS* locus seems to be pseudautosomal, and is not inactivated. It would obviously be of interest to look at the replication time and DNA modification status of the STS locus in humans and mouse. Genes which are shared by the X and Y chromosomes of some, but not all, eutherian species (see Table 2) suggest that recruitment into the X-inactivation system follows the divergence of the Y-linked allele. This X–Y differentiation is progressive, and the species differences in gene content of the PAR probably reflect gene order. The process may "skip" a gene, whose Y-linked allele remains functional.

Differences between the sex chromosomes of eutherian species in the gene content of the PAR (Disteche et al., 1992), the Y chromosome, and in the activity of X–Y-shared genes on the two X chromosomes in females indicate that the sex chromosomes are continuing to change in different lineages.

THE Y CHROMOSOME

The Y chromosome is almost devoid of functional genes. Only 10 Y-linked genes have been characterized in man and other mammals, and these all have related sequences on the X (Graves and Schmidt, 1992). One of these genes is responsible for testis determination, the first step of the sex-differentiation pathway. The role of several others is obscure, and they may represent the vagaries of evolution, rather than the bearers of important male-specific functions.

Sex Determination

The induction of testis formation is the initial step in the determination of male sexual differentiation. In man and other eutherian mammals, all other male-specific characteristics are induced by testicular hormones (Capel and Lovell Badge, 1992; Burgoyne, 1992).

The location of the testis-determining factor gene (TDF) on the Y chromosome was deduced by the analysis of the phenotypes of individuals with unusual sex chromosome complements: XO individuals being female, and XXY individuals male. Thus the numbers of X chromosomes are apparently not significant in eutherian sex determination, although they do affect growth and fertility (Burgoyne, 1992). However, in marsupials, XO and XXY animals are intersexes. The

Y chromosome determines testis, but the development of scrotum or mammary gland is a function of the number of X chromosomes, suggesting that the dosage of an X-linked gene independently determines the dimorphism (Sharman et al., 1990; Cooper et al., 1993).

The human *TDF* gene was eventually identified by positional cloning. Candidate genes, including HY antigen and *ZFY*, were excluded from a role in testis determination by mapping in mouse, man, and marsupial. The *SRY* gene was isolated from the human Y by deletion mapping of DNA from patients with only part of the Y chromosome (Sinclair et al., 1990) and its role in testis determination was confirmed when its mouse homologue, *Sry*, was shown to have a sex reversing effect in XX transgenic mice (Koopman et al., 1991). The translation product of the *Sry* gene shares a region of homology with several proteins of the high mobility groups (HMG), which contain a conserved DNA-binding domain of 79 amino acids (the HMG box) essential to SRY function (Harley et al., 1992; Nasrin et al., 1992). The *SRY* probe detects a number of other genes (*SOX* genes) which share this conserved box, but which have no function in sex determination. Giese et al. (1992) found that the HMG box of *SRY* binds specifically to a target sequence, producing bends in DNA that may control the tissue specific assembly of a functional chromatin structure.

Recently, homologues of the human *SRY* genes have been cloned and mapped to the Y chromosome in two marsupial species (Foster et al., 1992) implying that the *SRY* gene is sex-determining at least in all therian mammals. Surprisingly, homologous sequences were also detected on the marsupial X chromosome (Foster et al., 1993 submitted), and have since been found on the X in mouse and man (Stevanovich et al., 1993). Thus, even the sex-determining gene is not truly Y-specific and may have evolved, like *ZFX/ZFY*, from an ancestral gene with copies on X and Y (Foster et al., submitted). The poor homology between man, mouse, rabbit, and marsupial *SRY* within the conserved box region, and the total lack of homology outside this region, was also surprising. It suggested that rather than acting as an activator of transcription of testis determining genes, *SRY* may actually repress female determining genes. However, this now seems unlikely, as Cohen et al. (submitted) have demonstrated that *SRY* binds to a transcriptional activator.

Several of the *SRY*-related (*SOX*) genes identified in eutherian and marsupial species appear to be involved in a pathway of neutral development, while others are expressed in testis (Denny et al., 1992). These *SOX* genes are highly conserved in birds, reptiles, amphibians, and even insects, but none of them are sex-specific (Denny et al., 1992). One of those developmentally important genes may have been located on the proto-X and -Y chromsomes. Alleles at this locus diverged as the region in which they lay was isolated from recombination, and the Y-borne allele took on a testis-determining function (Foster et al., 1993, submitted). To date this event, experiments are currently being undertaken to determine whether any of the sequences detected in monotremes by human and mouse *SRY* gene probes are sex-specific.

Other genes which may be involved with sexual differentiation have also been identified on the sex chromosomes. The short stature and webbed neck characteristics of human XO females (and some XY deletions) ("Turner's Syndrome", Ford, 1959) has been ascribed to a single dose of a gene, probably on the PAR, and this has been recently confirmed (Ogata et al., 1992). There is also evidence from males with Y chromosome deletions for male-specific genes affecting spermatogenesis, meiosis, and sperm mobility (Bardoni et l., 1991; Burgoyne, 1992). However, *Ube ly* (which has been suggested to function in mouse sperm development) has no Y-borne homologue in humans, and *RPS4* which codes for a ribosomal protein (Fisher et al., 1990) and may effect early development, is absent from the mouse Y. *Zfy* apparently plays a role in spermatogenesis in mouse but not human (Koopman et al., 1990). Similarly, the analysis of the sexual phenotype of individuals with aberrant sex-chromosome constitutions suggest a role for genes on the X chromosome in early development and spermatogenesis in some species but not others (Burgoyne, 1992). Evidently, different species have evolved their own individual solutions to the genetic control of early development, spermatogenesis, and sperm function, and some of these solutions have apparently involved the evolution of differential functions for the X- and Y-borne loci of shared genes.

Y Chromosome Evolution

The conclusion that the mammalian X and Y were once homologous is logically inescapable, and is supported by the considerable cytological homology and pairing between the monotreme X and Y (Murtagh, 1977; Graves and Watson, 1991). All the genes on the human, mouse, and marsupial Y, including *SRY*, have a related sequence on the X. Similarly, the cloning and characterization of repeated sequences on the human Y has shown that, while some are apparently Y-specific and result from recent transposition, many are shared with the X.

The data of Foster et al. (1992) are consistent with gonadal sex being determined by the same Y-linked gene, *SRY*, in all therian mammals, and, together with the evidence for a marsupial Y-borne homologue of the mouse, Ube ly gene (Mitchell et al., 1992), provide support for the monophyletic origin of the therian Y. Similarly, the location of *UBE1* in the pairing region of the short arm of the platypus X chromosome (Wilcox et al., in preparation) suggests that there is also a Y-linked allele of *UBE1* in monotremes. This matter is currently being investigated using *in situ* hybridization to meiotic chromosome preparations in which the Y chromosome can be identified.

The genes and other sequences on the mammalian Y may be considered relics of an ancient autosome that bore the gene (not necessarily *SRY*) which became the original mammalian determination factor. Some of the isolated Y-linked genes took on new male specific functions (such as *SRY* in therians, and Ubely and *Zfy* in mouse). Others, such as *STS* have become only partially inactive, and most, like human *STS* and *AMGL*, are inactive or entirely missing.

CONCLUSION

The analysis of individuals with aberrant sex-chromosome complements and mutations of sex-determining genes in humans and mice have led to an understanding of the mechanism of mammalian sexual differentiation. The evolution of the mammalian system of gonadal sex determination has been approached by studying variants of sex chromosome form and function in other, more divergent mammal and animal species. Evidence from intra- and interspecies variation confirm that mammalian sex chromosomes evolved, and are still evolving from a homomorphic pair by progressive loss of active genes from the Y chromosome.

This differentiation has been accompanied by a number of changes: the repression of recombination, genetic inactivation, mutation and deletion from the Y chromosome; translocation of autosomal regions to sex chromsomes; and recruitment of genes on the X into a chromosome-wide system of inactivation. Mammalian sex determination may have originally involved dosage-dependent female determination, but this system has been overridden by *SRY*-induced testis determination.

Obviously, a very large number of genes interact to regulate the complex process of dimorphic sexual differentiation. Very basic similarities in dimorphic sexual characteristics between animal groups indicate that functionally similar genes are involved in sexual differentiation. However, the differences between the sex chromosomes of mammals and other vertebrates suggest that sex chromosomes have evolved independently several times, and that the gene(s) and the chromosome pair controlling the sex-determining pathways are also different. The particular choice of a specific controlling gene and chromosome pair to regulate sex determination has resulted in the variety of chromosomal systems observed in the animal kingdom. We suggest that the very elaborate system of Y chromosome directed testis determination and X chromosome inactivation in mammals has evolved from dose-dependent female determination to *SRY* specification of testis development.

ACKNOWLEDGMENTS

We gratefully acknowledge the contribution of the many colleagues and students with whom we have collaborated. Thanks are also due to Jo Anne La Rose for assistance with the preparation of the manuscript and artwork. JW is supported by a Queen Elizabeth II Fellowship and accommodated by the Department of Genetics and Human Variation, La Trobe University, Bundoora 3083, Australia.

ABBREVIATIONS

XCI	X chromosome inactivation
MyrBP	Million years before present
PAR	Pseudoautosomal region

XCR	X conserved region
XRA	X recently added region
XIC	X inactivation center
TDF	Testis determining factor
HMG	High mobility group

REFERENCES

Ashworth, A., Rastan, S., Lovell-Badge, R., & Kay, G. (1991). X-chromosome inactivation may explain the difference in viability of XO humans and mice. Nature 351, 406–408.

Ballabio, A. & Willard, H. F. (1992). Mammalian X—chromosome inactivation and the XIST gene. Current Opinion in Genetics and Development, 439–447.

Bardoni, B., Zuffardi, O., Guioli, S., Ballabio, A., Simi, P, Cavalli, P., Grimoldi, M. G., Fraccaro, M., & Camerino, G. (1991). A deletion map of the human Yq11 region: Implications for the evolution of the Y chromosome and tentative mapping of a locus involved in spermatogenesis. Genomics 11, 443–451.

Becak, W., Becak, M. L., Nazaneth, H. R. S., & Ohno, S. (1964). Close karyological kinship between reptilian suborder serpentes and the class Aves. Chromosoma 15, 606–617.

Brockdorff, N., Ashworth, A., Kay, G. F., Cooper, P., Smith, S., McCabe, V. M., Norris, D. P., Penny, G. D., Patel, D., & Rastan, S. (1991). Conservation of position and exclusive expression of mouse XIST from the inactive X chromosome. Nature 351, 329–331.

Brown, C. J., Lafreniere, R. G., Powers, V. E., Sebastio, G., Ballabio, A., Pettigrew, A. L., Ledbetter, D. H., Levy, E., Craig, I. W., & Willard, H. F. (1991). Localization of the X inactivation centre on the human X chromosome in Xq13. Nature 349, 82–84.

Brown, C. J. & Willard, H. F. (1990). Localization of a gene that escapes inactivation to the X chromosome proximal short arm: implications for X inactivation. Am. J. Hum. Genet. 46, 273–279.

Bull, J. (1983). Evolution of Sex Determining Mechanisms. Benjamin Cummings, Menlo Park.

Burgoyne, P. (1992). Y chromosome function in mammalian development. Adv. Dev. Biol. 1, 1–29.

Capel, B. & Lovell-Badge, R. (1992). The Sry gene and sex determination in mammals. Adv. Dev. Biol. (in press).

Chandra, H. S. (1985). Is human X chromosome inactivation a sex-determining device? Proc. Natl. Acad. Sci. USA 82, 6947–6949.

Chandra, H. S. (1991). How do heterogametic females survive without gene dosage compensation? J. Genet. 70, 137–146.

Chandra, H. S. & Nandjundiah, V. (1992). Dosage compensation from the viewpoint of natural selection. In: Molecular Genetics of Sex Determination (Caskey, C. T., Ed.). Blackwell Scientific, Oxford.

Coles, K., Mackenzie, M., Crolla, J., Harvey, J., Starr, J., Howard, F., & Jacobs, P. (1992). A complex rearrangement associated with sex reversal and the Wolf-Hirschorn syndrome: a cytogenetic and molecular study. J. Med. Genet. 29, 400–406.

Cooper, D. W. (1971). Directed genetic change model for X chromosome inactivation in eutherian mammals. Nature 230, 297–794.

Cooper, D. W., Johnston, P. G., Watson, J. M., & Graves, J. A. M. (1993). X activation in marsupials and monotremes. In: Seminars in Developmental Biology (Lucchesi, J. C., Ed.), Vol. 4. Academic Press, London.

Cooper, D. W., Johnston, P. G., Graves, J. A. M., & Watson, J. M. (1993). X-inactivation in marsupials and monotremes. In: Seminars in Developmental Biology: Dosage Compensation (Lucchesi, J. C., Ed.). Academic Press, London.

Dawson, G. W. & Graves, J. A. M. (1986). Gene mapping in marsupials and monotremes. III. Assignment of four genes to the X chromosome of the wallaroo and the euro (*Macropus robustus*). Cytogenet. Cell Genet. 42, 80–84.

Dawson, G. W. & Graves, J. A. M. (1984). Gene mapping in marsupials and monotremes I. The chromosomes of rodent—marsupial (*Macropus*) cell hybrids, and gene assignments to the X chromosome of the grey kangaroo. Chromosoma 91, 20–27.

Denny, P., Brand, N., Dabhade, N., Barton, P., & Ashwolth, A. (1992). A conserved family of genes related to the testis determining gene SRY. Nucleic Acid Res. 20, 2887.

Denny, P., Swift, S., Connor, F., & Ashworth, A. (1992). An SRY-related gene expressed during spermatogenesis in the mouse encodes a sequence-specific DNA-binding protein. EMBO J. 11, 3705–3712.

Disteche, C. M., Brannan, C. I., Larsen, A., Adler, D. A., Schorderet, D. F., Gearing, D., Copeland, N. G., Jenkins, N. A., & Park, L. S. (1992). Evolution of the pseudoautsomal region: The GM-CSF Receptor α subunit gene is on mouse chromosome 19. Nature Genet. (in press).

Dobovic, A. & Graves, J. A. M (1986). Gene mapping in marsupials and monotremes II. Assignment to the X chromosome of dasyurid marsupials. Cytogenet. Cell Genet. 41, 9–13.

Ewens, W. J., Griffiths, R. C., Ethier, S. N., Wilcox, S. A., & Marshall Graves, J. A. (1992). Statistical analysis of *in situ* hybridization data: Derivation and use of the Zmax test. Genomics 12, 675–682.

Fisher, E. M. C., Beer-Romero, P., Brown, L. G., Ridley, A., McNeil, J. A., Lawrence, J. B., Willard, H. F., Bieber, F. R., & Page, D. C. (1990). Homologous ribosomal protein genes on the human X and Y chromosomes: escape from X inactivation and possible implications for Turner's Syndrome. Cell 63, 1205–1218.

Fisher, E. M. C., Beer-Romero, P., Brown, L. G., Ridley, A., McNeil, J. A., Lawrence, J. B., Willard, H. F., Bieber, F. R., & Page, D. C. (1991). Cell 63, 1205–1218.

Ford, C. E., Jones, K. W., Polani, P. E., de Almeida, J. C. C., & Briggs, J. H. (1959). A sex chromosome anomaly in a case of gonadal dysgenesis (Turner's Syndrome) Lancet 1958i, 711.

Foster, J. W., Brennan, F. E., Hampikian, G. K., Goodfellow, P. N., Sinclair, A. H., Lovell-Badge, R., Selwood, L., Renfree, M. B., Cooper, D. W., & Graves, J. A. M. (1992). Evolution of sex determination and the Y chromosome: SRY-related sequences in marsupials. Nature 359, 531–533.

Foster, J. W., Goodfellow, P. N., Lovell-Badge, R., & Graves, J. A. M. (1993). Evolution of the mammalian testis determining gene SRY from an X-Y shared gene is suggested by a SOX3 homologue on the marsupial X. Nature Genetics (submitted).

Gartler, S. M., Dyer, K. A., Graves, J. A. M., & Rocchi, M. (1985). A two step model for mammalian-chromosome inactivation. In: Chemistry, Biochemistry and Biology of DNA Methylation (Cantoni, G. L. & Razin, A., Eds.), pp. 223–235. A. R. Liss, NY.

Giese, K., Cox, J., & Grosschedl, R. (1992). The HMG domain of lymphoid enhancer Factor 1 bends DNA and facilitates assembly of functional nucleoprotein structures. Cell 69, 185–195.

Graves, J. A. M. (1967). DNA synthesis in chromosomes of cultured leukocytes from two marsupial species. Exptl. Cell Res. 46, 37–57.

Graves, J. A. M. (1983). Inactivation and reaction of the mammalian X chromosome—on the threshold of molecular biology. In: Development in Mammals (Johnson, M. H., Ed.), pp. 251–295. Elsevier, Amsterdam.

Graves J. A. M., Cooper, D. W., McKenzie, L. M., Hope, R. M., & Watson, J. M. (1993). Gene maps of marsupial and monotreme mammals. In: Genetics Maps (O'Brien, S. J., Ed.), 6th ed. Cold Spring Harbor Laboratory Press, New York (in press).

Graves, J. A. M. (1987). The evolution of mammalian sex chromosomes and dosage compensation—clues from marsupials and monotremes. Trends Genet. 3, 252–256.

Graves, J. A. M. & Dawson, G. W. (1988). The relationship between position and expression of genes on the kangaroo X chromosome suggests a tissue specific spread of inactivation from a single control site Genet. Res. 51, 103–109.

Graves, J. A. M. & Schmidt, M. M. (1992). Mammalian sex chromosomes: design or accident? Current Opinion in Genetics and Development 2, 890–901.

Graves, J. A. M. & Watson, J. M. (1991). Mammalian sex chromosomes: Evolution or organisation and function. Chromosoma 101, 63–68.

Graves, J. A. M. & Short, R. V. (1990). Y or X which determines sex? Reprod. Fertil. Dev. 2, 729–735.

Harley, V. R., Jackson, D. I., Hextall, P. J., Hawkins, J. R., Berkovitz, G. D., Sockanathan, S., Lovell-Badge, R., & Goodfellow, P. N. (1992). DNA binding activity of recombinant SRY form normal males and XY females. Science 255, 453–456.

Harper, M. E. & Sauders, G. F. (1981). Localization of single copy DNA sequences on G-banded human chromosomes by *in situ* hybridization. Chromosoma 83, 431–439.

Hope, R. M., Cooper, S., & Wainwright, B. (1990). Globin macromolecular sequence in marsupials and monotremes. In: Mammals from Pouches and Eggs: Genetics, Breeding and Evolution of Marsupials and Monotremes (Graves, J. A. M., Hope, R. M., & Cooper, D. W., Eds.), pp. 147–172. CSIRO, Melbourne.

Hopson, J. A. (1970). The classification of nonetherian mammals. J. Mammal 51, 1–9.

Jones, K. W. & Singh, T. (1985). Snakes and the evolution of sex chromosomes. Trends Genet. 1, 55–61.

Johnston, P. G. & Robinson, E. S. (1987). X chromosome inactivation in female embryos of a marsupial mouse (*Antechinus stuartii*). Chromosoma 95, 419–423.

Kaslow, D. C. & Migeon, B. R. (1987). DNA methlation stabilizes X chromosome inactivation in eutherians but not in marsupials: Evidence for a multistep maintenance of mammalian X dosage compensation. Proc. Natl. Acad. Sci. USA 84, 6210–6214.

Kay, G. F., Penny, G. D., Patel, D., Ashworth, A., Brockdorff, W., & Rastan, S. (1993). Expression of Xist during mouse development suggests a role in the initiation of X chromosome inactivation. Cell 72, 171–182.

Kemp, T. S. (1982). Mammal like Reptiles and the Origin of Mammals. Academic Press, New York.

Kemp, T. S. (1983). The relationship of mammals. Zool. J. Linn. Soc. 77, 353–384.

Kirsch, J. A. W. & Calaby, J. M. (1977). The species of living marsupials: an annotated list. In: The Biology of Marsupials (Stonehouse, B., & Gilmore, D., Eds.). Macmillan, London.

Koopman, P., Gubbay, J., Vivian, N., Goodfellow, P., & Lovell-Badge, R. (1991). Male development of chromosomally female mice transgenic for Sry. Nature 351, 117–121.

Koopman, P., Munsterberg, A., Capel, B., Vivian, N., & Lovell-Badge, R. (1990). Expression of a candidate sex determining gene during mouse testis differentiation. Nature 348, 450–452.

Lau, Y-F.C., Yang-Feng, T. L., Elder, B., Fredga, K., & Wiberg, U. H. (1992). Unusual distribution of Zfy and Zfx sequences on the sex chromosomes of the wood lemming, a species exhibiting XY sex reversal. Cytogenet. Cell Genet. 60, 48–54.

Lyon, M. F. (1974a). Evolution of X chromosome inactivation in mammals. Nature 250, 651–653.

Lyon, M. F. (1974b). Mechanisms and evolutionary origins of variable X chromosome activity in mammals. Proc. R. Soc. Lond. B. Biol. Sci. 187, 243–268.

Lucchesi, J. C. (1978). Gene dosage compensation and the evolution of sex chromosomes. Science 202, 711–716.

Maaswinkel-Mooij, P. D., Van Zwietell, P., Mollervanger, P., Van Noort, E., & Beverstock, G. (1992). A girl with 71, XXXXY karyotype. Clin. Genet. 11, 96–99.

McKay, L. M., Wrigley, J. M., & Graves, J. A. M. (1987). Evolution of mammalian X-chromosome inactivation: Sex chromatin in monotremes and marsupials. Aust. J. Biol. Sci 40, 397–404.

Migeon, B. R., Shapiro, L. J., Norum, R. A., Mohandas, T. Axelman, J., & Dabora, R. L. (1982). Differential expression of steroid sulfatase locus on active and inactive human X chromosome. Nature. 299, 838–840.

Migeon, B. R., De Beur, S.J., & Axelman, J. (1989). Frequent de-repression of G6PD and HPRT on the marsupial inactive X chromosome associated with cell proliferation *in vitro*. Exptl. Cell Res. 182, 597–609.

Mitchell, M. J., Woods, D. R., Wilcox, S. A., Graves, J. A. M., & Bishop, C. E. (1992). The marsupial Y chromosome encodes a homologue of the mouse Y-linked candidate spermatogenesis gene Ubely. Nature 359, 528–531.

Murtagh, C. E. (1977). A unique cytogenetic system in monotremes. Chromosoma 65, 3757.

Nasrin, N., Buggs, C., Kong, X. F., Carnazza, J., Goebl, M., & Alexander-Bridges, M. (1992). DNA binding properties of the product of the testis-determining gene and a related protein. Nature 354, 317–320.

O'Brien, S. J. & Graves, J. A. M. (1991). Report of the committee on comparative gene mapping. Cytogenet. Cell Genet. 58, 1124–1151.

Ogata, T., Hawkins, J. R., Taylor, A., Matsuo, N., Hata, J., & Goodfellow, P. N. (1992). Sex reversal in a child with a 46,X,Yp+Karyotype: support for the existence of a gene located in distal Xp, involved in testis formation. J. Med. Genet. 29, 226–230.

Ohno, S. (1967). Sex Chromosomes and Sex Linked Genes. Springer-Verlag, Berlin.

Pfeifer, G. P., Tonguay, R. L., Steigerwald, S. D., & Riggs, A. D. (1990). *In vivo* footprint and methylation analysis by PCR-aided genomic sequencing comparison of active and inactive X chromosomal DNA at the CpG island and promoter of human PGK-1. Genes Development 4, 1277–1287.

Piper, A. A., Bennett, A. M., Noyce, L., Swanton, M. K., & Cooper, D. W. (1993). Isolation of a clone partially encoding hill kangaroo X linked hypoxanthine phosphoribosyl transferase: sex differences in methylation in the body of the gene. Somatic Cell Genet. (in press).

Riggs, A. D. (1990). Marsupials and the mechanism of X chromosome inactivation. In: Mammals from Pouches and Eggs: Genetics, Breeding and Evolution of Marsupials and Monotremes (Graves, J.A.M., Hope, R. M. & Cooper, D. W., Eds.), pp. 277–299. CSIRO, Melbourne.

Rofe, R. & Hayman, D. L. (1985). G-banding evidence for a conserved complement in the Marsupialia. Cytogenet Cell Genet. 39, 40–50.

Sharman, G. B., Hughes, R. L., & Cooper, D. W. (1990). The chromosomal basis of sex differentiation in marsupials. Aust. J. Zool. 37, 451–466.

Sharp, P. (1982). Sex chromosome pairing during male meiosis in marsupials. Chromosoma 86, 27–47.

Sinclair, A. H., Berta, P., Palmer, M. S., Hawken, J. R., Griffiths, B. L., Smith, M. J., Foster, J. W., Frischauf, A. M., Lovell-Badge, R., & Goodfellow, P. N. (1990). A gene from the human sex-determining region encoding a protein with homology to a conserved DNA-binding motif. Nature 346, 244–249.

Sinclair, A. H., Foster, J. W., Spencer, J. A., Page, D. C., Palmer, M., Goodfellow, P. N., & Graves, J. A. M. (1988). Sequences homologous to ZFY, a candidate human sex-determining gene, are autosomal in marsupials. Nature 336, 780–783.

Sinclair, A. H., Wrigley, J. M., & Graves, J. A. M. (1987). Autosomal assignment of OTC in marsupials and monotremes: Implications for the evolution of mammalian sex chromosomes. Genet. Res. 50, 131–136.

Singer-Sam, J., Grant, M., Le Bon, J. M., Okuyama, K., Chapman, V., Monk, M., & Riggs, A. D. (1990). Use of a HpaII-polymerase chain reaction assay to study DNA methylation in the PGK-1 CpG island of mouse embryos at the time of X-chromosome inactivation. Molec. Cell Biol. 10, 4987–4989.

Spencer, J. A., Sinclair, A. H., Watson, J. M., & Graves, J. A. M. (1991c). Genes on the short arm of the human X chromosome are not shared with the marsupial X. Genomics. 11, 339–345.

Spencer, J. A., Watson, J. M., & Graves, J. A. M. (1991a). The X chromosome of marsupials shared a highly conserved region with eutherians. Genomics 9, 598–604.

Spencer, J. A., Watson, J. M., Lubahn, D. B., Joseph, D. R., French, F. S., Wilson, E. M., & Graves, J. A. M. (1991b). The androgen receptor gene is located on a highly conserved region of the X chromosomes of marsupial and monotreme as well as eutherian mammals. J. Hered. 82, 131–139.

Stevanovic, M., Lovell-Badge, R., Collingnon, J. W., & Goodfellow, P. N. (1993). SOX3 is an X-linked gene related to SRY. Nature Genetics (submitted).

Thorne, M. H. & Sheldon, B. L. (1992). Cytological evidence of maternal meiotic errors in a line of chickens with a high incidence of triploidy. Cytogenet. Cell Genet. 57, 206–210.

Van Daal, A. M. (1986). The molecular basis of X-chromosome inactivation in kangaroos. Ph.D. thesis, Macquerie University, Sydney.

VandeBerg, J. L. & Cooper, D. W. (1978). Possible autosomal inheritance of erythrocyte phosphoglycerate kinase A in echidnas. Biochem. Genet. 16, 1031–1034.

Wareham, K. A., Lyon, M. F., Glenister, P. H., & Williams, E. D. (1987). Age related reactivation of an X-linked gene. Nature 327, 725–727.

Watson, J. M. (1990). Monotreme genetics and cytology and a model for sex chromosome evolution. Aust. J. Zool. 37, 385–406.

Watson, J. M. & Graves, J. A. M. (1988). Gene mapping in marsupials and monotremes V. Synteny between hypoxanthine phosphoribosyl transferase and phosphoglycerate kinase in the platypus. Aust. J. Biol. Sci. 41, 231–237.

Watson, J. M., Spencer, J. A., Riggs, A. D., & Graves, J. A. M. (1990). The X chromosome of monotremes shares a highly conserved region with the eutherian and marsupial X despite the absence of X inactivation. Proc. Natl. Acad. Sci. USA 87, 7125–7129.

Watson, J. M., Spencer, J. A., Riggs, A. D., & Graves, J. A. M. (1991). Sex chromosome evolution; platypus gene mapping suggests that part of the human X was originally autosomal. Proc. Natl. Acad. Sci. USA 88, 11256–11260.

Watson, J. M., Frost, C., Spencer, J. A., & Graves, J. A. M. (1992a). Sequences homologous to the human X- and Y-borne zinc finger protein genes (ZFX/Y) are autosomal in monotreme mammals. Genomics 15, 317–322.

Watson, J. M., Meyne, J., & Graves, J. A. M. (1992b). Chromosomal composition and position in the echidna meiotic translocation chain. In: Platypus and Echidnas (Augee, M. L., Ed.), pp. 53–63. The Royal Zoological Society of New South Wales, Sydney.

Watson, J. M., Riggs, A. D., & Graves, J. A. M. (1992c). Gene mapping studies confirm the homology between the platypus X and echinda X_1 and identify a conserved ancestral monotreme X chromosome. Chromosoma 101, 596–601.

Watson, J. M., Spencer, J. A., Graves, J. M. G., Snead, M. L., & Lau, E. C. (1992d). Autosomal localization of the amelogenin gene in monotremes and marsupials: implications for mammalian sex chromosome evolution. Genomics 14, 785–789.

Weinberg, J., Jauch, A., Stanyon, R., & Cremer, T. (1990). Molecular taxonomy of primates by chromosomal *in situ* suppression hybridization. Genomics 8, 347–350.

Wrigley, J. M. & Graves, J. A. M. (1988). Karyotypic conservation in the mammalian order Monotremata (subclass Prototheria) Chromosoma 96, 23?–247.

Wrigley, J. M. & Graves, J. A. M. (1988). Sex chromosome homology and incomplete, tissue-specific inactivation suggests that monotremes represent an intermediate stage of mammalian sex chromosome evolution. J. Hered. 79, 115–118.

RECOMMENDED READINGS

Background

Auge, M. L., (Ed.) (1992). Platypus and Echidnas. Royal Zoological Society of New South Wales, Sydney.

Griffiths, M. (Ed.) (1978). The Biology of the Monotremes. Academic Press, New York.

Calaby, J. H. & Tyndale-Biscoe, C. H. (Eds.) (1977). Reproduction and Evolution. Australian Academy of Science, Canberra.

Tyndale-Biscoe, C. H. (Ed.) (1973). Life of Marsupials. Edward Arnold Press, London.

Tyndale-Biscoe, C H. & Renfree, M. (Eds.) (1987). Reproductive Physiology of Marsupials. Cambridge University Press, Cambridge.

Specific

Graves, J. A. M., Hope, R. M., & Cooper, D. W. (Eds.) (1990). Mammals from Pouches and Eggs: Genetics, Breeding and Evolution of Marsupials and Monotremes. Reprinted from Australian Journal of Zoology 37, numbers 2, 3 and 4, 1990, CSIRO, Australia.

MacIntyre, R. J. (Ed.) (1985). Molecular Evolutionary Genetics. Plenum Press, New York.

O'Brien, S. J. (Ed.) (1993). Genetic Maps. Cold Spring Harbor (in press).

Ohno, S. (1967). Sex Chromosomes and Sex Linked Genes. Springer-Verlag, Berlin.

Reed, K. & Graves, J. A. M. (Eds.) (1993). Sex Chromosome and Sex Determination. Gordon and Breach, Melbourne (in press).

Chapter 5

The Evolution of Gametes

M. GOMENDIO and E. R. S. ROLDAN

INTRODUCTION

Throughout evolutionary history, natural selection has favored those traits that contribute to maximize the reproductive success of individuals. Such traits include a wide variety of morphological, physiological, and behavioral adaptations that influence not only the survival chances of individuals but also the number of descendants that they leave. The influence of natural selection can be detected at many different levels, but it will be more evident for traits that are directly associated with either survival or reproduction since these are decisive in evolu-

Principles of Medical Biology, Volume 1B
Evolutionary Biology, pages 115–151

ISBN: 1-55938-802-1

tionary terms. Because an individual's reproductive success ultimately rests with the success of its gametes at fertilization, these cells are likely to have been under intense selective pressure.

Gametes are cells that have become specialized to perform a reproductive role. When there is sexual reproduction, it generally involves two sexes: male and female. The most basic distinction between males and females is the type of gamete that they produce, although among higher animals males and females can normally be easily distinguished by their external appearance. Male and female gametes are fundamentally different and, as we shall see, these differences have far-reaching consequences for the reproductive strategies that males and females follow. Gametes are formed by meiosis and the fusion of a male and a female gamete constitutes the origin of a new individual which will carry genetic material from both parents (Figure 1).

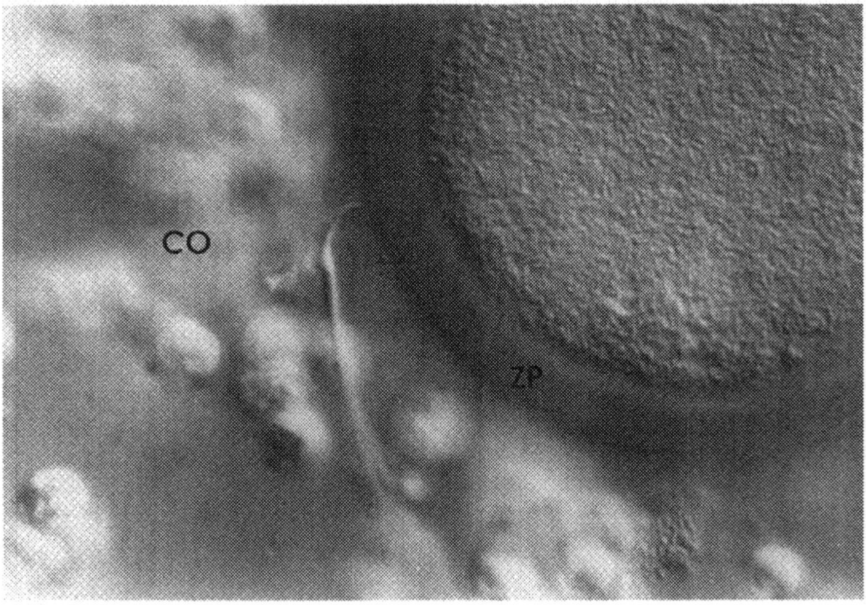

Figure 1. Spermatozoon and ovum from the Chinese hamster, *Cricetulus griseus,* at the time of fertilization. The spermatozoon has just traversed the cellular ovum coat, the *cumulus oophorus* (CO), and has just arrived at the surface of the acellular coat, the *zona pellucida* (ZP). Note the difference in size between the male and the female gametes.

MALE AND FEMALE GAMETES AND THEIR IMPLICATIONS FOR REPRODUCTIVE STRATEGIES

Males tend to produce numerous gametes known as spermatozoa that are small in size and are in most cases highly mobile. Spermatozoa contain the DNA that the male will contribute to the next generation and little more than the elements required to move towards the ova and achieve fertilization. Because males can produce vast numbers of tiny gametes rather quickly, and because they tend not to participate in parental care to the extent that females do (Clutton-Brock, 1991), male reproductive success is determined primarily by the number of females (or more specifically ova) that they can fertilize. Thus, a typical male's strategy is aimed at maximizing the number of matings or, strictly speaking, fertilizations. In many animal populations, the differences in reproductive success between males are huge, because a few males tend to monopolize mating access to the majority of the females, while most males do not reproduce at all. The great variance in male reproductive success implies that selective pressures have been strong, and as a result males have evolved features that either improve their competitive ability in male–male contests (i.e., intrasexual selection) or make them more attractive to females (intersexual selection) (Krebs and Davies, 1993).

Females produce gametes known as ova (singular: ovum) that are much larger in size, contain nutritive material to be used during the postfertilization period (i.e., by the developing zygote), and tend to be immotile. The greater investment in each gamete results in a smaller number of gametes being produced by females and in a slower production rate. This initial asymmetry in gamete production by the two sexes is further exacerbated by the fact that females tend to do most, if not all, the parental care. Thus, while a male's reproductive attempt may consist only of copulating with a female, for the female it involves a longer and more demanding process that starts when she becomes fertile and continues until the offspring are independent. All the available ova that a female has when she is sexually receptive can normally be fertilized by a single male. Thus, a female does not increase the number of offspring that she produces by increasing the number of matings like males do. Because males produce large numbers of spermatozoa rather quickly, males do not normally represent a limiting resource for females and, within populations, females tend to produce similar numbers of offspring. The main variable that explains the differences in reproductive success between females is the survival of offspring, which tends to be low in natural populations. Thus, the most efficient way in which a female can enhance her reproductive success is by obtaining enough food resources to ensure offspring survival. As a result, females invest more in offspring than males do and, rather than maximizing the number of matings, they tend to be choosy about who they mate with, particularly when males provide resources that influence offspring survival, or when offspring will benefit from the genetic quality of their fathers.

The divergent reproductive strategies of males and females may lead to conflicts of interest between the sexes which can also take place at the level of gametic interactions. As we have just seen, the fact that males and females produce different types of gametes has important implications for the reproductive strategies that the two sexes follow. It is important to realize that the interaction also works the other way round, so that the different reproductive strategies that males and females adopt may determine in part the kinds of selective pressures that have influenced the evolution of gametes.

In this chapter we will review the evolutionary changes that both ova and spermatozoa have undergone, and we will examine which selective forces are likely to have played a role. For many decades, the degree of interspecific differences in gamete morphology and physiology have puzzled scientists. Researchers have devoted a great deal of effort to describe gamete traits in different species but, despite all the interest generated, efforts to explain the adaptive value of these differences have often failed, leading many researchers to argue that these gamete traits are nonadaptive. In our view, given the crucial role that gametes play in reproduction, it seems unlikely that such major differences have no evolutionary meaning. Recent findings support this view as they suggest that many of the interspecific differences can be explained in terms of coevolution between male and female gametes, or as the result of sperm competition.

THE ORIGIN OF MALE AND FEMALE GAMETES

Several theoretical models have considered the issue of why and how did male and female gametes evolve. One of the best known is Parker's (Parker et al., 1972) who used a computer simulation to examine what would be the evolutionary stable strategy (ESS); that is, a strategy which when adopted by most members of the population cannot be beaten by any other strategy in the game. Let us imagine an ancestral population in which there is only one kind of gamete (isogamy), but individuals differ in the size of the gametes that they produce (Figure 2). Most individuals will produce gametes intermediate in size, but at one end of the spectrum there will be a small number of individuals that produce few large gametes with high prospects of survivorship as zygotes, and at the other end some individuals will produce many small gametes with poorer prospects of survival. The model assumes a trade-off between gamete size and number, and also a positive relationship between gamete size and zygote survival. Thus, individuals that produce many small gametes have the advantage that they maximize the number of fusions with other gametes, but zygotes will not survive unless they fuse with a large gamete that has enough material to ensure zygote survival. Given the numerical predominance of small gametes, large gametes are more likely to fuse with small ones, although they might enjoy greater prospects of zygote survival if they could fuse with another large gamete.

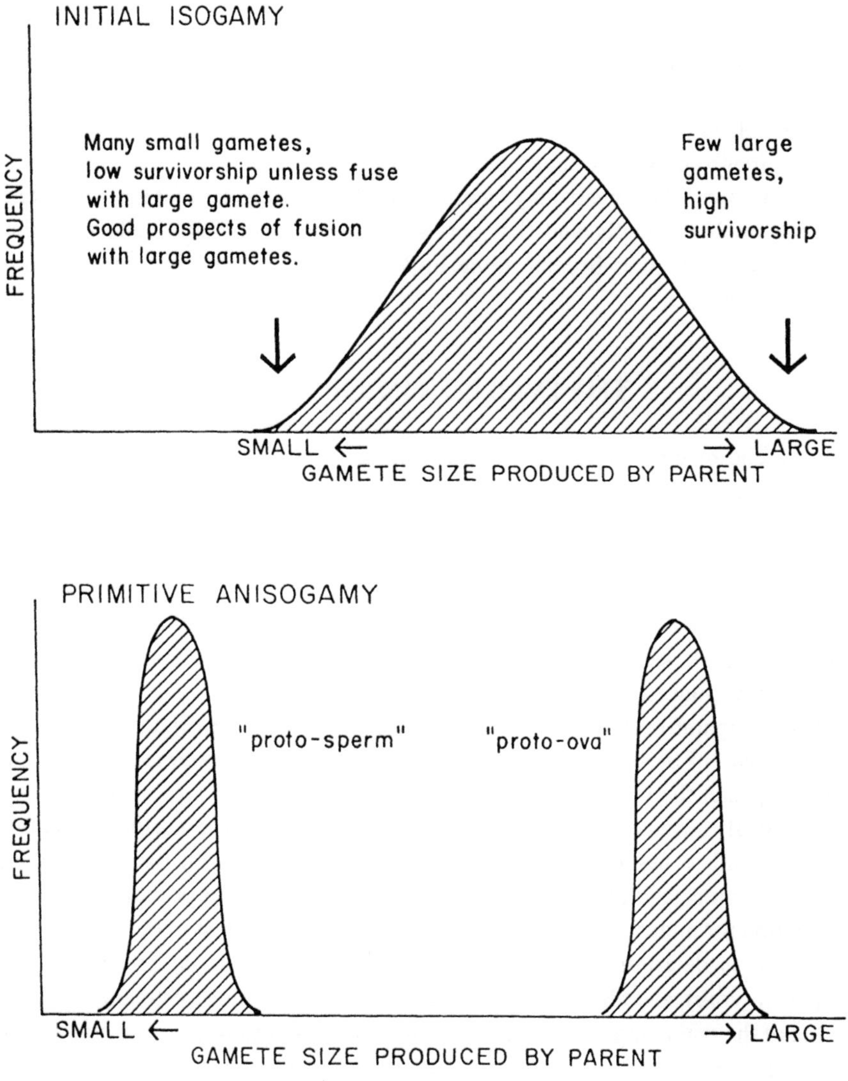

Figure 2. Evolution of anisogamy (two sexes) from a hypothetical ancestral population with isogamy (one sex) and a range of gamete size. [From Parker, G. A. (1984). In: Sperm Competition and the Evolution of Animal Mating Systems (Smith, R. L., Ed.). Academic Press, Orlando, with permission from the author and Academic Press].

The ESS solution to this problem is the coexistence of two morphs for gamete size ("proto-sperm" and "proto-ova"), and after several generations the population is essentially anisogamous (i.e., two types of gametes) as a result of disruptive selection. Female gametes would reach a size that maximizes the number of surviving offspring, while male gametes would be reduced in size to a minimum determined by survival before fusion, the ability to reach the ovum, and the ability to compete with other male gametes. Thus, male gametes can be seen as having become specialized in parasitizing the investment of female gametes in the zygote, and the whole system is driven by sperm competition which acts to maintain sperm small in order to maximize sperm productivity.

It is unknown when anisogamy emerged, but Parker suggests that gametes were probably small originally and that during the evolution of multicellularity the provisioning of zygotes became more important, thus setting the scene for the evolution of ova.

THE EVOLUTION OF OVA

Females tend to produce few large immotile gametes, with nutritive material to nourish the zygote. These gametes are known as ova. They are spherical in shape and contain a nucleus with DNA, a large cytoplasm with various organelles and nutritive material (yolk) which will be used by the zygote generated at fertilization, and surrounding "vestments" (Austin, 1982). It is important to highlight that only in very few animal species is the ovum nucleus ever truly haploid. Most ova are shed as primary or secondary oocytes, and complete the first and/or second meiotic division after (indeed, triggered by) fertilization.

Most comparative work has focused on differences in size (Figure 3), and the great majority of these studies have reported differences between taxa in the size of the eggs that the female lays rather than differences in the size of unfertilized ova, obviously because such measurements are much easier to obtain than the latter. It seems valid to assume that egg size will be correlated with ovum size and, therefore, we feel justified in reviewing the studies on egg size as providing relevant information for understanding the evolution of ovum size. Most of the variation in ovum size is determined by the amount of nutritive material present in the cytoplasm (yolk), and also by the thickness of the surrounding vestments.

One of the main variables that explains interspecific differences in ovum size is adult body weight, as there seems to be a clear allometric relationship at least in those taxa in which parents do not nourish the embryo throughout development, and offspring rely on the reserves contained in the egg (Blueweiss et al., 1978; reptiles: Kaplan and Salthe, 1979; birds: Rahn et al., 1975). Figure 4 shows the relationship between ovum size and body size for several species of salamanders. As can be seen, body size cannot explain all the variance in ovum size since the slope is rather different for species with different reproductive modes. In many other taxa, it is also the case that body size can explain some but not all the differences in ovum size between species; this is because mode of offspring development and ecological factors also play an important role.

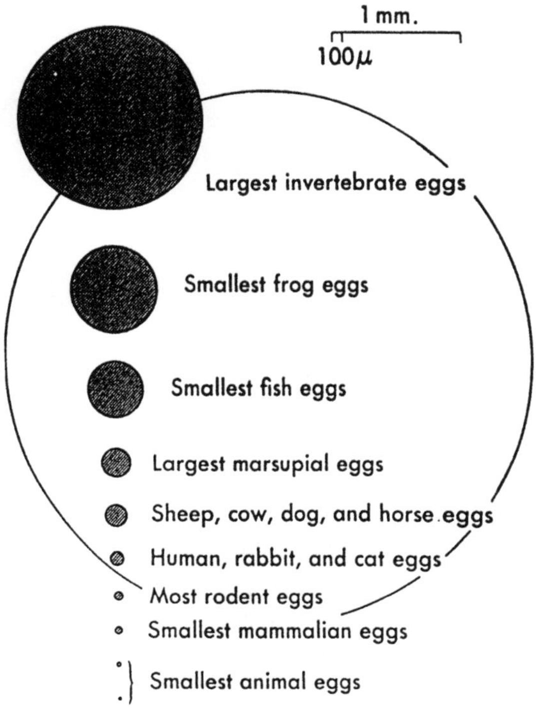

Figure 3. Schematic representation of different sizes of ova without vestments. The large circle represents the monotreme ovum. Some of the largest invertebrate ova are those of the squid *Loligo,* the gastropod *Busycon,* the starfish *Henricia,* and the crab *Libinia.* The largest marsupial ovum, that of the Australian native cat *Dasyurus,* is about the same size as the ovum of the sea-squirt *Amaroucium.* The sizes of most mammalian ova correspond to those of many echinoderms, tunicates, molluscs, polychaetes, nemertines, platyhelminths, and coelenterates. The smallest mammalian ovum, that of the field vole *Microtus agrestis,* corresponds in size to the ovum of the clam *Spisula.* The smallest animal ova include those of the bryozoan *Crisia* and the mesozoan *Dicyema.* [From Austin, C. R. (1965). Fertilization. Prentice-Hall, Englewood Cliffs, NJ, with permission from the author and Prentice-Hall].

In those taxa in which parents do not provide care for the offspring after fertilization, gamete production will be the major form of parental expenditure. In these cases, egg size will determine to a great extent the survival, growth, and reproductive success of offspring (Clutton-Brock, 1991). In *Daphnia magna,* females reared at high food availability produce larger eggs with more fat reserves, and the young are more resistant to starvation (Tessier et al,, 1983). Similarly, in the Coho salmon (*Oncorhynchus kisutch*) egg size is correlated with early growth, adult size, increased fecundity, and larger egg size (van den Berghe and Gross,

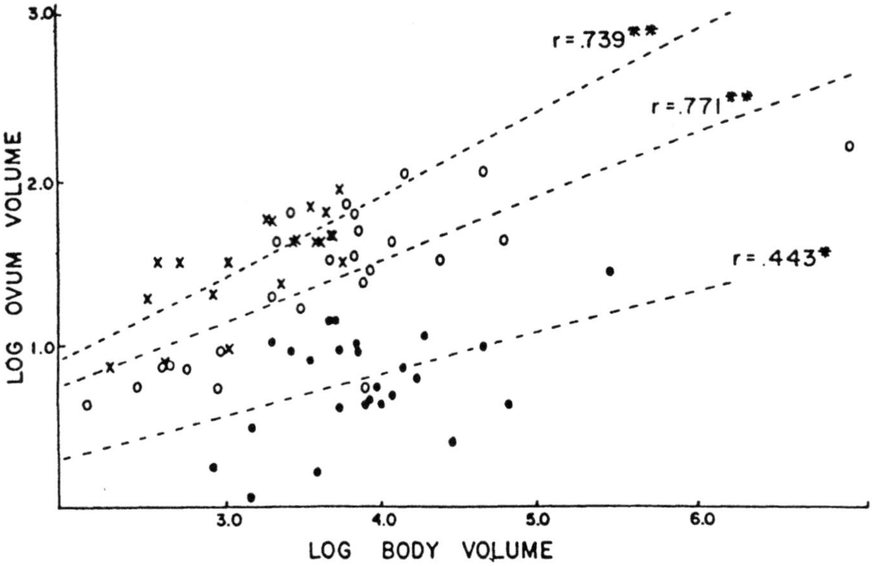

Figure 4. Relation between ovum size and body size among reptiles. Log mean ovum volume (mm^3) was plotted against log mean body volume (mm^3) for 74 salamander species with three different reproductive modes. *Closed circles* = pond breeders; *open circles* = stream breeders; *crosses* = direct developers; * = $P \leq 0.05$; ** = $P \leq 0.01$. (From Kaplan, R. H. and Salthe, S. N., 1979, Am. Nat. 113, 671–689, with permission from The University of Chicago Press).

1989). In these groups minor differences in egg size may become exacerbated throughout development, and have a major effect upon reproductive success.

While correlational evidence linking egg size with offspring survival strongly suggests that size is a key factor, experimental evidence is needed to determine causality. Recent experimental work on lizards has shown that egg size does determine to a great extent differences in offspring survival and has unveiled the mechanisms at work. Sinervo (1990) has studied a species of lizard (*Sceloporus occidentalis*) that shows a natural gradient in egg size and number. When different populations living in North America were compared, it was found that clutch size increases to the North and with elevation. Furthermore, there is a trade-off between egg size and number, so that when clutch size is large the eggs tend to be small (Figure 5). To determine whether size was directly responsible for differences in offspring survival and fitness, Sinervo extracted up to 50% of the yolk from a number of eggs and in this way managed to produce miniaturized hatchlings. These miniaturized hatchlings had slower sprint speed than normal-sized hatchings, and were therefore less successful at escaping from predators. Thus, lizard mothers produce a few large eggs in populations where predation pressure is strong (southern and low populations) and offspring survival will be enhanced by size-de-

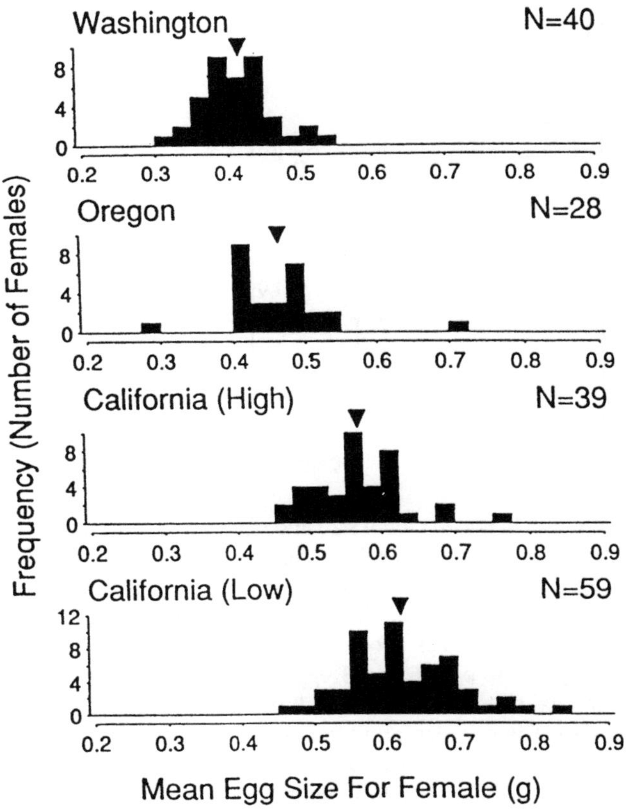

Figure 5. Frequency histograms of mean egg size for female's clutch (g) in a lizard, *Sceloporus occidentalis*, for each of four study populations, showing that it decreases to the North and with elevation. (From Sinervo, B., 1990, Evolution 44, 279–294, with permission from The Society for the Study of Evolution).

pendent performance traits, while they produce larger clutches of smaller eggs when predation pressure is low and they can safely trade offspring quality for numbers. Thus, the fecundity advantage of producing small offspring is balanced against the survival advantage of large offspring, and ecological conditions will determine to a great extent which is the optimal combination.

The presumed trade-off between the number of eggs that a female produces and their size is a fundamental tenet of life-history theory (Stearns, 1992). There is evidence of a negative correlation between number and size of eggs in fish (Wootton, 1979), insects (Hinton, 1981), and reptiles (Congdon et al., 1982), supporting the view that this trade-off may exist when there is no further parental provisioning. Sinervo and Licht (1991) have experimentally tested this assumption in the iguani lizard (*Uta stansburiana*) (Figure 6). They used a new approach to

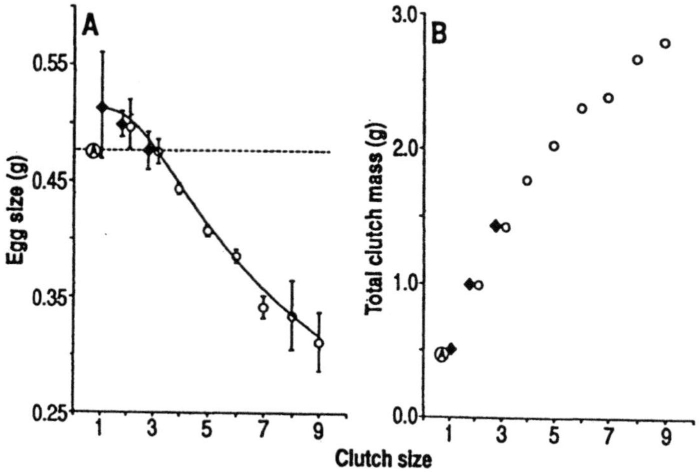

Figure 6. (**A**) Relation between clutch size and egg size (mean ± SE) for experimental (♦) and naturally laid eggs (○) obtained from female side-blotched lizards, *Uta stansburiana*. The effects of post-oviposition body size on egg size has been removed by ANCOVA of ln-transformed variables. Dashed horizontal line indicates the approximate threshold of egg sizes above which a large increase in the frequency of burst or oviducally bound eggs (i.e. ≤ 3 eggs or ≥ 0.475 g) were observed. (**B**) The relation between total clutch mass and clutch size was calculated by multiplying egg size by clutch size. The egg size and total clutch mass for *Anolis* lizards that only produce one-egg clutches is provided (Ⓐ) for comparison. (From Sinervo, B. and Licht, P., 1991, Science 252, 1300–1302, with permission from the author and The American Association for the Advancement of Science).

examine whether there is a specified energy budget allocated to reproduction which can be allocated either to a few large eggs or to many small ones. On the one hand, they produced gigantized eggs by surgically ablating a few young follicles early in vitellogenesis, but at a stage in which the female could no longer recruit replacements. This manipulation resulted in the production of fewer and larger young as predicted by the trade-off hypothesis. They also injected females with follicle-stimulating hormone, causing them to recruit extra follicles into vitellogenesis. Smaller eggs were produced, suggesting that a given amount of energy is distributed among more eggs, lending further support to the trade-off hypothesis.

It is important to bear in mind that the content of the eggs should be considered together with their size (Figure 7). In birds, egg quality varies widely, with the proportion of yolk ranging from 15 to 60% (O'Connor, 1984). Again, this variation can be explained by mode of offspring development and ecological factors since species with precocial young, or with young that can only be fed at long and irregular intervals (such as seabirds), produce large and yolky eggs (Lack, 1968).

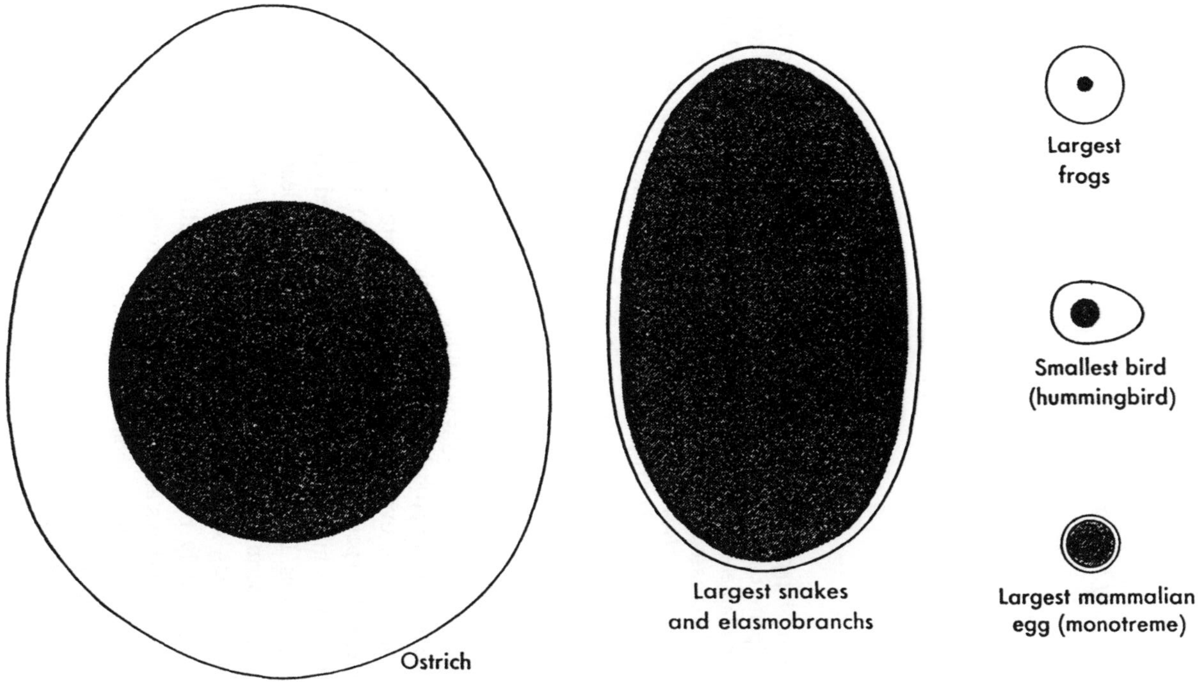

Figure 7. Over-all sizes of eggs showing the proportion of yolk in gray. [From Austin, C. R. (1965). Fertilization. Prentice-Hall, Englewood Cliffs, NJ, with permission from the author and Prentice-Hall].

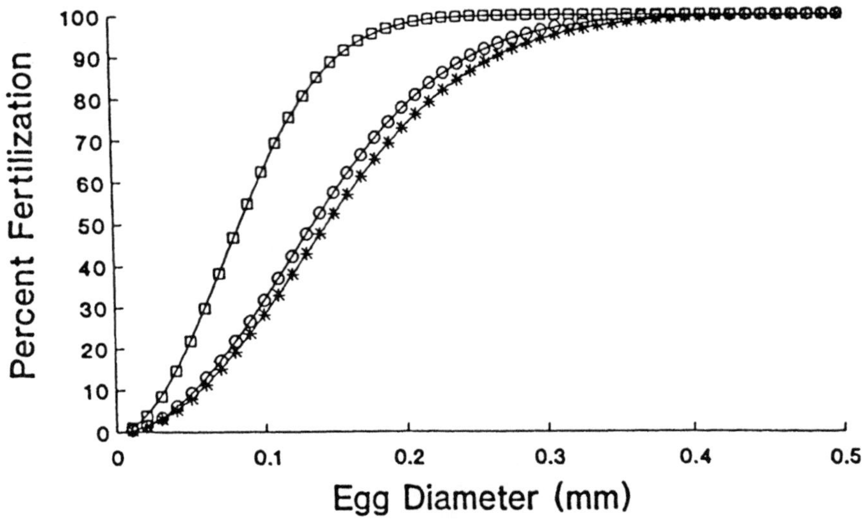

Figure 8. Relationship between ovum diameter and the percentage of ova fertilized in sea urchins. Sperm concentration is the concentration required to fertilize 50% of ova when ovum diameter is as measured empirically. Squares, circles and asterisks represent predictions based on fertilization kinetics of *Strongylocentrotus purpuratus, Strongylocentrotus franciscanus,* and *Strongylocentrotus droebachiensis,* respectively. Ovum concentration is 0.01 per μL. (From Levitan, D. R., 1993, Am. Nat. 141, 517–536, with permission from The University of Chicago Press).

As we have seen, variation in ovum size is normally explained by a trade-off between gamete quality and quantity, but alternative hypotheses are currently being developed. It has recently been suggested that, among marine invertebrates, ovum size may also be related to probability of fertilization, so that when there is a low concentration of gametes, ova tend to be large in order to maximize the chances of being encountered by sperm (Figure 8) (Levitan, 1993). On the contrary, when sperm concentration is high the risk of not being fertilized drops and ova tend to be small and numerous. Interestingly, this seems to be a clear case of coevolution between ovum and sperm traits, since when ova are small and numerous, sperm are fast at the cost of being short-lived in order to search a small volume of water vigorously for a rapidly diminishing pool of unfertilized ova. When gamete concentration is low and ova are large in order to increase target size, longer lived but slower sperm may be at an advantage. Thus, which is the optimal strategy will depend on the environmental conditions during spawning.

Ovum structure has been studied in more detail in some mammalian species than in any other taxon. It is, however, difficult to examine evolutionary trends, or to explain the adaptive value of the traits described, because these studies have been restricted to a few mammalian species (mostly domesticated or laboratory animals) and there is a lack of comparative data from other taxa. Despite this paucity of information, it is possible to make a number of generalizations and at least to point at potentially interesting aspects.

There are major differences between monotreme ova and the ova of therian mammals, which is hardly surprising given the striking differences in mode of offspring development (see reviews in Austin, 1982, Bedford, 1991). Monotreme females lay eggs and the ovum is large (2.5–4 cm) and rich in yolk, as is common in other egg-laying taxa where offspring rely on the reserves contained in the egg. The ovum is surrounded by a thin glycoprotein envelope—the zona pellucida—and, as in reptiles and birds, also by a shell and a shell membrane.

On the other hand, marsupial and eutherian ova are much smaller (120–250 μm, and 50–150 μm, respectively) because they contain a small amount of nutritive material and, among the latter, there seems to be no relationship with adult body weight. Obviously, the lack of nutritive material for the zygoye is related to the fact that the mother nourishes the offspring by other means. In marsupial ova, the zona pellucida (1–2 μm thick) is more prominent than in monotremes, and much thicker still in eutherian mammals (Figure 9). The extraordinary size of the eutherian zona pellucida (7–15 μm) makes the ovum much less fragile and more resilient. It is not clear why the zona pellucida has acquired such dimensions in eutherians, but it has

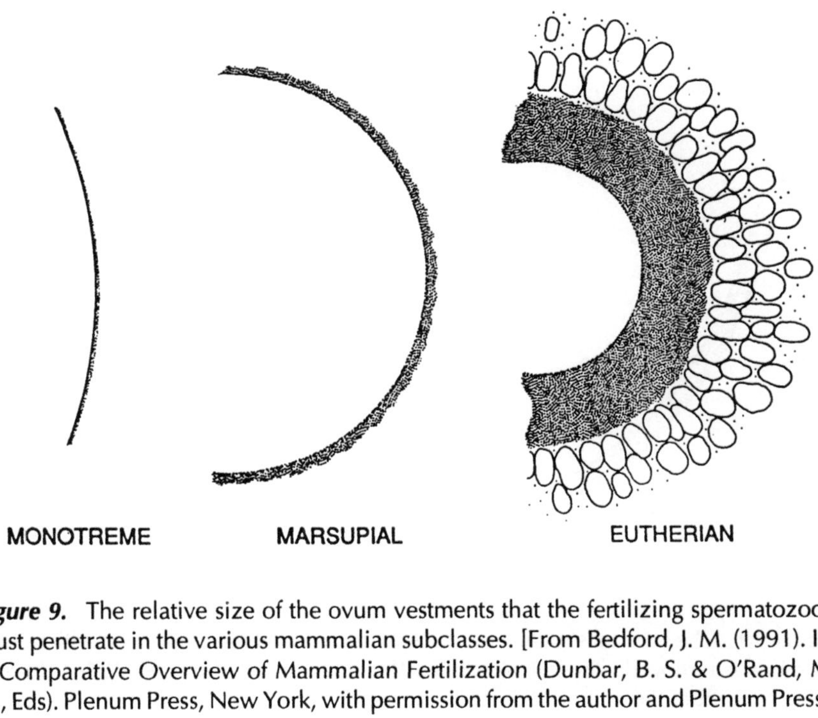

Figure 9. The relative size of the ovum vestments that the fertilizing spermatozoon must penetrate in the various mammalian subclasses. [From Bedford, J. M. (1991). In: A Comparative Overview of Mammalian Fertilization (Dunbar, B. S. & O'Rand, M. G., Eds). Plenum Press, New York, with permission from the author and Plenum Press].

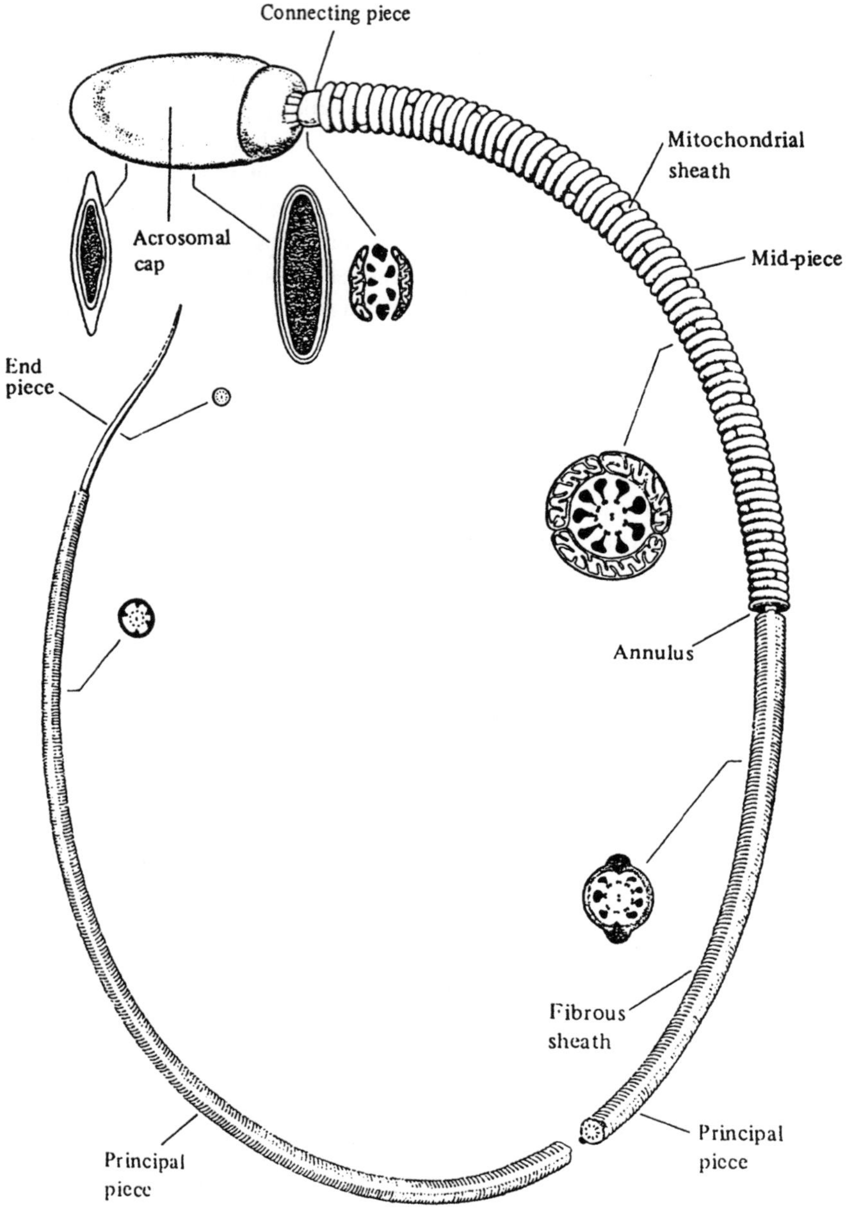

Figure 10. The internal structure of a typical mammalian spermatozoon with the cell membrane removed. (From Fawcett, D. W., 1975, Dev. Biol. 44, 394–436, with permission from the author and Academic Press).

been suggested (Bedford, 1991) that it may be related to the relatively long period of time that the ova must spend in the oviduct with no other protection but the zona. In addition, at the time of fertilization eutherian ova are surrounded by an additional vestment, the cumulus oophorus, which is a mass of follicle cells embedded in a proteoglycan matrix. The cumulus oophorus may act as a "trap" for spermatozoa and, in this way, may facilitate the meeting of unfertilized eggs with the very few spermatozoa available in the ampulla at the time of ovulation (Bedford and Kim, 1993).

THE EVOLUTION OF SPERMATOZOA

Males tend to produce vast numbers of gametes which contain no nutritive material for the zygote. These gametes are known as spermatozoa and their general structure responds to their basic function: to reach the ovum and to fertilize. The most common sperm design corresponds to a very compartmentalized cell with two main sections (Figure 10): (1) the head which contains a haploid nucleus and the acrosome (a vesicle with enzymes necessary to penetrate the ovum vestments); and (2) the tail with the mitochondria, which generate the energy necessary for movement, and the flagellum which propels the sperm as it beats (Bedford and Hoskins, 1990).

This basic design can be modified to a great extent giving rise to a striking variety of sperm morphs. There are spermatozoa with no tails and others with more than one. Some spermatozoa can be extremely large, exceeding the length of the male, while others are immobile dwarfs, and some spermatozoa even lack a genome and act as a sort of "vehicle" transporting other fertilizing spermatozoa (Figure 11). This bizarre gametic fauna has attracted much attention and stimulated comparative studies aimed at reconstructing the major evolutionary changes undergone by spermatozoa. These studies have either concentrated on lower taxa, or have outlined very general changes throughout the animal kingdom (e.g., Baccetti and Afzelius, 1976; Franzen, 1977; Baccetti, 1985; Jamieson, 1987, 1991) (Figure 12).

Species which show external fertilization and release spermatozoa into the water share a primitive type of spermatozoa which is rather simple in its morphology: a round head and a tail approximately 50 μm long with a 9 + 2 tubular structure. In a uniform environment like water, spermatozoa only have to swim until they reach the ova, and this simple design is well adapted to such conditions. Although the same basic design is present in all spermatozoa, a number of important changes have taken place throughout evolution. In terrestrial species with internal fertilization, spermatozoa have an elongated tail and experience a rearrangement of mitochondria. Mammals possess nine accessory fibers in the flagellum which are probably required to swim in the more viscous fluids encountered in the female tract.

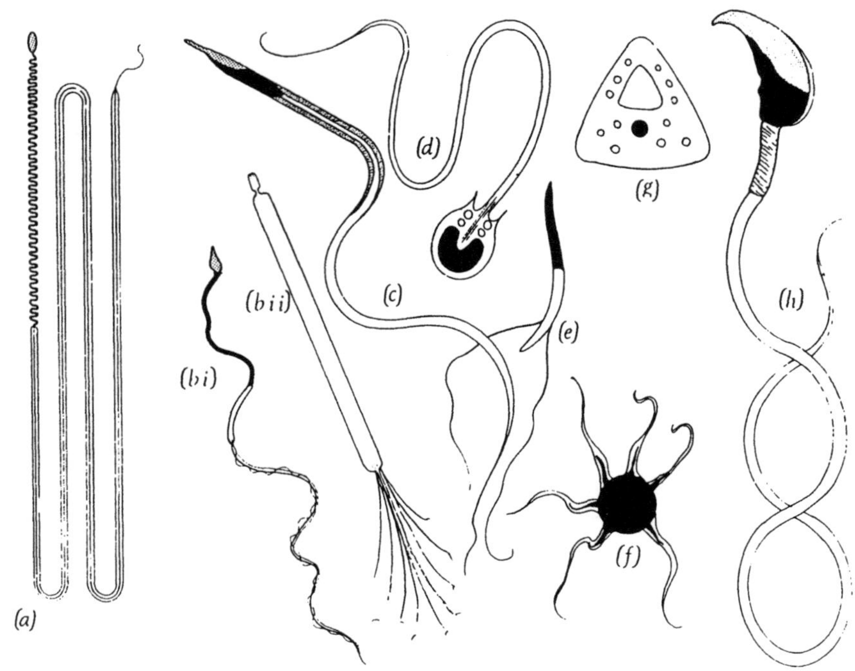

Figure 11. Variation in sperm form. The nuclei are shown solid black, the acrosome stippled. (**a**) *Lithobius* sperm [all centipede spermatozoa have this long head structure]; (**b**) *Paludina* sperm [(**bi**) is the fertilizing sperm (eupyrrhene), but (**bii**) is the oligopyrrhene form also found in the emitted semen of most snails]; (**c**) *Drosophila* sperm [some 2 mm long and with a complex midpiece structure, the nebenkern; many fly sperm are large and complex]; (**d**) *Tilapia* sperm [most teleost fish have tiny spermatozoa which lack acrosomes]; (**e**) *Polycelis* sperm [flatworms often have double-tailed sperm, frequently (e.g., *Fasciola*) with one very long tail]; (**f**) *Procambarus* sperm [many crustaceans have symmetric, immotile sperm]; (**g**) *Ascaris* sperm [nematode sperm are often triangular when inactive, amoeboid when fertilizing]; (**h**) mouse, *Mus musculus*, sperm [many rodents have hooked sperm heads]. [From Cohen, J. (1977). Reproduction, Butterworths, London, with permission from the author).

The range of variation in sperm morphology is such that spermatozoa from different species can in many cases be readily recognized. Thus, spermatozoa from different species may differ in their dimensions, their overall morphology, or both. Figure 13 illustrates the main sperm morphs found among mammals (Roldan et al., 1992). Monotremes are markedly different from other mammals since they have retained the vermiform spermatozoa which is typical of reptiles and birds. On the other hand, spermatozoa from marsupials and eutherian mammals have a similar ground plan with a round or oval head and a tail of varying length. Among therian mammals, the head shows the greatest variation in morphology, particularly among

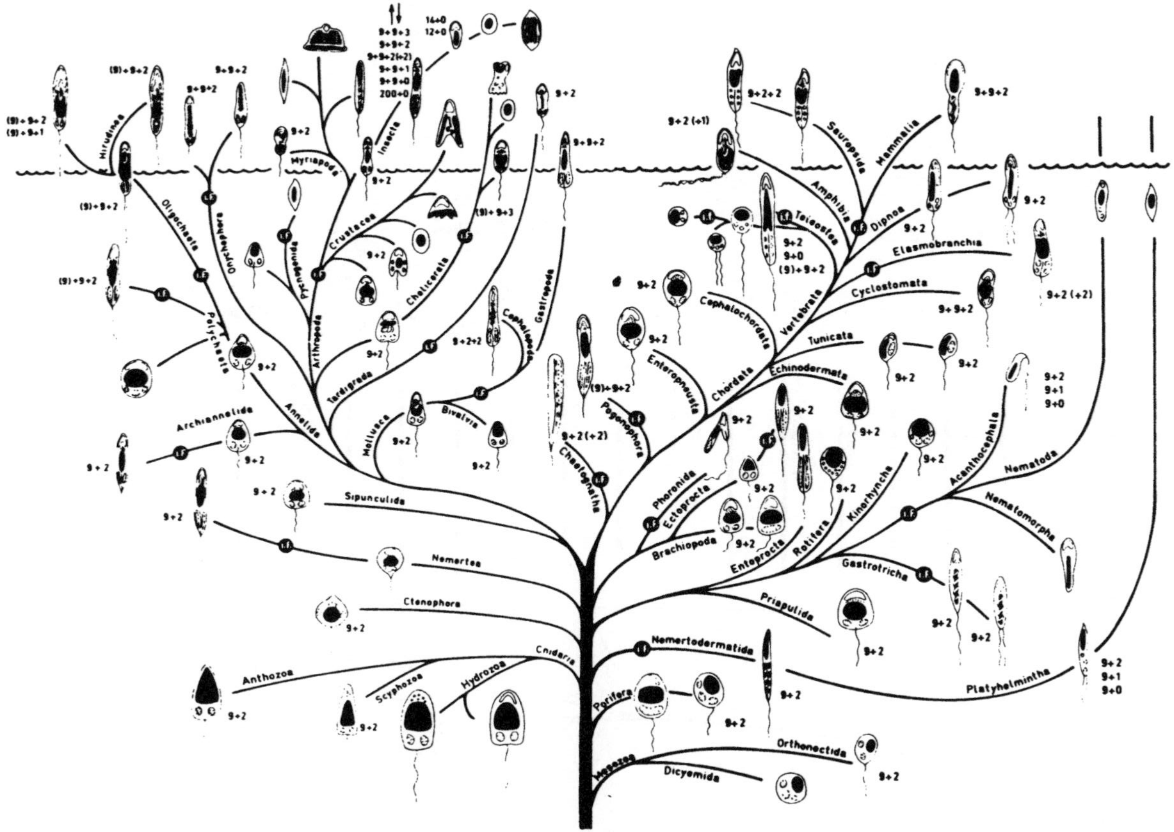

Figure 12. Schematic drawing of the evolution of sperm structure in the animal kingdom. i.f. = acquisition of internal fertilization. Numbers (e.g. 9 + 2) correspond to axonemal structure. (Courtesy of Professor Baccio Baccetti, University of Siena, Italy).

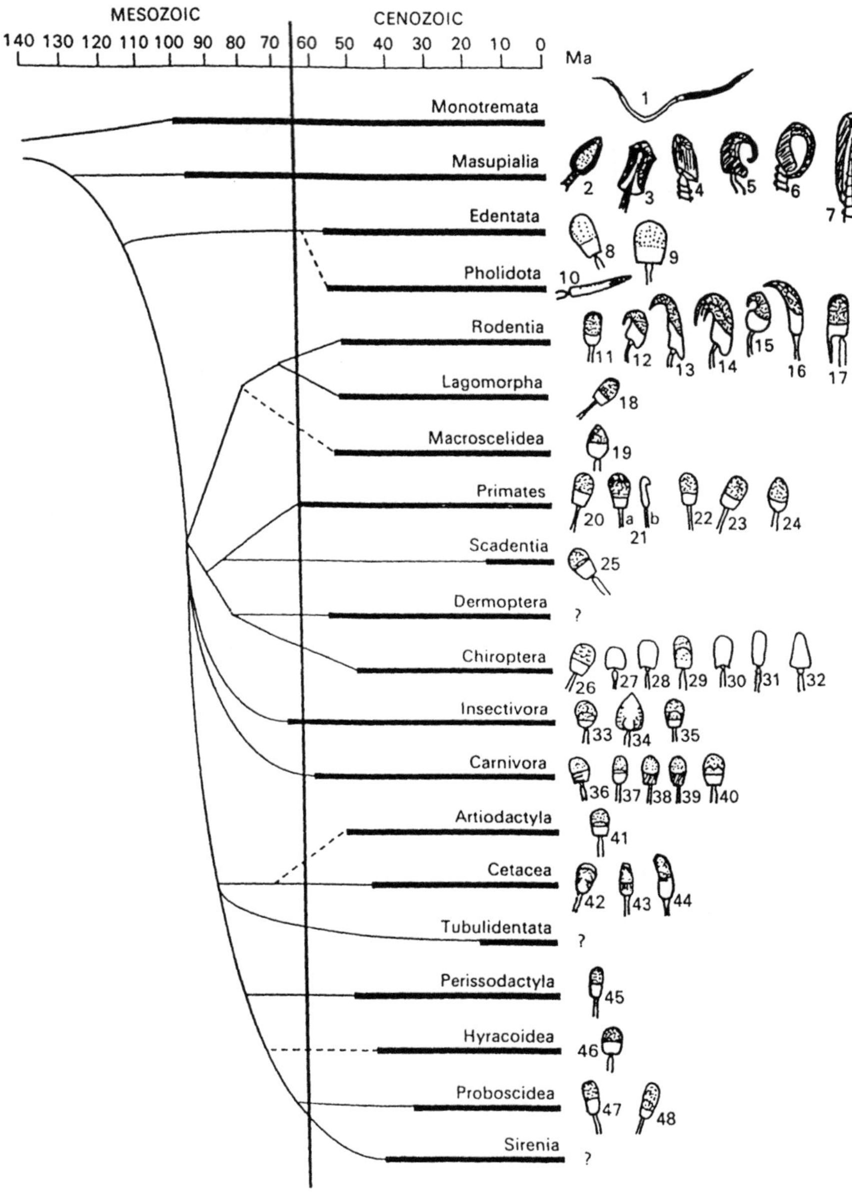

MESOZOIC
CENOZOIC
140 130 120 110 100 90 80 70 60 50 40 30 20 10 0
Ma
Monotremata
Masupialia
Edentata
Pholidota
Rodentia
Lagomorpha
Macroscelidea
Primates
Scadentia
Dermoptera
Chiroptera
Insectivora
Carnivora
Artiodactyla
Cetacea
Tubulidentata
Perissodactyla
Hyracoidea
Proboscidea
Sirenia

marsupials and rodents where heads with hooks in the apical region or with caudal extensions in the basal region can be found.

Among mammals, the sperm head shows little variation with respect to its dimensions; it is the length of the tail that varies markedly between species (Cummins and Woodall, 1985). Spermatozoa from marsupials and eutherian mammals show an average total length of 133.60 μm (N = 21 species), and 69.23 μm (N = 266 species), respectively, but this value masks the wide range in sperm lengths found between species: marsupials = 79–356 μM; eutherians = 33–258 μm (Figure 14) (Roldan et al., 1992). Rodents show the widest range of sperm lengths seen in any eutherian order ranging from 35 to 258 μm; that is, covering most of the total eutherian range. It is also noteworthy that some rodents have the longest spermatozoa so far described for eutherian mammals. Relatively large sperm length

---

Figure 13. A phylogenetic tree showing relationships among major extant mammalian clades and examples of sperm head morphologies seen in the various clades. The solid horizontal bars indicate the age range of the clade on the basis of dated first appearance in the fossil record. Solid lines indicate the branching sequence and dashed lines indicate relatively more ambiguous relationships. The sperm heads are not drawn to scale. Species = (1) *Tachyglossus aculeatus*; (2) *Pseudocheirus cupreus* (Pteuridae); (3) *Parameles nasuta* (Paramelidae); (4) *Macropus canguru* (Macropodidae); (5) *Phasolarctos cinereus* (Phasolarctidae); (6) *Vombatus ursinus* (Vombatidae); (7) *Sarcophilus harrisii* (Dasyuridae); (8) *Dasypus novemcinctus*; (9) *Cabassous unicinctus*; (10) *Manis longicaudata*; (11) *Calomys laucha* (Cricetidae: Sigmodontinae); (12) *Mus musculus* (Muridae: Murinae); (13) Rattus norvegicus (Muridae: Murinae); (14) *Pseudomys australis* (Muridae: Hydromyinae); (15) *Microtus ochrogaster* (Cricetidae: Arvicolinae); (16) *Cricetulus griseus* (Cricetidae: Cricetinae); (17) *Ctenomys maulinus* (Ctenomyidae); (18) *Oryctolagus cuniculus*; (19) *Elephuntulus edwardii*; (20) *Lemur fulvus* (Prosimii); (21) *Galago crassicaudatus* (Prosimii), a) planar view; b) lateral view; (22) *Saimiri sciureus* (Ceboidea); (23) *Macacca mulatta* (Cercopithecoidea); (24) *Homo sapiens* (Hominoidea; (25) *Tupaia glis*; (26) *Mormopterus planiceps* (Molossidae); (27) *Glossophaga soricina* (Phyllostomidae); (28) *Sturnira lilium* (Phyllostomidae); (29) *Pteropus conspiculatus* (Pteropodidae); (30) *Pipistrellus subflavus* (Vespertilionidae); (31) *Myotis sodalis* (Vespertilionidae); (32) *Plecotus townsendii* (Vespertilionidae); (33) *Talpa europaea* (Talpidae); (34) *Suncus murinus* (Soricidae); (35) *Sorex araneus* (Soricidae); (36) *Panthera leo* (Felidae); (37) *Acinomyx jubatus* (Felidae); (38) *Neofelis nebulosa* (Felidae); (39) *Canis familiaris* (Canidae); (40) *Mustela putorius* (Mustelidae); (41) *Bos taurus* (Bovidae); (42) *Globicephala melas* (Delphinidae); (43) *Tursiops truncatus* (Delphinidae); (44) *Physeter catodon* (Physeteridae); (45) *Equus caballus*; (46) *Procavia capensis*; (47) *Elephas maximus*; (48) *Loxodonta africana*. (From Roldan, E.R.S., Gomendio, M., and Vitullo, A.D., 1992, Biol. Rev. 67, 551–593, reprinted with the permission of Cambridge Philosophical Society and Cambridge University Press).

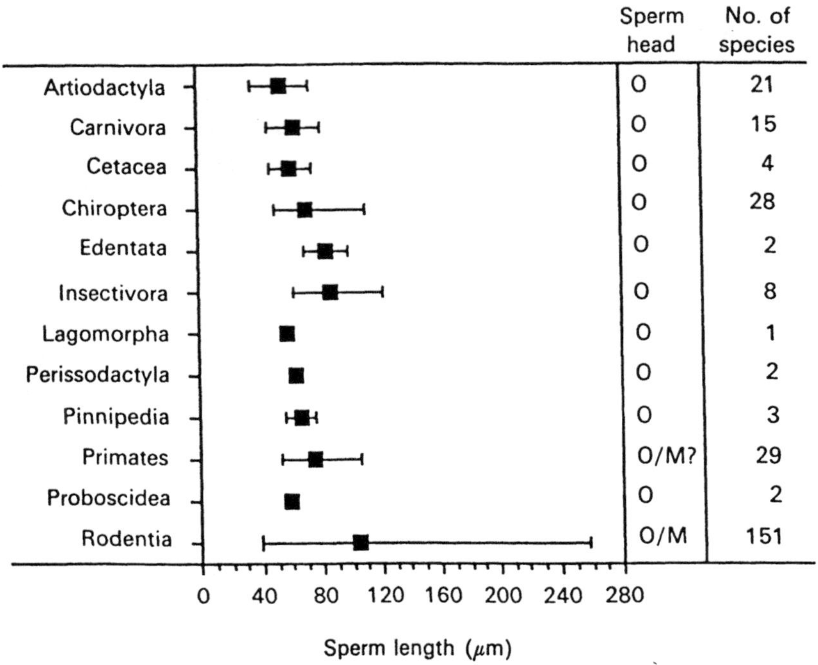

Figure 14. Sperm morphology and dimensions in several orders of eutherian mammals. Values given are means (■) and ranges. Sperm head shapes = o, oval or round; M = modified (either in the apical or the basal region of the head). (From Roldan, E.R.S., Gomendio, M. and Vitullo, A.D., 1992, Biol. Rev. 67, 551–593, reprinted with the permission of Cambridge Philosophical Society and Cambridge University Press).

ranges are also seen in *Chiroptera* (45–109 μm), *Insectivora* (64–121 μm) and *Primates* (50–105 μm).

Given the diversity of mammalian sperm morphology and dimensions, these features have been used as taxonomical tools in some analyses (e.g., Harding et al., 1987), as has been done with insect (reviewed in Jamieson, 1987) and fish

---

Figure 15. A model of the evolution of complexity in rodent sperm head organization and accompanying changes in total sperm length. Amu, Acr, and Aca = primitive type of sperm in Muridae, Cricetidae and Caviomorpha showing round or oval heads and lacking any appendices or protrusions at the basal or apical edges; the tail inserts centrally in the base of the head. Derived types show one or several hooks in the apical region of the head, modifications in the base of the head, displacements in tail insertion, and elongation of the flagellum (indicated by averages of total sperm length next to each head type). (From Roldan, E.R.S., Gomendio, M., and Vitullo, A.D., 1992, Biol. Rev. 67, 551–593, reprinted with the permission of Cambridge Philosophical Society and Cambridge University Press).

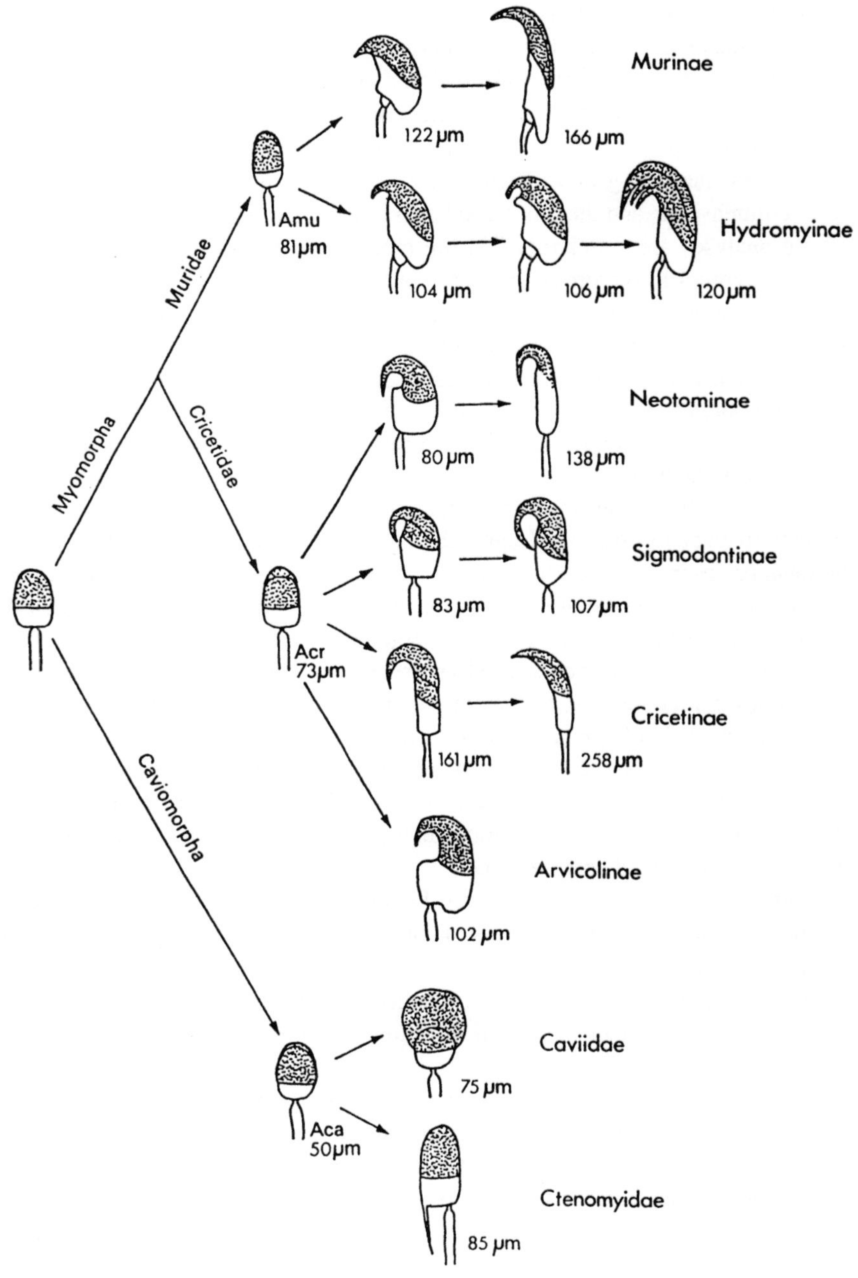
Murinae
122 µm
166 µm
Amu
81µm
Hydromyinae
104 µm
106 µm
120µm
Muridae
Myomorpha
Cricetidae
Neotominae
80 µm
138 µm
Sigmodontinae
83 µm
107 µm
Acr
73µm
Cricetinae
161 µm
258 µm
Caviomorpha
Arvicolinae
102 µm
Caviidae
75 µm
Aca
50µm
Ctenomyidae
85 µm

spermatozoa (reviewed in Jamieson, 1991). Moreover, some eutherian species which cannot be easily distinguished by their external phenotype have been shown to have very different spermatozoa.

To trace the evolutionary changes undergone by eutherian spermatozoa, Roldan et al. (1992) superimposed the information available on sperm morphology and dimensions onto phylogenetic trees based on biogeographical, morphological, chromosomal, and genic data. These authors examined evolutionary trends of rodent sperm since this group shows the greatest variation in both sperm morphology and dimensions, and there is a good deal of information available. The results of these analyses are summarized in the model depicted in Figure 15. Several ancestral myomorph (peromyscines, arvicolids, conilurines) and caviomorph rodents (ctenomyines, phyllotines) from South and North America and Australia have spermatozoa that are both short and with simple head morphology. These are not isolated examples of rodent species with short sperm and simple head morphology. In myomorphs, round/oval hookless sperm heads and short tails have also been found in species of South American Oryzomyini and Akodontini (Cricetidae), European, Southeast Asian, and African Murinea, and Sulawesian Muridae, and North-African Dipodidae. In caviomorphs, most species show this basic sperm head pattern and short flagella. Regrettably, a lack of information on phylogenetic relationships among the latter rodent taxa precludes further analyses. In any case, the widespread occurrence of spermatozoa with these traits can be taken as a clear indication that the rodent ancestors that diverged very early from the main eutherian branch (see Figure 13) have conserved a sperm type which is both short and simple. This implies that an increase in the complexity of rodent head morphology (e.g., the appearance of hooks) could have occurred repeatedly in different lineages in different continents. The fact that no two major rodent lineages present the same overall head design (Figure 15) could be taken as further evidence in favor of this hypothesis. It follows that modifications in head morphology may confer a highly adaptive value to rodent spermatozoa.

In summary, trends in rodent sperm evolution involve: (1) changes in head organization leading to higher degrees of complexity, which result in different head patterns in each taxon, and (2) an elongation of the tail (head length is fairly constant across taxa). In fact, there is a clear association between the appearance of hooks, head elongation, displacement of tail insertion, and elongation of the tail (Figure 15) (Roldan et al., 1992).

The simple rodent spermatozoon, consisting of a round/oval hookless head and a short tail, may thus be considered as the ancestral rodent form and, since it is similar to the type of spermatozoon that is found in several eutherian orders (overall length about 65 μm), it has been proposed as the basic ancestral form for all eutherians (Roldan et al., 1992). Interestingly, this sperm type seems to conform to the ground plan recognized as ancestral and primitive for other metazoans, where spermatozoa with round heads and a tail of about 50 μm have widespread occurrence (Franzen, 1977). Whereas in most eutherian species spermatozoa have retained the ancestral form, the reason why vastly modified heads with hooks and appendices, and long tails, have evolved in rodents is not known, especially why they have appeared repeatedly, and independently, in separate lineages.

SELECTIVE FORCES

Coevolution between Male Gametes and Female Reproductive Biology

In order to achieve fertilization, spermatozoa must be adapted to the variety of environments in which it may take place. When fertilization is external, spermatozoa swim in water until they reach the ova. This rather homogeneous environment has led to the evolution of spermatozoa that are rather simple in design (see Baccetti, 1985). On the contrary, when fertilization is internal, females from different species show a wide range of internal environments and of genitalia, and this has led to the evolution of a surprising variety of sperm morphologies (Sivinski, 1984). This relationship suggests that some sperm features may represent adaptations to the environment that they encounter within the female reproductive system.

It is surprising how few studies have examined whether sperm morphology and dimensions can be explained as adaptations to the female reproductive tract. In insects and birds, females possess storage organs where sperm can survive for years in the case of insects, or weeks in the case of birds (Parker, 1970; Birkhead and Hunter, 1990; Birkhead and Møller, 1992). Little is known about how sperm survival is enhanced in these organs, but it is clear that a male will benefit from having his sperm inside the sperm storage organs. In a study of featherwing beetles, where a low number of long spermatozoa can fit in the spermatheca, sperm length was found to positively correlate with the length of the spermathecal lumen and the spermathecal duct (Figure 16) (Dybas and Dybas, 1981).

A recent study on passerine birds has found a positive correlation between sperm length and the length of the female storage tubules, and a negative correlation between sperm length and the number of these tubules (Briskie and Montgomerie, 1992; see also Birkhead and Møller, 1992). It is possible that males benefit from producing sperm that can fill the storage organs, because in this way they may prevent sperm from rival males from getting in. If this were the case, sperm competition would select for sperm that are long enough to block the storage organ. It would thus be a case of sperm length being determined by the joint action of female selection and sperm competition.

Female mammals differ from insects and birds in that they do not possess proper storage organs for spermatozoa (reviews in Ginsberg and Huck, 1989; Gomendio and Roldan, 1993a). This difference may be related to the fact that in birds and insects it is in the female's interest to make sure that they have enough spermatozoa ready to fertilize during the rather long period of time in which the ova are released and fertilized in sequence; female mammals, on the contrary, release all the ova

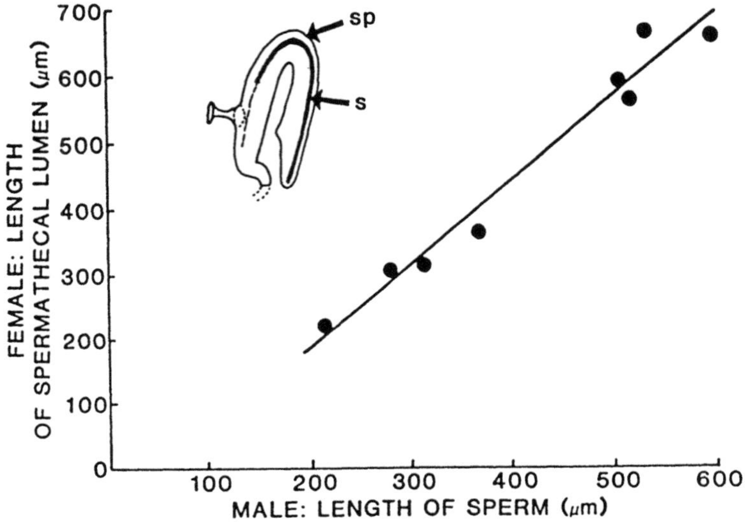

Figure 16. Relation between sperm length and the length of the female's sperm storage organ. The close relation between sperm length and spermathecal length is shown for a number of different species of beetles. Drawing: Sperm (s) of a conspecific featherwing beetle as it lies within the spermatheca (sp) of a conspecific female. (From Dybas, L.K. and Dybas, H.S., 1981, Evolution 35, 168–174, with permission from The Society for the Study of Evolution).

simultaneously in each cycle and these are all fertilized synchronously. Thus, female mammals may have no need to have a sperm population "in reserve." Perhaps as a consequence of the lack of sperm storage organs in mammals, spermatozoa rarely survive more than 48 hours in the female tract (Austin, 1975), although there are a few exceptions.

Among mammals, the female tract represents a formidable barrier for spermatozoa. Out of the millions of spermatozoa contained within an ejaculate, most die or are phagocytized before they reach the vicinity of the ova (Austin, 1965; Hunter, 1988). In most of the species which have been studied so far, millions of spermatozoa are ejaculated into the female tract, but only of the order of thousands reach the isthums, and only 2 to 20 reach the ampulla (Suarez et al., 1990). This drastic reduction in sperm numbers implies a very strong selection indeed within the female tract.

Among mammals, testes size and the number of sperm contained in each ejaculate are known to increase with body size (Harcourt et al., 1981; Kenagy and Trombulak, 1986), and it has been suggested that by increasing sperm numbers males counteract the dilution effects of large female tracts (Short, 1981). Sperm numbers also seem to be related to the likelihood that spermatozoa will survive within the female tract. In those exceptional cases in which female mammals do have storage crypts where sperm survival is enhanced (e.g., some marsupials), males inseminate low sperm numbers (Bedford et al., 1984).

It should also be mentioned that Cummins and Woodall (1985) reported a negative correlation between sperm size and body size in an interspecific study within mammals. This relationship has been interpreted as being the result of a trade-off between sperm numbers and sperm size: large mammals producing high numbers of small spermatozoa, and small mammals producing few large spermatozoa. Although the idea of a trade-off between sperm numbers and sperm size has gained widespread acceptance, it still remains to be tested. It should be noted that in Cummins and Woodall's (1985) analyses the negative relationship between sperm size and body size is only significant in two (Cetacea and Rodentia) out of the six mammalian orders considered. Furthermore, there is one order (Chiroptera) where the relationship is positive. Other studies on primates, muroid rodents, bats, and dogs have not found any relationship between sperm size and body size (reviewed in Gomendio and Roldan, 1993b). Thus, the apparent relationship between sperm size and body size could well be due to a few extreme outliers: some small rodents having large spermatozoa and some large cetaceans having small spermatozoa. If this is the case, the reported relationship between sperm size and body weight may not be truly meaningful.

In most mammalian species, spermatozoa retain their fertilizing ability within the female tract for up to 48 hours, except in bats where spermatozoa may survive for months (reviewed in Gomendio and Roldan, 1993a,b). On the other hand, mammalian females are sexually receptive for limited periods of time known as "estrus." The length of the estrous period varies considerably between species: in some cases, females are receptive for a few hours in each sexual cycle (e.g., hamster, cattle), whereas in others females are receptive for several days (e.g., dog, horse), or throughout the whole cycle (e.g., marmosets, humans). Among mammals, all ova mature synchronously and are fertile for only a few hours in each sexual cycle (irrespective of whether one ovum or several are ovulated). In addition, most species show no clear signs of the exact timing of ovulation. Thus, when a male mates with an estrous female he has little, if any, information about when ovulation will take place. Because of the short life spans of mammalian spermatozoa, males run the risk of not having any of their sperm still capable of fertilizing by the time the female ovulates. Parker (1984) originally suggested that, in order to maximize the chances of fertilizing the ova, sperm life span should be positively related to the duration of estrus. In support of this hypothesis, Parker found a positive correlation between sperm life-span and duration of estrus when comparing nine mammalian species. This relationship, however, does not hold when phylogenetic effects are removed by using new statistical techniques, but the analysis can be refined by taking into account that the maximum period of time that spermatozoa have to remain fertile is that between the onset of estrus and ovulation. Although estrus usually continues after ovulation takes place, the time period between

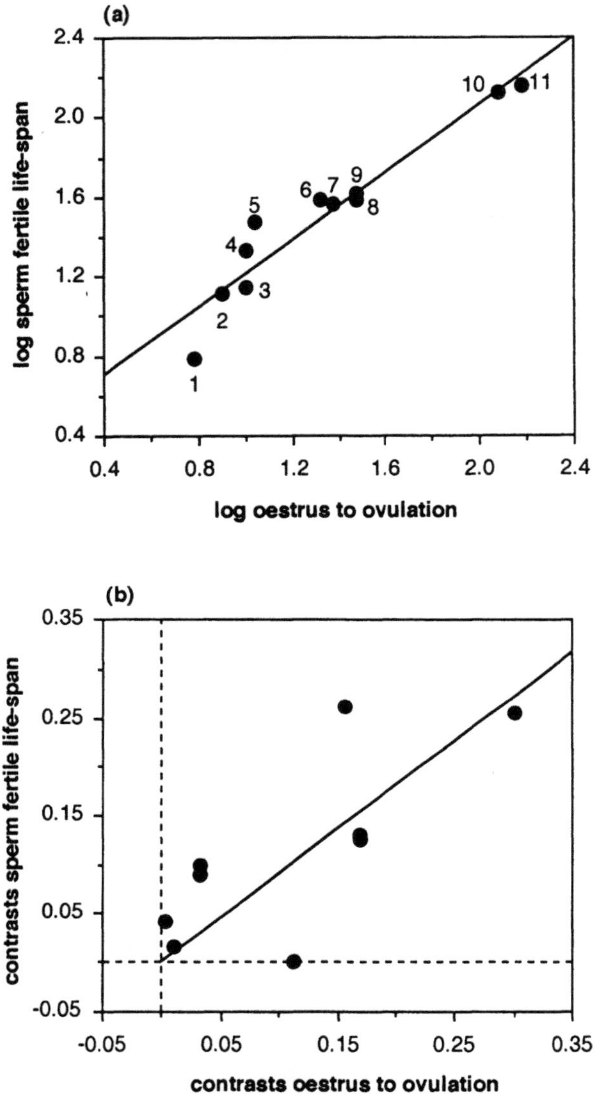

Figure 17. Relation between the interval between the onset of estrus to ovulation and sperm fertile lifespan in mammals. (**a**) Log values for each species: 1, *Mus musculus*; 2, *Mesocricetus auratus*; 3, *Rattus norvegicus*; 4, *Cavia porcellus*; 5, *Oryctolagus cuniculus*; 6, *Ovis aries*; 7, *Sus scrofa*; 8, *Bos taurus*; 9, *Mustela putorius*; 10, *Canis familiaris*; 11, *Equus caballus*. (**b**) Data points are contrasts for each variable. Contrasts are standardized linear differences in a trait between taxa at each node in the phylogeny. The use of contrasts removes phylogenetic effects in comparative studies. The regression line was forced through the origin. (From Gomendio, M. and Roldan, E.R.S., 1993, Proc. R. Soc. Lond. B 252, 7–12, with permission from The Royal Society).

ovulation and the end of estrus is irrelevant in this context. Thus, spermatozoa should have been selected to remain fertile until ovulation takes place, rather than for the whole estrous period. There is, in fact, a strong relationship between sperm fertile life span and the interval between the onset of estrus and ovulation when comparing 11 mammalian genera, which remains highly significant after controlling for phylogenetic effects (Figure 17) (Gomendio and Roldan, 1993b).

An analogous finding has been reported for birds (Birkhead and Møller, 1992). Among birds, the ova are ovulated and fertilized sequentially, and sperm storage duration in the female tract is related to the spread of laying, which is the maximum time span needed to ensure that all ova are fertilized. At present, there is not enough information to test whether sperm storage duration in birds is related to the duration of the female's fertile period.

Eutherian spermatozoa are unique in that they need to become capacitated before they can fertilize the ova, and this process can take several hours (Yanagimachi, 1988). Since sperm fertile life span is so short, it may constrain the time available for capacitation. This seems to be the case indeed: when sperm fertile life span is long, capacitation takes longer than when it is short (Gomendio and Roldan, 1993b). Interestingly, experiments in which spermatozoa have been placed in direct competition, and where there was no differential sperm transport, strongly suggest that spermatozoa which capacitate more quickly are more successful at fertilizing the ova, particularly when inseminated around or after ovulation (Krzanowska, 1986).

Differences in sperm fertile life span between different species may be related to differences in sperm size. Intergeneric comparisons have revealed a negative relationship between sperm size and sperm life span, and showed that when females are receptive for long periods before ovulation takes place, males produce short spermatozoa which may live for longer (Gomendio and Roldan, 1993b). The obvious question to arise from these results is why should sperm size be related to sperm life span. Tentatively, these authors have suggested an energetic argument based on Cummins and Woodall's (1985) finding that increases in total sperm length are achieved mainly through increases in the length of the principal piece of the tail which are accompanied by relatively smaller increases in midpiece length. The midpiece contains the mitochondria which are the source of energy for the sperm cell. Gomendio and Roldan (1993b) have proposed that large spermatozoa may have relatively less energy to utilize, and that this energy may be consumed more quickly due to the higher requirements of larger cells and to their faster swimming speed. Alternatively, it may be that sperm fertile life span is determined through the interactions that take place between spermatozoa and the walls of the female tract. In fact, it has been shown that during the period of residence in the lower isthmus of the oviduct only those spermatozoa that associate to the oviductal epithelial cells survive and, furthermore, that the duration of this period of residence depends on the timing of ovulation (e.g., Smith and Yanagimachi, 1990).

Mammalian males can deposit their ejaculates in the vagina, the cervix, or directly into the uterus. In addition, female mammals vary greatly in the length of

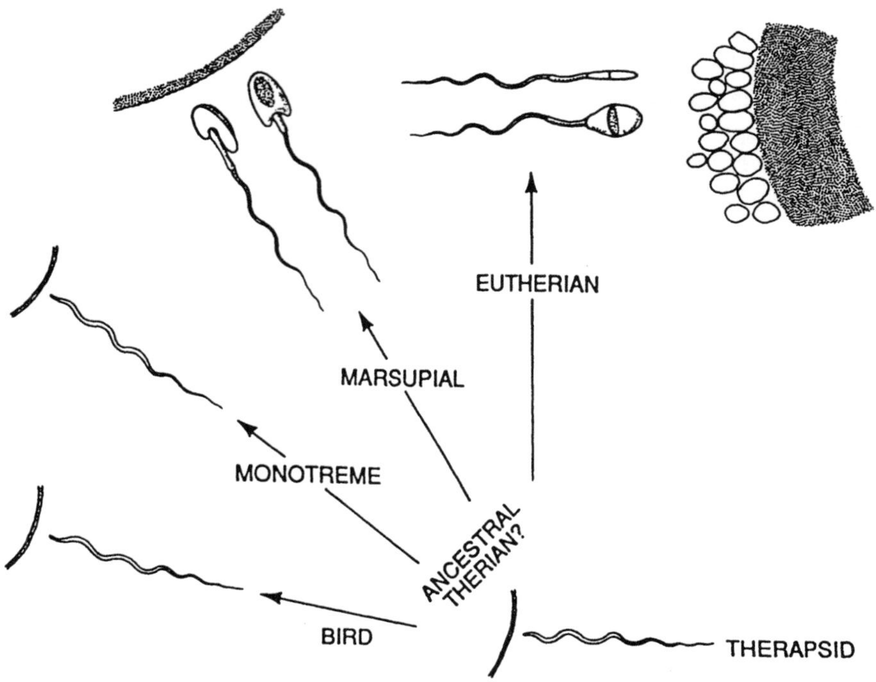

Figure 18. Schematic representation of the evolutionary radiation of the three subclasses of mammals and accompanying development of certain group-specific characteristics in both the male and female gametes. [From Bedford, J.M. (1991). In: A Comparative Overview of Mammalian Fertilization (Dunbar, B. S. and O'Rand, M. G., Eds.). Plenum Press, New York, with permission from the author and Plenum Press).

the reproductive tract. Thus, spermatozoa from different species will have to travel very different distances in order to reach the ova. Once in the uterus, spermatozoa are first transported passively by muscular movements of the female tract, but must then swim actively to go through the utero-tubal junction and along the oviduct (Suarez et al., 1990). It may be advantageous for males to produce relatively greater numbers of sperm when the distance they have to swim is long to compensate for the greater losses incurred. It has been found that males do indeed produce greater number of spermatozoa in relation to body size when females have long oviducts in relation to their body size (Gomendio and Roldan, 1993b).

Within mammals, Cohen (1977) found an inverse relationship between the complexity of flagellar fiber arrangements and the distance the gamete must swim, which was just the opposite of what he had predicted. However, as Sivinski (1984) has pointed out, the original prediction may have been misguided, for no matter how short the race is, the object is to win.

Sperm traits may also be related to the thickness of the ovum vestments that sperm have to penetrate in order to achieve fertilization. Bedford (1991) has argued that the evolution of the different types of spermatozoa found in the three mammalian groups can be explained in terms of differences in ovum vestments (Figure 18). At the time of fertilization, the monotreme ovum is surrounded by a rather thin zona pellucida and, in keeping with this minimal coat, monotreme spermatozoa have remained primitive with a rather fragile head. Among marsupials, the ovum is surrounded by a more elaborated zona pellucida and spermatozoa, although still rather fragile, have evolved an unusual sperm head design which appears to favor the sperm's ability to apply the full area of the acrosome to the zona pellucida. Thus, marsupial spermatozoa would seem to have responded to an increase in the thickness of the ovum vestments by enhancing the spermatozoon's ability to penetrate them via enzymatic activity. Finally, the eutherian ovum has developed an exceptionally thick zona pellucida and, in addition, it is generally surrounded by the cumulus oophorus at the time of fertilization. In response to these changes, mammalian spermatozoa have evolved a number of traits which improve the physical thrust needed to penetrate the ovum vestments. These sperm traits include: enhanced stability of the sperm head, a larger acrosome with a stable inner acrosomal membrane, and the presence of hyperactivated motility. Rather than enhancing the enzymatic content of the acrosome, these traits lend a stability and shape appropriate to the adoption of a strategy of physical thrust by the spermatozoon.

Sperm penetrate the ova vestments by thrusting movements of the flagella and by lateral oscillations of the head (Bedford, 1991). It has been suggested that sperm size might determine to some extent the ability of sperm to penetrate the ova vestments, because sperm with long flagella are able to generate greater forces. If this argument is correct, then sperm size should increase as the thickness of the zona pellucida increases. Contrary to this prediction, sperm size does not seem to be related to the thickness of the zona pellucida (Gomendio and Roldan, 1993b). Thus, the forces generated by the flagellum when the spermatozoon penetrates the zona pellucida do not seem to have played an important role in selecting for sperm size. The lack of association between sperm size and the thickness of the zona pellucida suggests that either sperm penetration may be achieved primarily by enzymatic digestion of this ovum vestment, or that it involves synergy between enzyme activity and sperm forces. Moreover, it is likely that both the physical properties of the ova vestments and the enzyme contents of the acrosome vary between species.

It has also been proposed that the presence of hooks in the sperm head may be related to the ability to penetrate the ovum vestments. However, in a comparative study between three species of Australian conilurine rodents, McGregor et al. (1989) failed to find a positive relationship between the complexity of the sperm head (i.e., presence of either one or three hooks) and the thickness of the zona pellucida or the structure of the surrounding vestments.

Given the magnitude of the barriers that spermatozoa encounter within the female tract, it is worth asking why these have evolved (reviewed in Roldan et al., 1992). These barriers may enable the females to avoid fertilizations which would result in nonviable embryos, such as those resulting from interspecific fertilizations (Roldan and Yanagimachi, 1989) or polyspermy, and to prevent abnormal sperm from reaching the ova. The need for active swimming seems to be an important barrier since abnormal sperm swim more slowly and are soon left behind. It is possible that females may select on a finer scale spermatozoa within a given ejaculate. The active participation of females in the transport of sperm along the female tract, and the complexity of the interactions between spermatozoa and the oviductal walls and ova vestments, suggest that there may be many ways in which this selection could be achieved. In this context, it could be argued that the biases in sex ratio at birth found in several mammalian species (see review in Clutton-Brock, 1991) could be the result of differential female selection for X- or Y-bearing sperm; it would be highly adaptive for females to be able to exercise such choice given the considerable reproductive consequences derived from producing male or female offspring, at least under some circumstances.

Finally, when spermatozoa from different males compete within the female tract (see below), females may be able to select spermatozoa from particular males (reviewed in Birkhead and Møller, 1993; Gomendio and Roldan, 1993a). Even if direct selection on the part of the female were not possible, the female tract may provide a competitive arena where the "abilities" of different ejaculates would be tested. Selection between ejaculates from different males would be highly adaptive if females could pass on the competitive traits of the successful ejaculates to their sons. On highly speculative grounds, it has also been suggested that females would benefit from choosing the best male genotypes on the basis of ejaculate features. The possibility that females may favor ejaculates from particular males, or choose certain kinds of sperm within a given ejaculate, highlights the role that conflicts of interest between males and females may have played in the evolution of reproductive traits.

Sperm Competition

Sperm competition takes place when sperm from rival males compete to fertilize the limited number of ova available. It has been known for a long time that sperm competition is common among insects, partly due to the long periods of time that females can store sperm from several matings in the storage organs (Parker, 1970). When there is sperm competition, male insects seem to compete primarily for access to the spermatheca, and in some species males have developed penile structures that remove the sperm from previous matings.

Because birds are "monogamous," sperm competition was thought to be rare in this group. However, recent work has shown that members of these monogamous pairs commonly engage in the so-called "extra-pair copulations" and thus that

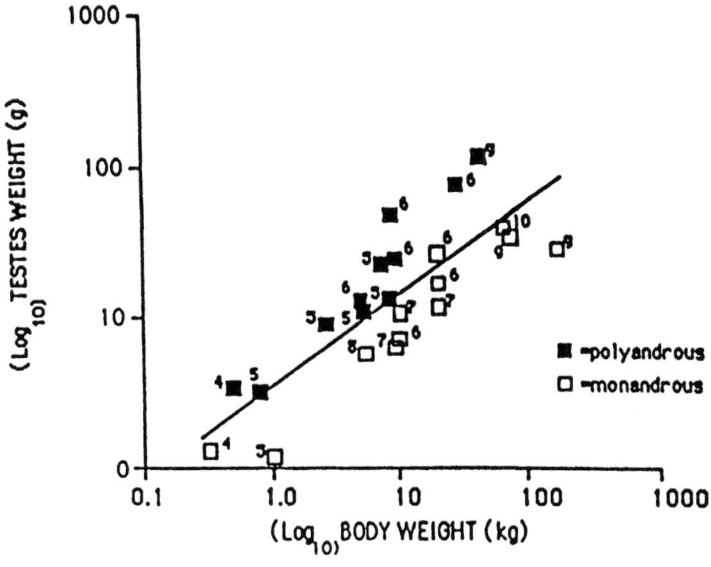

Figure 19. Relation between testes weight and body weight and the effect of mating system among primates. Testes weight (g) relative to body weight (kg)(both $\log_{10}$) were plotted for 23 primate genera, comparing monandrous (N = 12) with polyandrous (N = 11) genera. Empty symbols = monandrous genera; filled symbols = polyandrous genera. Numbers = families: 4, Callitrichidae; 5, Cebidae; 6, Cercopithecinae; 7, Colobinae; 8; Hylobatidae; 9, Pongidae; 10, Hominidae. (From Harcourt, A. H.. 1991, Evolution 45, 314–328, with permission from The Society for the Study of Evolution).

sperm competition is rather widespread (Birkhead and Møller, 1992). Since female birds also store sperm in storage organs there is competition between sperm from rival males to occupy the storage tubules. This competition could take place in two different ways: stratification or displacement (Birkhead and Hunter, 1990). Although our understanding of these mechanisms is still rather poor, preliminary evidence seems to suggest that displacement of previous spermatozoa by spermatozoa from the next male to mate is the most likely option.

By contrast, the lack of sperm storage organs in mammals precludes the possibility of physical displacement or stratification between ejaculates from rival males. Once ejaculated, mammalian spermatozoa disperse and are rapidly transported towards the upper sections of the tract. It has therefore been suggested that, among mammals, competition between ejaculates from rival males resembles a raffle with the male who delivers more spermatozoa being more likely to fertilize the ova (the so-called "raffle" principle) (Parker, 1984). For quite a long time the best supporting evidence that sperm competition had favored the evolution of high sperm numbers was the fact that mammalian males in polyandrous species (i.e., females mate with more than one male in each sexual cycle) have larger testes in relation to their body size than males in monandrous species (i.e., females mate with one male in each sexual cycle) (Figure 19) (Harcourt et al., 1981; Kenagy and Trombulak, 1986). However, it remained to be shown that relatively large testes do indeed produce more sperm. Recent studies have addressed this issue and shown that relatively large testes have a higher daily production rate, generate larger sperm reserves, and lead to more sperm per ejaculate (Møller, 1989).

Parker (1982, 1984) originally suggested that the evolution of high sperm numbers under sperm competition may have been achieved at the expense of a reduction in the size of each male gamete. The assumption underlying this proposal is that all spermatozoa have equal chances of fertilizing the ova and, therefore, that the best possible strategy is to produce as many spermatozoa as possible no matter how small. A trade-off between sperm numbers and sperm size has been assumed ever since, but has remained untested. A recent study has shown that such a trade-off does not exist in mammals since, contrary to the prediction, polyandrous species have longer spermatozoa than monandrous species (Figure 20) (Gomendio and Roldan, 1991). The differences in sperm size are largely explained by differences in tail length, which is the most variable portion of mammalian sperm. The adaptive value of long-tailed spermatozoa under sperm competition seems to lie in the faster swimming speed that they achieve. Thus, when competing against spermatozoa from rival males to fertilize the ova, long-tailed spermatozoa are able to swim faster and reach the ova sooner.

It is still not known in which parts of the mammalian female tract do rival spermatozoa compete against each other, but physiological evidence points to the portion between the lower isthmus of the oviduct and the oviductal ampulla (the site of fertilization) as a potentially important one. This is because, after the period of residence in the lower isthmus of the oviduct, spermatozoa seem to detach from the oviductal epithelial cells in response to a "signal" released around the time of ovulation (Hunter, 1988) and start to swim actively towards the ova; the spermatozoa that reach the ova first are the ones most likely to achieve fertilization.

The advantages conferred by longer spermatozoa to male mammals may not be directly applicable to other taxa because of the differences in the physiological mechanisms underlying sperm competition. In birds, sperm competition seems to have favored the evolution of long (perhaps, also in this case, faster-swimming) spermatozoa when sperm storage tubules are in short supply, and of spermatozoa that are long enough to fill the sperm storage tubules thus preventing access to them by rival spermatozoa (Briskie and Montgomerie, 1992).

An alternative view holds that the way spermatozoa from rival males compete within the female tract resembles more a battle than a race. The "kamikaze sperm hypothesis" proposes that not all spermatozoa within an ejaculate have a fertilizing function, but rather that certain sperm morphs are designed to block spermatozoa from rival males (e.g., by forming plugs), or to directly attack and destroy other spermatozoa (Baker and Bellis, 1988, 1989). This hypothesis has attracted some

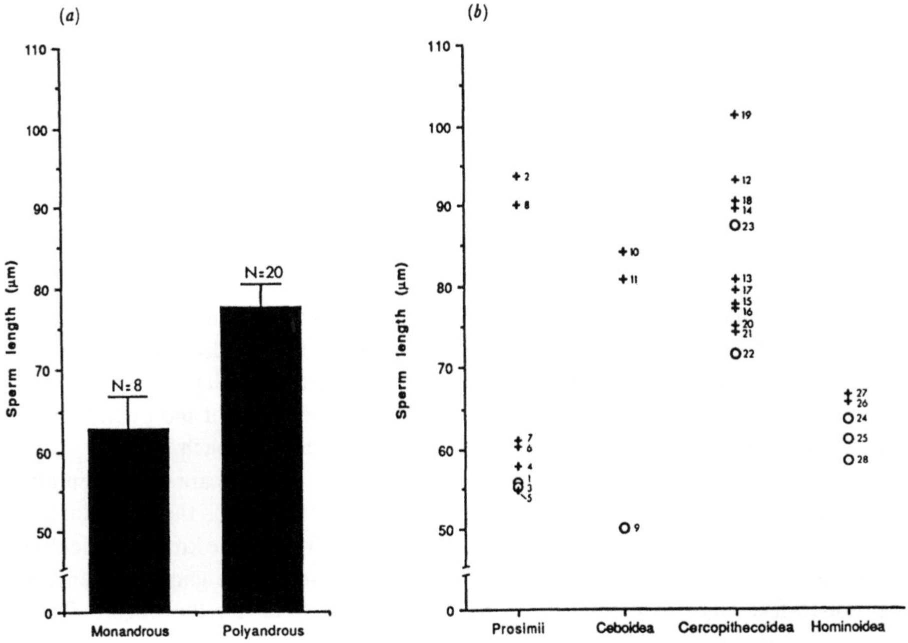

Figure 20. Differences in total sperm lengths between monandrous and polyandrous primate species. (**a**) Mean sperm lengths in monandrous versus polyandrous species (N = 28; *P* = 0.01); (**b**) Sperm length within different taxa; circles = monandrous species; crosses = polyandrous species (2-factor ANOVA, mating system: *P* = 0.005; taxon: *P* = 0.009, AB: not significant). Prosimii: (1) *Daubentonia madagascarensis*; (2) *Microcebus murinus*; (3) *Lemur mongoz*; (4) *Lemur catta*; (5) *Lemur macaco*; (6) *Galago crassicaudatus*; (7) *Galago senegalensis*; (8) *Nycticebus coucang*. Ceboidea: (9) *Callithrix jacchus*; (10) *Cebus apella*; (11) *Saimiri sciureus*. Cercopithecoidea: (12) *Cercocebus galeritus*; (13) *Cercopithecus aethiops*; (14) *Erythrocebus patas*; (15) *Macaca arctoides*; (16) *Macaca fascicularis*; (17) *Macaca mulatta*; (18) *Macaca nemestrina*; (19) *Macaca speciosa*; (20) *Papio anubis*; (21) *Papio cynocephalus*; (22) *Papio sphinx*; (23) *Theropithecus gelada*. Hominoidea: (24) *Hylobates lar*; (25) *Gorilla gorilla*; (26) *Pan troglodytes*; (27) *Pongo pygmaeus*; (28) *Homo sapiens*. (From Gomendio, M. and Roldan, E.R.S., 1991, Proc. R. Soc. Lond. B 243, 181–185, with permission from The Royal Society).

attention, but so far there is little evidence to support it. Perhaps the most serious drawback is the lack of specific predictions concerning which sperm morphs perform the various functions that have been proposed. Harcourt (1991) has derived a number of predictions from the original hypothesis and concluded that they are not supported by the available evidence. Harcourt's argument relies mainly on the fact that monandrous and polyandrous primates do not differ in the proportion of normal spermatozoa, proportion of motile spermatozoa, or the coefficient of variation of sperm size in their ejaculates.

Gametes can therefore be seen as the "ultimate tool" that males and females have to maximize their reproductive success. Under this light, the diversity of gametic morphology and behavior acquires an entirely new dimension which we hope will stimulate at least some of those reading this chapter to work towards improving our understanding of these highly specialized cells.

SUMMARY

The evolutionary changes undergone by gametes have drawn much attention, but most work has remained at a rather descriptive level given the difficulties encountered when efforts have been made to explain gamete traits as adaptations. Our aim in this chapter has been to try to advance our understanding of the evolutionary meaning of gamete traits by adopting a multidisciplinary approach.

In species that reproduce sexually, male and female gametes are fundamentally different. These differences have far-reaching implications for the reproductive strategies followed by the two sexes which, in turn, influence the kinds of selective pressures that have modeled the evolution of male and female gametes. Females produce few large gametes that contain nutritive material for the zygote. Their reproductive strategy is aimed at maximizing offspring survival; one of the ways in which this can be achieved is by allocating the right amount of nutrients to the ovum. Thus, many of the differences in ovum size and content can be explained as adaptations aimed at maximizing offspring survival under given ecological conditions.

Males, on the contrary, produce vast numbers of small gametes and their main reproductive strategy is aimed at maximizing the number of fertilizations that they achieve. Thus, the main selective forces which have driven the evolution of spermatozoa are the need to be adapted to the environment in which fertilization will take place (in many cases this will be the female's internal environment) and competition with sperm from rival males to fertilize the ova.

ACKNOWLEDGMENTS

We wish to thank Alfredo Vitullo for many stimulating discussions over a number of years. We also thank authors and publishers who have allowed us to reproduce illustrations. Our work is funded, in part, by the Agricultural and Food Research Council (U.K).

REFERENCES

Austin, C. R. (1965). Fertilization. Prentice-Hall, New Jersey.

Austin, C. R. (1975). Sperm fertility, viability and persistence in the female tract. J. Reprod. Fert., Suppl. 22, 75–89.

Austin, C. R. (1982). In: Reproduction in Mammals (Austin, C. R. & Short R. V., Eds.), Vol. 1, 2nd ed. Cambridge University Press, Cambridge.

Baccetti, B. (1985). In: Biology of Fertilization (Metz, C. B. & Monroy, A., Eds.), Vol. 2. Academic Press, Orlando.

Baccetti, B. & Afzelius, B. A. (1976). The Biology of the Sperm Cell. S. Karger, Basel.

Baker, R. R. & Bellis, M. A. (1988). 'Kamikaze' sperm in mammals ? Anim. Behav. 36, 936–939.

Baker, R. R. & Bellis, M. A. (1989). Elaboration of the kamikaze sperm hypothesis: A Reply to Harcourt. Anim. Behav. 37, 865–867.

Bedford, J. M. (1991). In: A Comparative Overview of Mammalian Fertilization (Dunbar, B. S. & O'Rand, M. G., Eds.). Plenum Press, New York.

Befdord, J. M. & Hoskins, D. D. (1990). In: Marshall's Physiology of Reproduction (Lamming, G. E., Ed.), Vol. 2. Churchill Livingstone, London.

Bedford, J. M. & Kim, H. H. (1993). Cumulus oophorus as a sperm sequestering device, *in vivo*. J. Exp. Zool. 265, 321–328.

Bedford, J. M., Rodger, J. C., & Breed, W. G. (1984). Why so many spermatozoa—a clue from marsupials? Proc. R. Soc. Lond. B. 221, 221–233.

Birkhead, T. R. & Hunter, F. M. (1990). Mechanisms of sperm competition. Trends Ecol. Evol. 5, 48–52.

Birkhead, T. R. & Møller, A. P. (1992). Sperm Competition in Birds. Academic Press, London.

Birkhead, T. & Møller, A. (1993). Female control of paternity. Trends Ecol. Evol. 8, 100–104.

Blueweiss, L., Fox, H., Kudzma, V., Nakashima, D., Peters, R. & Sams, S. (1978). Relationships between body size and some life-history parameters. Oecologia 37, 257–272.

Briskie, J. V. & Montgomerie, R. (1992). Sperm size and sperm competition in birds. Proc. R. Soc. Lond. B. 247, 89–95.

Clutton-Brock, T. H. (1991). The Evolution of Parental Care. Princeton University Press, Princeton.

Cohen, J. (1977). Reproduction. Butterworths, London.

Congdon, J. D., Dunham, A. E., & Tinkle, D. W. (1982). In: Biology of Reptilia (Gans, C. & Billett, F., Eds.). Academic Press, New York.

Cummins, J. M. & Woodall, P. F. (1985). On mammalian sperm dimensions. J. Reprod. Fert. 75, 153–175.

Dybas, L. K. & Dybas, H. S. (1981). Coadaptation and taxonomic differentiation of sperm and spermathecae in featherwing beetles. Evolution 35, 168–174.

Franzen, A. (1977). Sperm structure with regard to fertilization biology and phylogenetics. Verhand. Deutsch. Zool. Gessellsch. 1977, 123–138.

Ginsberg, J. R. & Huck, U. W. (1989). Sperm competition in mammals. Trends Ecol. Evol. 4, 74–79.

Gomendio, M. & Roldan, E. R. S. (1991). Sperm competition influences sperm size in mammals. Proc. R. Soc. Lond. B. 243, 181–185.

Gomendio, M. & Roldan, E. R. S. (1993a). Mechanisms of sperm competition: Linking physiology and behavioural ecology. Trends Ecol. Evol. 8, 95–100.

Gomendio, M. & Roldan, E. R. S. (1993b). Coevolution between male ejaculates and female reproductive biology in eutherian mammals. Proc. R. Soc. Lond. B. 252, 7–12.

Harcourt, A. H. (1991). Sperm competition and the evolution of nonfertilizing sperm in mammals. Evolution 45, 314–328.

Harcourt, A. H., Harvey, P. H., Larson, S. G. & Short, R. V. (1981). Testis weight, body weight and breeding system in primates. Nature (London) 293, 55–57.

Harding, H. R., Aplin, K., & Shorey, C. D. (1987). In: New Horizons in Sperm Cell Research. (Mohri, H., Ed.). Japan Scientific Societies Press, Tokyo.

Hinton, H. E. (1981). Biology of Insect Eggs. Pergamon Press, Oxford.

Hunter, R. H. F. (1988). The Fallopian Tubes. Their Role in Fertility and Infertility. Springer-Verlag, Berlin.

Jamieson, B. G. M. (1987). The Ultrastructure and Phylogeny of Insect Spermatozoa. Cambridge University Press, Cambridge.

Jamieson, B. G. M. (1991). Fish Evolution and Systematics: Evidence from Spermatozoa. Cambridge University Press, Cambridge.

Kaplan, R. H. & Salthe, S. N. (1979). The allometry of reproduction; An empirical view in salamanders. Am. Nat. 113, 671–689.

Kenagy, G. J. & Trombulak, S. C. (1986). Size and function of mammalian testes in relation to body size. J. Mammal. 67, 1–22.

Krebs, J. R. & Davies, N. B. (1993). An Introduction to Behavioural Ecology (3rd. Edition). Blackwell Scientific Publications, Oxford.

Krzanowska, H. (1986). Interstrain competition amongst mouse spermatozoa inseminated in various proportions, as affected by the genotype of the Y chromosome. J. Reprod. Fert. 77, 265–270.

Lack, D. (1968). Ecological Adaptations for Breeding in Birds. Methuen & Co., London.

Levitan, D. R. (1993). The importance of sperm limitation to the evolution of egg size in marine invertebrates. Am. Nat. 141, 517–536.

McGregor, L., Flaherty, S. P., & Breed, W. G. (1989). Structure of the zona pellucida and cumulus oophorus in three species of native Australian rodents. Gamete Res. 23, 279–287.

Møller, A. P. (1989). Ejaculate quality, testes size and sperm production in mammals. Funct. Ecol. 3, 91–96.

O'Connor, R. (1984). The Growth and Development of Birds. Wiley, Chichester.

Parker, G. A. (1970). Sperm competition and its evolutionary consequences in insects. Biol. Rev. 45, 525–567.

Parker, G. A. (1982). Why so many tiny sperm? The maintenance of two sexes with internal fertilization. J. Theor. Biol. 96, 281–294.

Parker, G. A. (1984). In: Sperm Competition and the Evolution of Animal Mating Systems (Smith, R.L., Ed.). Academic Press, Orlando.

Parker, G. A., Baker, R. R. & Smith, V. G. F. (1972). The origin and evolution of gamete dimorphism and the male-female phenomenon. J. Theor. Biol. 36, 529–553.

Rahn, H., Pagnelli, C..V., & Ar, A. (1975). Relation of avian egg weight to body weight. Auk 92, 750–765.

Roldan, E. R. S., Gomendio, M., & Vitullo, A. D. (1992). The evolution of eutherian spermatozoa and underlying selective forces: Female selection and sperm competition. Biol. Rev. 67, 551–593.

Roldan, E. R. S. & Yanagimachi, R. (1989). Cross-fertilization between Syrian and Chinese hamsters. J. Exp. Zool. 250, 321–328.

Short, R. V. (1981). In: Reproductive Biology of the Great Apes (Graham, C.E., Ed.). Academic Press, New York.

Sinervo, B. (1990). The evolution of maternal investment in lizards: An experimental and comparative analysis of egg size and its effects on offspring performance. Evolution 44, 279–294.

Sinervo, B. & Licht (1991). Proximate constraints on the evolution of egg size, number, and total clutch mass in lizards. Science 252, 1300–1302.

Sivinski, J. (1984). In: Sperm Competition and the Evolution of Animal Mating Systems (Smith, R. L., Ed.). Academic Press, Orlando.

Smith, T. T. & Yanagimachi, R. (1990). The viability of hamster spermatozoa stored in the isthmus of the oviduct: The importance of sperm-epithelium contact for sperm survival. Biol. Reprod. 42, 450–457.

Stearns, S. C. (1992). The Evolution of Life Histories. Oxford University Press, Oxford.

Suarez, S., Drost, M., Redfern, K., & Gottlieb, W. (1990). In: Fertilization in Mammals (Bavister, B. D., Cummins, J. & Roldan, E. R. S., Eds.). Serono Symposia, Norwell, MA.

Tessier, A., Henry, L. L., Goulden, C. E., & Durand, M. W. (1983). Starvation in *Daphnia*: Energy reserves and reproductive allocation. Limnol. Oceanogr. 28, 667–676.

van den Berghe, E. P. & Gross, M. R. (1989). Natural selection resulting from female breeding competition in a pacific salmon (Coho: *Oncorhynchus kisutch*). Evolution, 43, 125–140.

Wootton, R. J. (1979). Energy costs of egg production and environmental determinants of fecundity in teleost fishes. Symp. Zool. Soc. Lond. 44, 133–159.
Yanagimachi, R. (1988). In: The Physiology of Reproduction (Knobil, E. & Neill, J., Eds.). Raven Press, New York.

RECOMMENDED READINGS

Austin, C. R. & Short, R. V. (1982). Reproduction in Mammals, Vol. 1: Germ Cells and Fertilization (2nd ed.). Cambridge University Press, Cambridge.
Bedford, J. M. (1991). In: A Comparative Overview of Mammalian Fertilization (Dunbar, B. S. & O'Rand, M. G., Eds.). Plenum Press, New York.
Bedford, J. M. & Hoskins, D. D. (1990). In: Marshall's Physiology of Reproduction (Lamming, G. E., Ed.), Vol. 2. Churchill Livingstone, London.
Birkhead, T. R. & Møller, A. P. (1992). Sperm Competition in Birds. Academic Press, London.
Birkhead, T. & Møller, A. (1993). Female control of paternity. Trends Ecol. Evol. 8, 100–104.
Gomendio, M. & Roldan, E. R. S. (1993). Mechanisms of sperm competition: Linking physiology and behavioural ecology. Trends Ecol. Evol. 8, 95–100.
Roldan, E. R. S., Gomendio, M., & Vitullo, A. D. (1992). The evolution of eutherian spermatozoa and underlying selective forces: Female selection and sperm competition. Biol. Rev. 67, 551–593.
Smith, R. L. (Ed.) (1984). Sperm Competition and the Evolution of Animal Mating Systems. Academic Press, Orlando.

Chapter 6

The Role of Mutations in Evolution

JAMES F. CROW

Principles of Medical Biology, Volume 1B
Evolutionary Biology, pages 153–182

ISBN: 1-55938-802-1

INTRODUCTION

In many ways Darwin said it all in 1859. Although innumerable details have been added, especially with the growth of molecular genetics, the principle of natural selection stands essentially unchanged. But Darwin's generation had no understanding of heredity; the rules of inheritance were supplied by the rediscovery of Mendel's laws in 1900. Neither Darwin nor Mendel knew the ultimate source of variability, mutation. This came in 1901 from the Dutch geneticist, de Vries, who found sudden phenotypic changes in his evening primroses. The combined ideas of Darwin, Mendel, and de Vries constitute the basic premise of evolution as we now understand it. (For this early history see Sturtevant, 1965.)

From one viewpoint, mutation is the most basic process of evolution; for without mutation, evolution would never have occurred. At the same time, without natural selection, adaptive evolution would not have occurred. Mutation continuously presents new candidate genes. Most of these are rejected by natural selection; however, a few are preserved and become the normal genes of the future. Paraphrasing the Medieval Monk, "Mutation proposes, but selection disposes." The role of Mendelian segregation and recombination is to scramble the genome every generation so that each gene is tested in combination with many others; those that prevail are the ones that interact favorably with a large number of others.

This is a review of the mechanisms of biological evolution. For prebiotic evolution—evolution of self-reproduction (e.g., did protein or RNA come first?), origin of cells, and similar questions—see Bittar (1991). The emphasis of this chapter, as the title implies, is on the role of mutation in evolution, especially the contrasting role of mutation in molecular and organismic evolution.

MUTATION

Mutation is a sudden, stable change in the genetic material; that is to say, the altered state is as stable as the original one before the mutation occurred. Mutations can occur in any cell in the body, and the effects range from a local effect, such as a cell death, to a major bodily impairment, depending on the stage of occurrence and the tissue affected. From the standpoint of evolution, the mutations that matter are the ones that occur in the germ line and hence may be transmitted to future generations. Therefore, only these mutations will be considered.

Mutations and Chromosome Aberrations

In classical genetics, genomic changes are divided into two classes: chromosome aberrations and gene mutations. Chromosome aberrations are visible by light microscopy. They can be changes in chromosome number or rearrangements caused by chromosome breakage. Gene mutations are the residue—changes too small to be visible microscopically. Thus mutation is a catch-all category comprising all changes not classified as cytological. Noncytological mutations, since they are local, follow the rules of Mendelian inheritance. These were often called "point mutations" since they could be localized at a point in the chromosome. This classification, like most others, breaks down at the borders. Small chromosome changes may be indistinguishable from gene mutations. The size of a "point" depends on the sensitivity of the system. In the giant salivary gland chromosomes of *Drosophila*, minute cytological changes can be detected—changes that would be classified as point mutations in most other species. (For a review of transmission genetics, mutation, and chromosome rearrangements, see Strickberger, 1985.)

Molecular genetics has greatly refined the meaning of point mutation. Mutations can now be studied in minute detail and include a continuum from the alteration of a single base pair to rearrangements involving a million nucleotides.

Kinds of Mutational Changes

The smallest, most elementary mutation is the change of a single nucleotide pair. Because of the complementary nature of the two strands of DNA, a change in one strand is soon accompanied by a corresponding change in the other. If A is changed to G, its partner will change from T to C by the Watson–Crick pairing rules. It is convenient to classify single base changes into *transitions* (changes from one purine to another, or a pyrimidine into another) and *transversions* (changes from a purine to a pyrimidine, or vice versa). Of the three possible changes of a nucleotide, two are transversions and one is a transition; for example, A can change to G (transition), or to C or T (transversions). Although transversions can occur in twice as many ways, transitions occur far more frequently. This is reasonable because most mutations involve the misincorporation of a base, often at the time of replication, and such an error more often involves incorporation of a base of the same size.

There are numerous repair and proofreading processes that detect errors and set them right. As a result, mutations are much rarer than would be expected from ordinary chemical principles (e.g., from the frequency of tautomeric shifts which cause misincorporation). The errors that persist are those that elude the team of efficient repair enzymes. A typical mutation rate per base pair is 10^{-8} per generation. The rate per DNA replication is still less because of the large number of cell divisions between zygote and gamete. Thus, the error rate in higher eukaryotes is of the order 10^{-10} per DNA replication. One of evolution's finest accomplishments is the remarkable accuracy of DNA replication.

Although single base substitutions predominate in molecular evolution, they are by no means the only changes. There are many mechanisms for removing, adding, or transposing pieces of DNA ranging in size from a single base to several genes.

Another way in which DNA is changed is by errors in crossing over. Ordinarily, the crossover process is remarkably exact; homologous segments are exchanged precisely, leaving no gaps or duplications. But the process occasionally makes errors so that one recombinant product has extra material and the other a corresponding deficiency. If such a change adds or subtracts an amino acid, it may make little difference. But if the change is not an even multiple of three, it throws the translation system out of register, usually rendering the gene nonfunctional. The duplication or deficiency may also be larger, involving a whole gene or many genes.

Once a duplication occurs it predisposes to additional ones, for the meiotic chromosomes are then much more likely to synapse out of register. Because of this, the exchange produces further duplications or deficiencies. Sometimes there are mechanisms that preserve increases in preference to decreases. By a continuation of this process, long sequences of repeated units can build up. Many of these are known, some involving thousands or even millions of repeated units.

Another process that can cause changes larger than one or a few base pairs is gene conversion. In the process of repair of chromosome breaks that arise, often during crossing over, the repair process may erroneously introduce bases from the homologous chromosome or even from nonhomologous regions. Such changes are often indistinguishable from orthodox mutations.

Still another process that produces mutational changes is movement of a piece of chromosome by a *transposable element*. Such elements exist in almost all species and are of a variety of types. What they have in common is the capacity to leave one chromosomal location and insert themselves at another. If an element is inserted into a functional gene, it is very likely to disrupt the gene function, perhaps mildly but more often drastically. It is thus a mutation, and differs from other mutations only in that the new form is likely to be comparatively unstable and to revert to the original form if the element is excised. If the excision is imprecise, the gene may be permanently mutated. Most of the DNA in higher eukaryotes, and mammals in particular, has no known function. If an element is inserted into this "junk DNA" it may have no observable consequence. It may eventually lose its capacity to move and simply add to the amount of (presumably) functionless DNA.

Although all these mechanisms play a role in evolution, which of them are the most important is not known. Single base substitutions predominate numerically in molecular evolution, but they are not necessarily more important for this reason. Various other changes also are found in evolutionary phylogenies: duplications, deletions, conversions, and transpositions. All have played a part in generating the variability on which evolution depends.

Cytogenetic Changes

The mechanisms so far mentioned can generate changes of various sizes, from single bases to those involving large segments of the chromosome. Thus they merge into classical cytogenetic changes of the kind that have been known for many years.

An *inversion* is produced if a chromosome breaks in two places and the interstitial piece is reversed and reattached. This may have no phenotypic effect if both breaks fall between genes, although such a change may lead to meiotic complications when an inverted chromosome tries to pair with a partner chromosome having the original sequence. If one of the breaks is within an important region, the function is disrupted. Such inversions frequently have a drastic or lethal effect, and are quickly eliminated from the population. The main consequence of inversion for long-time evolution is the rearrangement of gene order. Gene sequences of related species often show traces of inversions in their history, so that the gene orders of even closely related species are not necessarily the same.

If the interstitial piece between the two breaks is lost rather than inverted the chromosome has a *deficiency*. Large deficiencies are usually lethal.

Two nonhomologous chromosomes may both break and reattach in new ways, producing a *translocation*. If the broken ends join in such a way that each new chromosome has a centromere, the translocated chromosomes go through mitosis normally. Meiosis, however, is upset if a pair of translocated chromosomes inherited from one parent synapse with a normal set from the other. This frequently leads to the formation of gametes with too many or too few chromosome parts, and typically these lead to lethality. If the translocated chromosomes become homozygous, they behave like normal chromosomes in meiosis, and the consequence is a strain with genes in a different order. Translocations do occur in evolution, but rarely. Humans differ from chimpanzees by a translocation. Actually the "mutations" that deVries reported at the turn of the century turned out to be segregants from complex chromosome rearrangements.

Even normal chromosomes can make errors during meiosis (or during somatic mitoses, but our interest is in those that are transmitted to the next generation). An individual with too many or too few chromosomes or chromosome parts is called *aneuploid*. Usually the gain or loss of a chromosome is lethal, so most of these changes have no evolutionary future. Occasionally, however, there are exceptions, and gaining a chromosome in this way has had a permanent effect in the evolution of some species.

A larger and more significant chromosome error is the duplication of a whole set of chromosomes—a *ploidy* change. Just as gametes are haploid and zygotes are diploid in most animals (humans for example), there can be *triploids* (one extra chromosome set), *tetraploids* (two extra sets), and so on. The generic adjective for individuals with more than the regulation number of chromosome sets is *polyploid*.

The most important generalization is that polyploidy is usually less harmful than aneuploidy. A proper balance among the relative numbers of different chromosomes is more important than the absolute numbers. Therefore, polyploidy does less harm than aneuploidy; frequently none. Triploids usually have no evolutionary future, since an odd number of chromosomes complicates meiosis and leads to sterility. However, they go through mitosis normally. For this reason, triploids are essentially nonexistent in sexually reproducing species, but are relatively common in those that reproduce asexually. Several domestic plants that are propagated by cuttings (bananas, for example) have an odd number of chromosomes, whereas their wild relatives that reproduce by seeds have an even number.

The great evolutionary importance of polyploidy is that it provides for a new species in one step, or can render fertile a sterile interspecies hybrid. The significance of these for evolution will be discussed later.

Cytoplasmic Mutations

Some cytoplasmic organelles have their own DNA and are self-reproducing. Mitochondria occur in almost all eukaryotes (some protozoa provide interesting exceptions), and green plants have chloroplasts in addition. Mutations in cytoplasmic organelles are essentially the same as those of chromosomal genes. Their transmission is maternal, since the sperm has very little cytoplasm. Mitochondrial genes typically have much higher mutation rates than chromosomal genes; one reason is the absence of some repair enzymes. Mitochondria and plastids were derived in the distant evolutionary past from microorganisms that were once free-living, but have been domesticated and incorporated into the cellular machinery.

Although mitochondria and plastids have genes that carry out many of their functions more or less independently of the cell in which they occur, organelles are not responsible for all their own functions. Some of their enzymes are controlled by nuclear genes. This makes sense: If the organelle were completely autonomous it would determine its own replication cycle, which would soon get out of phase with the cell nucleus. Hence, the nuclear genotype has cleverly maintained enough control to assure synchrony.

Although mitochondrial DNA is an insignificant fraction of the total DNA—and therefore relatively unimportant in evolution—its rapid mutation rate and maternal inheritance have made it a favorite subject for evolutionary research. The controversial "Eve" hypothesis (that the present female lineage can be traced to a single African woman) is based on mitochondrial studies. Mitochondrial changes provide a convenient way for tracing the ancestry of closely related forms, such as human races.

Induced Mutations

There are a number of ways of influencing the mutation process. Not surprising for a chemical process, the mutation rate is temperature-dependent with the rate increasing sharply with temperature. In general, there is little specificity; all genes mutate more rapidly as the temperature rises, although this subject has been explored only sporadically.

In the 1920s, two geneticists, H. J. Muller and L. J. Stadler, independently discovered that ionizing radiation increases the mutation rate. In general, the effect is to increase the rate, but as with temperature the increase is not specific for particular genes or particular mutant alleles. It is true, however, that radiation increases chromosome breakage much more than it increases point mutations. Indeed, the efficiency of radiation in producing mutations by processes other than chromosome breakage is very low. Ultraviolet radiation is also an effective mutagen, although little UV reaches the germ cells in higher animals.

During World War II, Charlotte Auerbach and J. M. Robson discovered that mustard gas produces mutations (for the interesting history of this discovery and the complications of wartime secrecy, see Beale, 1993). This opened the way for the discovery of a large number of mutagenic (and some antimutagenic) chemicals. At first, the search for such mutagens was rather haphazard and there were few guides as to how to predict mutability of a chemical, although many were alkylating agents. The breakthrough came with the discovery that the gene is DNA. This opened mutation to rational analysis, based on the chemistry of nucleic acids.

The subject is now very sophisticated. It is now possible to produce specific base changes in specific genes—site-directed mutagenesis—and this is now one of the most useful tools for genetic research. Another tool is a "knockout" mutation, which inactivates the gene function. Much of what is known about development comes from noticing what goes wrong when a normal gene mutates.

Useful as induced mutation is as a subject for genetic research, it has had little impact on our knowledge of evolution. There have possibly been times when a changed mutation rate caused by an environmental agent had an evolutionary consequence, but this is highly speculative. The most likely candidate is ultraviolet radiation in the early stages of life when atmospheric oxygen and UV-opaque ozone were much less abundant than now.

Epigenetic Changes

Modern research has turned up a large number of transient DNA changes. One of these is *imprinting*, whereby the expression of a gene differs in accordance to whether it came from the mother or the father. Frequently, the mechanism involves methylation of some nucleotides, which is known to regulate gene action. Differences between maternal and paternal inheritance constitute unusual, but interesting exceptions to the usual rules of Mendelian inheritance, and are an active subject of research. The evolutionary consequences are minor, however, for the effects are evanescent, persisting only a few generations (usually only one). They therefore have little or no lasting influence on evolution (although how such processes have evolved is of great interest, but totally unknown).

Mutant Phenotypes

From an evolutionary standpoint, the importance of a mutation is its phenotypic effect. These effects range from trivial to lethal; they involve every structure, every process, and every stage of development. Everything that is gene-influenced—and virtually everything in biology to some extent—is mutation-influenced. A mutation that kills or sterilizes its host is of course eliminated from the population and has no evolutionary consequence. Evolution depends on those with milder effects, among which some may be beneficial and become the normal genes of a later time.

There is some correlation between the size of the region of the chromosome affected by a mutation and its phenotypic effect. Individual base substitutions often have minor, even negligible effects. But there are many exceptions; for example, a single base substitution changes a certain glutamic acid to valine and thereby changes normal hemoglobin A to sickle cell hemoglobin S. A chain-terminating ("nonsense") mutation usually incapacitates a protein. Another drastic example is the mutation causing the devastating Huntington's disease. In this case, a CAG triplet, normally duplicated a dozen or so times, increases several fold, and this change somehow produces the disease. In contrast, tetraploids are often normal plants, sometimes superior to their diploid parents. So the correlation between the size of the genomic change and the phenotypic effect is weak.

In contrast, there is a strong correlation between the magnitude of the phenotypic effect and the degree of harmfulness. In the first place, the overwhelming majority of mutations, if they have an effect large enough to be detected at all, are harmful. This is not surprising, for mutations are random changes in the genetic material; random at least with respect to phenotypic consequences. Random tinkering with a functioning machine, such as the human body, is far more likely to do harm than good. As one would expect, a tiny change is much less likely to do harm than a large one. To use the familiar watch analogy, a tiny movement of the timing screw may even make the watch more accurate, whereas a large change is almost certain to make it worse, and a hammer blow will incapacitate it.

One would expect that as the size of the mutation effect decreases, the ratio of beneficial to deleterious effects becomes higher and approaches 1/2 as the mutation effect approaches zero. There is good reason, as will be discussed later, to believe that most of the base substitutions that occur in DNA evolution are so small in their effect as to be selectively neutral.

Mutation Rates

The most direct and obvious way to measure the human mutation rate is to see how often two normal parents produce a child with a disease caused by a dominant gene. In a slightly less direct way, the rate of mutation of severe X-linked recessives can be measured. Estimates of human mutation rates made in this manner vary rather widely but most are in the range of 10^{-6} to 10^{-5} per gene per generation.

(Such a procedure does not, however, work for autosomal recessives; the mutation observed as a homozygote may have occurred many generations earlier, during which time its frequency could have increased or decreased, thus rendering any estimate nearly useless.)

Similar rates per generation are obtained for individual gene mutations that produce conspicuous phenotypes in *Drosophila*. This leads to a first conclusion: mutation rates in different species are much more nearly equal when measured as the rate per generation than when measured as the rate per absolute time unit. Clearly, evolution has adjusted the average mutation rate to the length of the reproductive life cycle.

A striking fact emerging from the study of human dominant diseases is that the probability of having a mutant child rises sharply with the age of the father, and not with the age of the mother (except in so far as husband and wife's ages are correlated). The probability of a mutation increases by about fourfold as the age of the father goes from 30 to 45. This is consistent with the fact that there are many more cell divisions between zygote and sperm than between zygote and egg, and that the number prior to spermatogenesis increases with age. This makes sense if mutations occur in connection with DNA replication. The rise in mutation rate, however, is too steep to be explained by a linear increase in number of cell divisions (Crow, 1993). Presumably, older men have less effective repair mechanisms. As expected, the age discrepancy is much less in the mouse, where males do not have so many cell divisions, and is nonexistent in *Drosophila* where the number of premeiotic divisions is roughly the same in the two sexes.

A major technical breakthrough was made many years ago by Muller, who devised techniques for measuring the mutation rate of an entire chromosome rather than a single gene. The typical experiment used a special X chromosome (called *ClB* by Muller) that permitted detection of all the recessive lethal mutations on the chromosome. An extension of the Muller method was made by Mukai et al. (1972). By accumulating mutations over several generations in a sheltered chromosome and by using the variance between replications as a means of distinguishing between many small and few large mutations, they were able to estimate the total rate of mutations causing small effects on viability and fertility, even though such mutations are not individually observable. The mutation rate turned out to be some 30 times as high as for lethals, the estimated genomic rate being about one per zygote. The rate could well be higher, for the assumptions inherent in the method make this a minimum estimate.

Estimates of the human mutation rate are about 2×10^{-8} per nucleotide per generation. This is based partly on direct measurements—quite uncertain because of paucity of data—and indirect estimates from evolutionary data to be discussed later. With 6×10^{9} base pairs per cell, this implies that a new human zygote has some 10^2 new mutations. If any substantial fraction of these are deleterious, we have a problem.

These values are averages. The mutation rate is not constant for all base pairs. In particular there are certain sequences, such as CG, that predispose to mutation, sometimes leading to a mutational hot spot.

Correlation of Mutation Rate and Genome Size

Drake (1991) has noted that among small haploid organisms, from DNA viruses to Neurospora, mutation rates per nucleotide vary enormously—over a factor of 6500. Yet the genomic rates vary by a factor of only about 2.5. The mean genomic rate is about 0.003 mutations per replication per genome. This remarkable constancy argues strongly that the total mutation rate has been adjusted to a roughly constant amount, presumably somewhere near an optimum balance between a too low rate that would not produce enough mutations to keep up with a changing environment and a too-high rate that would produce too many lethals.

Yet, this neat generalization breaks down completely for higher eukaryotes. As stated above, the genomic rate in mammals is very much higher. Perhaps the mammalian genome has enlarged too rapidly for mutation rate adjustment to keep up. Of course, most of the DNA is junk, so most of this plethora of mutations are presumably neutral.

THE DYNAMICS OF EVOLUTION

Although Darwin got the basic idea right, the detailed mechanisms are now much better understood, thanks first to Mendel for providing the rules of inheritance and to deVries for discovering mutation, and then to the whole field of genetics for providing a much deeper understanding of the processes. I believe the best way to describe current views of evolution is to follow Sewall Wright (1988) and classify evolution as happening in four modes: (1) evolution at the DNA level; (2) evolution by incorporation of favorable mutations; (3) evolution of complex, multifactorial traits; and (4) evolution of novel gene combinations. As I mentioned earlier, the ultimate source of genetic variability is mutation. Yet, as we advance from Mode 1 to Mode 4, there is a decreasing dependence on mutation as the driving mechanism.

The rise of molecular biology as a tool for studying evolution at the DNA level has greatly changed our outlook. It is now possible to study directly the most elementary process of evolution; evolution at the DNA level. These findings have reinforced many traditional ideas of evolution, but have revealed a whole new set of phenomena, and some surprises.

Contrasts between Molecular and Organismic Evolution

Molecular evolution follows a quite different set of rules from those long observed in evolution at the organismic level. The study of molecular evolution, primarily individual base substitutions, permits the following generalizations (Kimura, 1983).

- The rate of evolution of a particular DNA region or particular gene is quite similar among different species. For example, amino acids in hemoglobin have evolved at about the same rate in the various mammals studied, and even in fish.
- The rate of molecular evolution is quite constant over long time periods. This is not to say that the rate is absolutely constant. The variability is considerably higher than would be expected from a Poisson process, but it is far less variable than phenotypic evolution.
- The rate of molecular evolution is quite independent of external, or even internal environment. For example, the number of amino acid substitutions between human hemoglobin α and β is essentially the same as that between fish α and human β.
- The least important parts of the DNA or of parts of protein molecules generally evolve most rapidly. For example, synonymous codon changes are substituted about three times as fast as those producing an amino acid replacement. Remarkably, about the same ratio is found in RNA viruses, which mutate and evolve at a rate perhaps a million times as fast.
- There is little if any correlation between the rate of molecular evolution and the rate of change of phenotype of the organisms involved. Fish, which have changed relatively little, have changed their DNA and most proteins at a rate comparable to mammals, whose morphology and physiology have changed much more. Furthermore, species that have hardly evolved at all phenotypically for very long periods ("living fossils") still evolve molecularly at the same rate as normally evolving lineages.

In contrast, phenotypic evolution rates differ enormously from one lineage to another. Most insects have changed relatively little over long periods of time, whereas mammals have changed much more rapidly. Phenotypic evolution goes by fits and starts, greatly uneven in its rate. This has been dramatized in the words "punctuated equilibrium"—the idea that there are long periods of stasis punctuated by sudden rapid spurts (Gould and Eldredge, 1977). Whether the changes are this extreme is doubtful, but that the rate is enormously variable is clear. In another contrast, the rate of morphological evolution depends on such things as population structure, environment, and ecological opportunity. For example, species invading a previously uninhabited island evolve rapidly and quickly fill the various niches.

These provide enough contrasts and examples to make clear that molecular and molar evolution are marching to different drummers. Let us now examine the probable reasons for this contrast.

There is an elaborate theory of evolution by mutation, migration, selection, and random genetic drift, worked out over the past 75 years, mainly by three pioneers, R. A. Fisher and J. B. S. Haldane in England and Sewall Wright in the United States, and more recently by Motoo Kimura in Japan. The subject is now called population genetics (for a short textbook, see Crow 1986). But in one way population genetics has been frustrating. It deals mainly with changes at the gene level, whereas most of the observations have been at the phenotypic level and it has usually not been possible to study the most important traits by the elegant methods of population genetics. For example, we have no idea how many gene changes were required to go from a fox-sized animal to the modern horse. Molecular evolution stands in sharp contrast. Here is a subject that can be studied at the gene or DNA level, and that provides an opportunity for gainful employment of the large body of mathematical theory that has been lying in wait for just such an opportunity.

Mode 1: Evolution at the Genomic Level

The Neutral Theory: Evolution Driven by Mutation

The clock-like regularity of molecular evolution and its independence of environmental conditions forced evolutionists to look for a new mechanism. This led Motoo Kimura in Japan, and Jack King and Tom Jukes in the United States (for a general summary, see Kimura, 1983) to propose in 1968 and 1969 that molecular evolution consists mainly of neutral DNA substitutions brought about by mutation and random genetic drift. This theory, as developed primarily by Kimura, is discussed in detail in Chapter 7 and will be touched on here only briefly. The kinetics of evolution of a completely neutral gene, however, are remarkably simple.

In a diploid population the number of new mutations arising at a specific site in a single generation is $2N\mu$, where μ is the mutation rate per site (usually a nucleotide), and N is the population number (counted as adults of reproducing age). The factor 2 comes from diploidy; there are two representatives of each site, either of which may have mutated with probability μ. In each generation some mutations are lost by chance while others increase in frequency. If we wait a very long time, random effects will insure that all the nucleotides at this site will be descended from a single one. The probability that the prevailing nucleotide is the mutant, and not some other one at this site, is simply the frequency of the mutant in the population, which is $1/2N$. Thus, the probability of a mutation arising and replacing all the others is $(2N\mu)(1/2N) = \mu$.

If we view the process over a long time period, the number of mutant substitutions per generation is simply the mutation rate, μ. How long is a "long time period"? The average time required for a mutation to sweep through the population and replace the others is not easily derived (Crow and Kimura, 1970) but is easily stated: it is about $4N$. So in counting mutant substitutions, we must view the process over a number of generations that is many times the population number.

This purely random process might seem much too slow and too chancy to serve as a basis for understanding molecular evolution. But we must remember that very long time periods are involved. The time of radiation of the mammalian orders was some 500 million years ago. A nucleotide that mutated at a rate of 10^{-9} per generation, supposing one generation per year, would have about a 50% chance of being replaced during that time. About 40% of the 3×10^9 nucleotides in the mammalian genome would have changed one or more times during these 500 million years.

This is not the place to discuss the evidence for or against the neutral theory (see Chapter 8). I shall mention here only what is emerging as the consensus among evolutionists. It is convenient to divide the theory roughly into two forms: weak and strong. The weak form asserts that most of the DNA in higher eukaryotes evolves under neutral kinetics. Since the great bulk of the DNA seems functionless, the weak form says that functionless DNA evolves neutrally. The strong form asserts that most amino acid substitutions are also neutral.

Molecular evolutionists generally accept the weak form of the theory. There is less unanimity about the strong form. There is no doubt that many amino acid changes follow the neutral paradigm; probably most do. There is also no doubt that some are selected; a few have been demonstrated (see Chapter 6 for an example). I believe that the evidence for most being neutral outweighs the evidence against it. The problem is to determine the proportions.

How small must the average selective difference between a nucleotide and its mutant replacement be for the process to be neutral? The answer turns out to be that a mutant is effectively neutral if its selective advantage or disadvantage, usually designated by s, is smaller than $1/4N$. If the value of s is much larger, selection dominates; if the value is lower, chance dominates.

Of course, mutations don't suddenly change from selected to neutral as their selective advantage or disadvantage, s, becomes less than $1/4N$. Figure 1 shows the evolution rate for mutations in a narrow range of s values on either side of zero. A mutation whose s value fluctuates between $+1/4N$ and $-1/4N$ will evolve at about the neutral rate, μ. At values around $\pm 1/N$, selection and random drift interact.

In summary, the majority of nucleotide substitutions are so nearly neutral that their fate has been determined by random drift. The driving force, then, is mutation and the nucleotide substitution rate is equal to the mutation rate.

Selfish DNA

The business of DNA is to make more DNA. One might even say—and several have—that a human being is DNA's way of making more DNA (modifying Samuel Butler's quip that a hen is an egg's way of making another egg). Some transposable

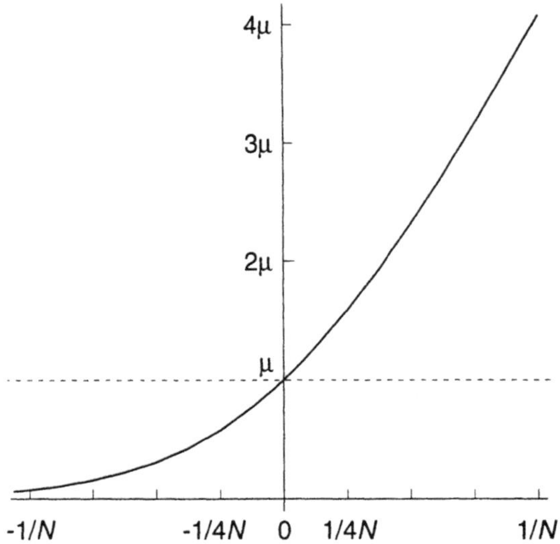

Figure 1. Evolution in the near-neutral range. The evolution rate (rate of substitution per generation in units of the mutation rate, μ) is plotted as a function of the selective advantage of the mutation, s. When $s = 0$ (neutrality) the evolution rate is equal to the mutation rate. If s fluctuates in the range $\pm 1/4N$ the rate is essentially the neutral rate.

elements have the property of being excised from the chromosome, but leaving behind a copy of themselves, and then inserting at a new site. This leads to a net increase of the copy number. If the element moves to an inert region of the chromosome, it does no harm, and the process can be repeated. So the net tendency for transposons to increase the amount of DNA goes on, and there is little the organism can do about it. Unequal crossing-over with some sort of built-in bias in favor of retaining those products that increase the length will produce the same increase, but this time the duplicated material is localized to one area instead of being scattered.

There is little doubt that the amount of DNA in most multicellular eukaryotes has been steadily increasing, perhaps for hundreds of millions of years, reaching enormous amounts in mammals and some amphibia and plants. As I have emphasized earlier, probably most of the increase is in nonfunctional DNA and has little consequence. At most it slows down cell division and adds to the energy costs of the cell, which must copy increasing amounts of DNA when the cell divides. But most multicellular organisms are not in a hurry to develop, so this is a cost that can be afforded. It is noteworthy that those organisms whose ability to survive depend on rapid reproduction (e.g., bacteria) have almost no excess DNA. Copying extra DNA takes time, so their DNA contains only the essentials.

In any case, the excess DNA in multicellular eukaryotes, with its relentless propensity to increase its amount still further, is hard for the organism to combat.

The only mechanism is sterility or death of those individuals with the greatest excess; not very inviting. Thus most species must wage a losing battle against their own excess DNA. It is ironical that the very stuff on which our life and evolution depend is also, because of its ruthless selfishness, our enemy. We have parasites at the DNA level, just as we do on our skin, or in our blood and intestinal tracts.

Cytogenetic Evolution

Gross chromosome changes are also a part of genomic evolution. Here, however, there is little reason to suspect strict neutrality. Heterozygous inversions lead to aneuploid products and thus to inviability and sterility. As a result, they tend to be eliminated from the population in the same way as deleterious mutations. New translocations usually suffer the same fate.

Yet, translocations sometimes persist and become a normal part of the species. It is not surprising that those cytogenetic changes that persist in evolution are those that cause the least sterility and inviability in heterozygotes. Thus, they have a greater chance of spreading. The problem is that cytogenetic changes are selected against when they are rare, although they are positively selected when they become common. The usual explanation is that a chromosome change occurs when the population is very small. The rearrangement can sometimes increase by chance even if it is mildly harmful. Once it has attained frequency higher than 50%, it becomes the prevailing type and is now advantageous. If this population expands the change expands along with it. Thus, related species that had a common ancestor sometimes differ by one or more chromosome rearrangements. Presumably this happened by one of the populations being very small following the split, permitting random drift to carry the frequency beyond the 50% hurdle.

Thus molecular evolution and cytogenetic evolution are both strongly influenced by random drift. The difference is that nucleotide changes are so nearly neutral that random drift is important even in very large populations, whereas cytogenetic changes occur by random drift in populations small enough that deleterious changes can sometimes be fixed.

As I noted earlier, when related species are compared they often have changes in gene sequence. Most of these are probably random. But in rare instances a specific rearrangement carries with it some improved function.

Mode 2: Evolution by Incorporation of Favorable Mutations

A second mode of evolution is the incorporation of favorable mutations, one (or a very few) at a time. Examples abound. In England during the industrial revolution, a dark colored mutant strain of the peppered moth gradually replaced its light colored progenitors. The reason was that smoke from factories darkened the tree trunks, partly by depositing soot and partly by killing light colored lichens. Therefore, the light colored moths were no longer protectively colored and were picked off by birds while the dark mutants survived and multiplied. Another example is the incorporation of mutant genes causing insecticide resistance in flies and mosquitoes.

Although the best known examples are evolution under man-made changes, natural evolution has been taking place all along, although usually at much lower rates. One example is successive mutations causing resistance to fungus infections in plants, or to parasite infections in animals. At the same time there are mutations in parasites making them better able to attack their hosts.

What is the role of mutation in this mode of evolution? Obviously, without the mutations, such evolution would be impossible. So this evolution clearly depends on mutation. As first shown by Haldane, the probability of fixation of a favorable mutation is about $2s$, where s is the selective advantage of the mutant heterozygote (Crow and Kimura, 1970). Multiplying this by $2N\mu$, as we did earlier for neutral mutations, the rate of substitution is $4N\mu s$. Not surprisingly, the rate depends on the mutation rate, μ, and the selective advantage, s. But it is not so obvious that it depends on the population number, N. Yet this is reasonable, since the larger the population the greater the probability of a particular mutation occurring. So clearly, evolution by incorporation of favorable mutations is mutation-rate dependent.

But this is not the whole story. Whether a mutation is beneficial depends on the circumstance. In the case of DDT resistance, the mutations had been occurring all along, but were selectively favored only when the presence of the insecticide made them advantageous. Surely the same kind of thing has happened repeatedly in evolution. Thus, evolution in this mode is driven partly by mutation and partly by environmental changes that render previously deleterious mutations favorable.

Some of the most rapid evolution has been in viruses, such as influenza and HIV. The very high mutation rate means that these viruses rapidly produce new strains capable of varying their mode of attack. Yet, if one examines these for nucleotide changes, synonymous changes predominate just as in higher organisms. As rapid as virulence evolution is, it still involves a minority of the nucleotide substitutions. Both processes, Mode 1 and Mode 2, are occurring.

Two things are obvious. One is that without mutation evolution would eventually come to a standstill, so a minimal mutation rate is necessary. We have little idea what this minimum rate is. It is also obvious that a too-high mutation rate would be fatal. The mutation rate is at least roughly adapted to the life cycle of the species. But between the extremes of a too-low and a too-high mutation rate lies a great deal of variability.

Mode 3: Evolution of Complex Multifactorial Traits

Although there are numerous instances of single genes that are important, most of the traits of importance to a species such as humans—size, blood pressure, intelligence—depend on multiple genes, and on the environment as well.

The genes affecting such traits work more or less cumulatively. Evolution occurs, not as much by incorporation of new mutations as by shuffling existing genes, making some more common and others less so. For most quantitative traits there is an intermediate phenotype that is optimum; too-high and too-low blood pressure are both harmful. But there is usually a great deal of variability, of which some is due to genetic differences. If the environment changes, alteration of gene frequencies can adjust to the new phenotypic optimum.

A great deal of evolution thus consists of changing the frequencies of genes, each with a very minor effect, but which collectively determine the quantitative trait. By having a number of genes with such small effects, natural selection can fine-tune the phenotype to produce the exquisite adaptations that are so fascinating to the biologist.

This third mode of evolution is the way in which each species becomes fitted to its particular environment. The combination of many genes, shuffled every generation by the Mendelian mechanism, permits each gene to be tried in many different combinations; those that work best with other genes and in the prevailing environment are retained.

Genetic Variability

The rate at which a population evolves depends on its genetic variability. Before Mendel, inheritance was generally thought to be analogous to a fluid. We still have words like "blood" in our vocabulary, as a synonym for heredity. A consequence of such a view is that every time there is a mating between individuals of contrasting character, the distinction is forever lost, just as after mixing white and red paint it is impossible to obtain pure white and red again. As a result of such a system, genetic variability decays very rapidly. It was necessary to postulate sources that would generate a high level of new variability. Much of Darwin's later work was devoted to inventing mechanisms for generating such variability, all of which have turned out to be not only wrong but unnecessary. In Mendelian heredity, variability is conserved. If Darwin had only known of Mendel!

In a Mendelian population, genetic variability is lost primarily by two mechanisms. The first is random drift, the same process I mentioned earlier to account for neutral evolution. The population eventually becomes homogeneous. But the process is exceedingly slow, except in very small populations. To quantify this we measure variability in the units in which selection acts, namely the average of the squared deviations from the average, called the variance. In a diploid population of size N, the variance is reduced by a fraction $1/2N$ per generation. This means that in a population of even moderate size, say 10,000, the decay of variability is very slow and is easily compensated by a small mutation rate.

The second mechanism is selection. Selection reduces variation by increasing the frequency of some genes while eliminating others, doing the same thing in a directional way as genetic drift does in a random way. How much selection reduces variability depends on how intense the selection is and on how many genes are involved in the selected trait. If the number of genes is large, the reduction in variability for any one gene and therefore for the total is very slight.

The conclusion is that sexually reproducing populations, except for very small ones, have a great deal of genetic variability. It is unlikely that they ever run out of genetic variability unless selection is very intense. This is borne out by artificial selection in laboratory animals and plants. Populations selected strongly for a specific trait, such as size in mice, oil content in maize, or bristle number in *Drosophila*, all show large and continuous change in phenotype over many generations. If a plateau is reached it usually has an explanation other than running out of genetic variability. Selection in a downward direction has an obvious limit at zero. Selection in an upward direction may reach a state where further change in the selected trait is incompatible with some other function. For example, selection for large size in one mouse experiment eventually produced very large mice with reduced fertility. Loss of genetic variability itself is hardly ever a problem.

The rate at which selection changes a population depends on: (1) the intensity of selection; (2) the amount of variability in the population; and (3) the mode of inheritance. Selection is most effective when each gene acts additively on the trait so that heterozygotes are intermediate between their respective homozygotes. The proportion of the variance that is as responsive to selection as additively acting genes is called the *heritability*. Factors that reduce heritability are dominance, epistasis (interaction of nonallelic genes), and environment. Heritability is a statistical measure that averages out these various effects. Its value varies from trait to trait, and for many is low, say 0.2. This does not mean, however, that selection is ineffective; only that it is slower. And evolution has had lots of time.

In any moderately large population there is sufficient genetic variability to respond to ordinary changes in the environment. Evolutionary adaptation consists of many changes of allele frequencies, cumulatively adding up to better and better organisms. Of course, as each species improves its genetic makeup by selection, other species are doing the same thing. Thus, although each is getting better with respect to what it was in the past, so are its competitors which are making its environment worse. It's as Lewis Carroll said about the treadmill: "You have to run awfully fast to stay in the same place."

The Darwinian paradigm is that most evolution depends on a large number of genes with individually small effects. This permits the Mendelian shuffle to work at its greatest efficiency and allows evolutionary fine tuning. In some ways the neutral theory stands in contrast to the Darwinian theory, but in one way Darwinism is reinforced by neutralism. For, if there are neutral genes there are also very weakly selected genes, and these are the stuff of Darwinian evolution.

Molecular analysis shows that related species differ in a very small proportion of nucleotides in coding regions (except for synonymous changes). How do such different phenotypes as chimpanzee and *Homo* occur. The answer probably lies in the fact that we do not differ so much qualitatively, as quantitatively. We have the same proteins, but in different amounts—more hair in chimps, more brains in humans. This argues that much of the fine-tuning that I have been discussing is done by changes in regulatory regions rather than in structural genes. An exciting field for the near future is the quantifying of such changes, as more sequence information becomes available.

Sexual versus Asexual Reproduction

A species that reproduces asexually, such as a dandelion, can respond rapidly to selection, for each offspring has exactly the same genotype as its parent. Furthermore, it does not produce inferior genotypes by segregation and recombination. But the other side of the coin is that the future amount of change in the population is limited by the phenotype of the most extreme individual in the population. Usually mutation is too slow to make up for the decreased variability. Thus, the effectiveness of selection in an asexual species is limited.

A sexual species, with its free Mendelian segregation and recombination, has no such limitation. Recombination among the genes within a population can generate types that are far beyond the range of any individual in the current population. Thus, a change in environment may lead to very large phenotypic changes. It is the ability to scramble genes and try new combinations that gives sexual reproduction such an evolutionary advantage. This is borne out by the paleontological data. Many asexual species are highly successful (think of dandelions again). But asexual species are relatively short-lived when long geological periods are considered, and contemporary asexual species had sexual ancestors.

The conclusion from all this regarding the role of mutation in organismic evolution is that in sexual species mutation plays an essential role in providing the ultimate source of genetic variability, but that the rate of evolution is not coupled to the rate of mutation. All that is required is a bare minimum. Selection and recombination play a more decisive quantitative role. If mutation were to stop entirely, the population would retain variability and be able to respond to selection for many generations. Unless selection were very intense, this could be hundreds or even thousands of generations.

Mode 4: Evolution of Novel Gene Combinations

Evolution in the third mode has one drawback. There is no way in a large, randomly mating population to evolve a favorable combination of genes if the intermediate stages are harmful. If *aa bb* is favorable and *Aa Bb* is still more favorable, but *Aa bb* and *aa Bb* are harmful, the population cannot evolve from *aa bb* to *Aa Bb* by natural selection alone. If the *Aa Bb* combination does occur, it is quickly lost by recombination.

One way to get around this problem was proposed by Sewall Wright (for his last, summarizing paper, see Wright 1988). In his "shifting balance" theory, evolution proceeds best if a large population is broken up into small subpopulations within which random drift occurs. Occasionally, one of these will drift into a happy gene combination. This subpopulation, being more fit than the others, will expand, sending migrants to other subpopulations. In this way the entire population is upgraded to the status of the fortunate subpopulation.

Wright's theory has been the subject of much discussion among evolutionists. Its strongest support comes from the observation that major changes in evolution often leave poor fossil records. Thus, the populations must have been sparse, which favors the subpopulation structure that Wright's theory requires. Another argument is the correlation between morphological and cytological changes; for example both are rapid in mammals. As I said earlier, the most plausible way for cytogenetic changes to occur is by random drift in small populations; thus, this supports Wright's view. But whether this is an important mode of evolution is still not clear. To work, it requires a rather delicate balance between migration rates and subpopulation sizes. It is also possible that it is not needed; perhaps mass selection, although it may fail to create some specific genotypes, can always find a way to increase the fitness of the population.

In the Wright theory mutation plays a very minor role. The variability is introduced mainly by migrants from other subpopulations. Although mutation, as always, remains the ultimate source of variation, it is very loosely coupled to the evolution process.

Although the first three modes of evolution are well established, the fourth remains uncertain. It has been very popular among those biologists who emphasize gene interactions and the interrelatedness of developmental processes. But as to its importance in evolution, the jury is still out.

Evolution of Cooperation and Altruism

It is easy to understand the evolution of survival and reproductive traits by natural selection. Those individuals that compete most successfully and are the most successful at finding mates leave the most progeny and their genes increase. Cutthroat competition would be expected, and indeed is found throughout nature.

But how did what we like to regard as the finer human traits—generosity, cooperation, even self-sacrifice—evolve. The answer probably lies in *kin selection*. An individual shares 50 percent of his genes with his brother or sister. From the standpoint of gene transmission, saving his brother's life is half as good as saving his own. Thus, brothers should not only cooperate, but one should be willing to run a substantial risk (theoretically, 50 percent) to himself to save the life of a sib. Thus, natural selection would be expected to lead individuals to behave altruistically toward relatives in rough proportion to the fraction of genes that they share.

The altruism of parents toward their children is apparent throughout the mammals. This is expected, for a parent approaching the end of the reproductive period, can best perpetuate its genes by caring for the progeny sharing 50 percent of those genes.

It is not likely that animals recognize kinship beyond very close relatives. But another factor comes into play in many organisms including humans. Our ancestors in the not too distant past had a distinct tribal structure. Individual tribes were small enough that all individuals within were related. Thus, on kin selection principles, they would be cooperative and altruistic with members of their group. The other side of the coin is that they would not be so with other groups. Thus, the biological foundations of within-group cooperation and between-group xenophobia may well lie in the structure of human populations in the past when our behavioral genes were evolving. Needless to say, social influences are of great importance and disentangling these from biological effects in determining human behavior is no easy task.

Speciation

If a species is split into two or more parts, as for example by groups finding themselves on opposite sides of a river or mountain range, then each of them evolves independently of the other. They will then accumulate different genes, especially if they are in different environments, but even if the two environments are identical. Since there are many genetic ways of doing the same job, the same result may be produced in different ways. An illustration is given by crosses between two closely related Drosophila species, *D. melanogaster* and *D. simulans*. Both have identical bristle patterns, but the hybrids have irregular numbers of misshapen bristles; each species produces the same end result in somewhat different ways that cause confusion in the hybrid.

Eventually two separated populations accumulate enough differences that they are no longer cross-fertile. They may produce inviable or sterile hybrids, or they may simply no longer mate. The result, then, is that even if the two groups come to inhabit the same region, they remain distinct.

Evolutionists believe that this is the major process by which new species are formed, first by some form of separation, usually geographical, followed by separate evolution to the extent that they become incompatible. There is considerable debate as to how often, if ever, species diverge while both are in the same locality. One can invent mechanisms, but how important they are remains unsettled. Speciation by geographical separation, however, is well established.

The role of mutation in speciation is not clear. If the traits causing species isolation are multifactorial, then the mutation rate is unimportant. Yet, recent researches have isolated individual genes with major species-differentiating effects, so perhaps the mutation rate plays a larger role than previously thought.

Ordinarily, as I have said, speciation is a gradual process involving some form of separation for many generations. There is an important exception to this rule. If a diploid individual becomes tetraploid, called *autotetraploidy*, it and its ancestors are no longer compatible for they produce sterile triploid progeny. If the new tetraploid can find a tetraploid mate, or is capable of self-fertilization, a new species is immediately created, since it is forever isolated from its diploid forbears.

Another form of polyploidy that is important in evolution is *allopolyploidy*. As mentioned earlier, related species often differ in cytogenetics. Inversions and translocations, or different chromosome numbers can cause the hybrids between the species to be sterile, for the chromosomes from the two parents fail to find pairing partners. If the chromosomes are doubled, however, each chromosome now has a mate and meiosis proceeds smoothly. Such allopolyploids have been of great value in agriculture, permitting the production of fertile hybrids between widely different plants. This also has happened in nature.

Both allo- and autopolyploids are common in the plant kingdom. The reasons why animals have not availed themselves of this opportunity are not fully understood. For one thing, plants seem to tolerate chromosome changes better than animals. Also polyploidy is sometimes incompatible with the sex-determining mechanism. What sex is an animal with 3 X chromosomes and 1 Y chromosome, and is it fertile? Nevertheless, it is likely that there have been a few polyploidization events in the ancestry of contemporary vertebrates. The evidence is particularly strong in fish.

Evolution of Complexity: Where Do New Genes Come From?

We mammals are more complex than a fungus or nematode, partly because we have more genes and a greater diversity of gene functions. It takes more genes to carry out complicated functions, such as the human brain, and the new genes must have come from somewhere.

It is difficult to imagine genes being created *de novo* (although this must have happened at least once in the early history of life). A much better way to get a new gene is to modify an old one. The problem is that when a gene mutates to produce a new, possibly useful function it must lose the old, also useful, one. The obvious answer is to duplicate the gene. Then the organism can eat its cake and still have it. One of the duplicates can evolve to produce a new function while the other remains to carry out the old one. This is a fine idea; does evolution actually take advantage of it?

The answer is an emphatic *yes*. Time and again we find tandem clusters of genes with each one carrying out a somewhat different function. A classic example is provided by the globin genes. The human β globin region consists of 5 genes each with a slightly different function, producing different hemoglobins at different times—embryo, fetus, and adult. In addition, there is one gene that is nonfunctional.

This is exactly as would be expected. If a gene duplicates, perhaps by unequal crossing-over, and one of them mutates there is a good chance that the mutation will be harmful rather than leading to a new useful function. The nonfunctional gene in the beta cluster has done this. After duplication, it mutated into inactivity.

Usually this is caused by a chain-terminating (nonsense) mutation. Such "dead" genes are called *pseudogenes*. They no longer do anything, but are simply carried along as one more piece of excess baggage.

Even though most duplications lead nowhere, once in a while one of them mutates to produce a new and useful function, as the hemoglobins illustrate. Gene duplication provides a plausible way for the organism to become increasingly complex by a series of orderly steps in which new genes—really modified old genes—are added one by one, with compensatory adjustments along the way. To make all the changes at once would invite catastrophe.

THE STATICS OF EVOLUTION

So far, I have treated evolution as a dynamic process, whether the change is brought about by mutation and random drift, or by selection, or by all together. There are also genetic processes that, instead of leading to continuous change, lead to a stable equilibrium.

Mutation-Selection Balance

As stated before, the great majority of mutations that affect function in any way are harmful. They are eliminated by natural selection rather quickly in terms of evolutionary time, and make no permanent contribution to the evolution of the species. They do, however, affect the fitness of the individuals of the population at any particular time.

A harmful mutation persists in the population until all its descendants are eliminated by natural selection or chance. How long the mutation persists depends on how harmful it is and how it interacts with other genes, especially on whether it is dominant or recessive. A harmful dominant mutation is eliminated rather quickly. If it reduces the viability or fertility of its carrier by 10%, it will persist in the population on the average for 10 generations. In contrast, a recessive mutation may persist for hundreds or thousands of generations before there have been enough homozygous descendants to eliminate the mutation from the population.

Actually, however, completely recessive genes are rare. Almost always the gene, even if seemingly recessive, has some effect in the heterozygous state. Usually it is these heterozygous effects that cause the gene to be eliminated from the population. The reason is that homozygotes are so rare that even a small selective disadvantage of the much more common heterozygotes has a bigger net effect. In *Drosophila*, the only species where the answer to this question is known, the average recessive mutation has enough heterozygous expression to persist in the population about 50 generations (Crow, 1993).

If the evolutionary process continues for a long time, the population will approach an equilibrium in which the rate of occurrence of new deleterious mutations is balanced by the elimination of old ones. Roughly the equilibrium frequency (proportion) of mutant genes is $q = \mu/s$, in which μ is the mutation rate per gene per generation and s is the selective disadvantage of the mutant heterozygote. We can safely ignore rare homozygotes.

Balancing Selection

There are other causes of stable equilibria. The most important is balancing selection. This can happen in several ways of which two are most important.

First is overdominance for fitness; that is to say, the heterozygote is more fit than homozygotes for either of its components. The classical example of this is sickle-cell hemoglobin. Designating the sickling form by *S* and the normal form by *A*, *S/S* homozygotes have severe sickle-cell anemia and *A/A* homozygotes are susceptible to falciparum malaria, while *A/S* heterozygotes have normal hemoglobin but are resistant to malaria. Therefore, in areas where malaria is a major cause of illness and death, heterozygotes are more fit than either homozygote. This tends to keep both genes in the population, and their ratio is determined by the relative decrease in fitness caused by anemia in *S/S* genotypes versus that from malaria in *A/A*.

In this case, the mutation rate is largely irrelevant. The mutation had to occur only once, or enough times to overcome the chances of getting lost by chance when it was rare. The selective forces are so great relative to mutation rates that new mutations make almost no impact on the gene frequency. As a mechanism for protecting the population against malaria, this is an expensive one, for the price is a high incidence of anemia. Furthermore, the *A/A* type is much more common than *A/S*, so only a fraction of the population is protected. One wonders why an unequal crossover has not put the *A* and *S* genes on the same chromosome, so that an *AS/AS* population could have malaria resistance without anemia. There must be a reason, but so far it is not evident.

Although overdominance is an obvious mechanism for maintaining variability in the population, instances like hemoglobin *S* are unusual. It is possible that some much milder genes are held in balance by this mechanism, but that remains conjectural. In particular, it is often impossible to distinguish true overdominance from closely linked combinations like *Ab/aB*, where *b* and *a* are both deleterious as homozygotes. Whether overdominance plays an important role in maintaining variability of quantitative traits remains an open question, although most of the evidence is against it.

Another mechanism for maintaining a selective polymorphism occurs when a genotype is favored when rare, but not when common. An example is a gene in plants that causes pollen grains not to function on a stigma that carries the same gene. Such a gene is favored when rare because its pollen can function on almost all flowers, but as it becomes common it encounters more plants on which its pollen cannot function. Such a system generates many alleles, each rare. Similar mechanisms have been proposed for maintaining variability of pathogen-combating systems, such as MHC.

When two or more alleles are maintained at a locus, this is called *polymorphism*. The word originally meant, as is clear from its roots, having more than one structure, and it was applied to organisms such as bears where black and cinnamon variants coexist. The word has been extended (some would say corrupted) by being used in situations where the root "morph" is not strictly applicable.

Polymorphisms, whether maintained by overdominance or by selection for rarity, are stable and in some cases have been maintained over very long time periods. Some MHC polymorphisms are very old. For example, in mice this antedates the separation of different species. It is usually impossible by population genetic studies to distinguish between overdominance and selection for rarity, and that is the case with MHC. Discriminatory evidence will have to come from other sources. This system is discussed in detail in Chapter 6.

The Mutation Load

The impact of deleterious mutation on the welfare of the population depends almost entirely on the mutation rate and not on the deleteriousness of the individual mutation. One way to understand this is to imagine a new mutation that kills or sterilizes its bearer. The mutation is eliminated and one individual suffers a "genetic death." Alternatively, consider a mutation that causes a 1% risk of death or sterility. Such a mutation will persist in the population an average of 100 generations. On a system of mutation cost-accounting that equates one individual exposed to a 100% risk of death or sterility to 100 individuals exposed to a 1% risk the two mutations are equivalent. Each causes one genetic death. Thus, a dominant or partially dominant mutation causes an average of one genetic death regardless of its degree of impairment. A recessive mutation causes only half as much damage, since two mutations are eliminated with each genetic extinction.

This is a rough description of the principle first enunciated by Haldane (1937). It says that the reduction of mean fitness of a population by a dominant or partially dominant mutation is equal to the zygotic mutation rate (i.e., the number of new mutations per zygote). Since mutation rates are usually measured per gene or per gamete, the principle says that the *mutation load*, as it is now called, is equal to twice the mutation rate for a dominant gene and the mutation rate for a recessive gene.

If the mutations act independently the combined load is roughly the sum of the mutation rates; more exactly, it is $1 - e^{-2U}$ where U is the total mutation rate for all loci (per haploid genome). On this principle, a deleterious mutation rate of 20% per gamete would cause an average fitness reduction of $1 - e^{-0.4}$ or 33%. How can this load be lessened? The most obvious way is if several mutant genes can be eliminated with one genetic death. If genes act in such a way that several cooperate to produce a deleterious effect, this can materially reduce the load.

The most general statement of the load principle is this: The mutation load is twice the gametic mutation rate divided by the difference in the number of mutant genes per individual eliminated by selection and the number before selection (Kondrashov and Crow, 1988). Thus if the population somehow contrives to make this difference equal to 10, the load is decreased by a factor of 10.

Reducing the Mutation Load

Now that deleterious mutation rates have been shown to be very high, at least in *Drosophila* and in all probability in other species as well, the question of how the population gets rid of all these mutations without suffering a substantial loss of fitness is one of the major unsolved questions in evolutionary genetics.

Geneticists are not clear as to how the mutation load is reduced, but they are sure that it happens somehow. The most likely hypothesis is rank-order, or *truncation* selection, whereby all individuals above a certain number of mutations are eliminated. This is illustrated by a hypothetical example in Figure 2. Suppose that the mean number of mutant genes in the group eliminated by selection, n_S, is 50 while the number in the population before selection, n_P, is 40. The mutation load is thus reduced by a factor of 50 – 40, or 10. If 10% of the population is eliminated by selection, each carrying an excess of 10 deleterious mutations, an average of

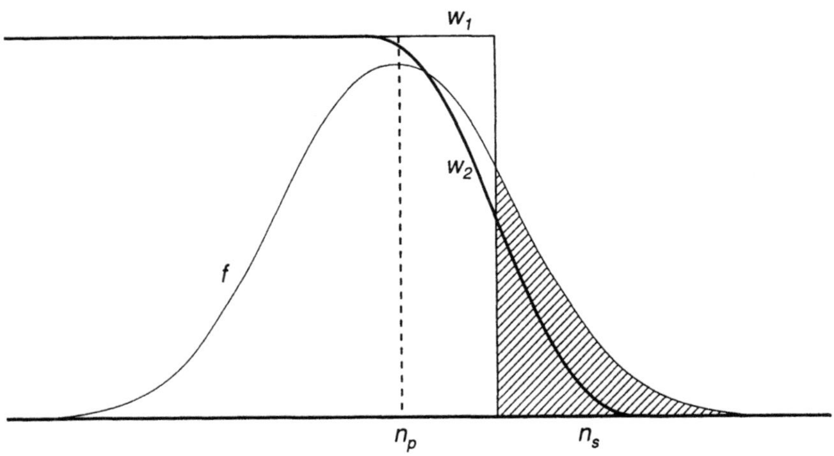

Figure 2. Fitness as a function of the number of deleterious mutations, *n*. The bell-shaped curve, *f*, is the distribution of the number of mutations per individual. With fitness function W_1 (strict truncation) individuals to the right of the truncation value (shaded area) are eliminated each generation. The mean number of mutations in the population is n_p and n_s is the number in the eliminated group. The S-shaped curve, W_2, shows a less extreme and more realistic fitness function.

one mutant is eliminated from the population each generation, enough to balance a mutation rate of 0.5 per gamete.

Nobody thinks that nature truncates precisely, by eliminating strictly those in the hatched area in Figure 2. But it has been shown (Crow and Kimura, 1979) that approximate or *quasi-truncation*, as shown by the S-shaped curve is almost as effective, and this does seem realistic.

One more note: Getting rid of deleterious mutations in groups requires genetic recombination, so it is not possible in an asexual species. I mentioned one advantage of sexual reproduction earlier; this is another.

MUTATION AND HUMAN WELFARE

The Total Human Mutation Rate

Mutation, as has already been emphasized, is a necessary condition for evolution. Yet it has its price, for the great majority of mutations, unless they are neutral, are deleterious. Even amino acid sites that evolve neutrally through replacement by similar amino acids can mutate to produce dissimilar, harmful amino acids; in fact, these are the majority. As just discussed, species in nature have ways of getting rid of mutations efficiently without an excessive mutation load. But what about the human population?

We can use the neutral theory to get an idea of the human mutation rate. Pseudogenes, being functionless, must evolve by neutral kinetics. We can also use, with somewhat less assurance, the rate of evolution in synonymous codon positions. The rate of synonymous divergence of human and chimpanzee is about 1.3×10^{-9} per year, or about 2×10^{-8} per 15 to 20-year reproductive generation. This is the average rate of mutation in the ancestry of humans and chimpanzees. Given 3×10^{-9} nucleotide pairs per genome, this implies a mutation rate of 60 per gamete, or 120 new mutations per newborn individual, an astonishing number. Roughly this same value is given by direct measurements of the mutation rate from isozyme mutations, but the numbers are small enough to make an exact answer impossible. The evolutionary rate of pseudogenes suggests an even higher rate.

The Human Mutation Load

How can the human population cope with such a high mutation rate? Surely the great majority of these changes are in nonfunctional DNA and are neutral. But if only 5% are selectively disadvantageous, this means 6 new deleterious mutations per generation. I believe the best answer lies in a combination of neutrality of most nucleotide mutations and quasi-truncation selection for those mutations that are selectively disadvantageous. But as yet we have very little basis for a quantitative understanding.

Clearly, the human mutation rate is a problem. We do not need mutations for future evolution, at least for the foreseeable future. There is already a large amount of genetic variability. If we could reduce the mutation rate to zero (which I think would be a good idea if we only knew how to do it), we would notice this only by a decrease in severe dominant diseases. Increasing molecular technology is making possible the prenatal detection of more and more types of deleterious gene mutations and cytogenetic abnormalities. One way to reduce the mutation rate substantially is for males to reproduce at an earlier age. Society may some day decide that semen should be collected shortly after puberty and stored for possible future use. This would reduce greatly the rate of those mutations associated with DNA replication. Finally, we have extensive programs to detect potential mutagens before they have a chance to affect the public, and this will help keep the mutation rate from getting higher.

Relaxed Selection in Recent Times

In countries with advanced technology there is very little natural selection. A newborn child has an excellent chance of surviving through the age of reproduction. If the population in the past was somewhere near equilibrium between mutation and natural selection, it is not likely to be so at the present time. The relaxation of natural selection almost certainly means that deleterious mutations are accumulating faster than they are being eliminated.

We do not notice this by such things as increased illness and death rate because the genetic weaknesses are compensated by improved environment. At the moment, environmental changes are in the lead, for health, survival, and life expectancy continue to increase. The question is: Can we keep improving the environment enough to stay ahead of the deleterious effects of an increased accumulation of mutations? The answer is not obvious. But one thing is clear: If for some reason we had to go back and live the way our ancestors did 10,00 years ago, we would have a miserable time. We would suffer all the ills that they did, plus the weakening effects of mutations that have accumulated in the intervening years as selection was relaxing.

It is very likely that accumulated deleterious mutations will eventually be a serious problem for mankind. Fortunately, any such change will be slow. If we can judge from *Drosophila*, the changes from our present high standard of living and relaxed selection will gradually manifest themselves over a period of some 50 generations. This is about 1500 years and perhaps time enough for humanity to learn more efficient ways of coping with the problem, provided we are not overwhelmed by a runaway population increase.

SUMMARY

Mutation is the elementary process of evolution; without it, evolution could not happen. But the quantitative importance of mutation differs greatly in the different modes of evolution, which are as follows:

Mode 1: Genomic evolution. Evolution at the DNA level is largely driven by the mutation rate. The base substitutions, which predominate in molecular evolution, are mainly neutral and the survival of any one is determined by random drift. The average evolutionary substitution rate is equal to the mutation rate.

Mode 2: Evolution by incorporation of favorable mutations. This mode depends on the rate of occurrence of favorable mutations, but also depends strongly on environmental conditions. A changed environment can change a previously deleterious mutation into a favorable one.

Mode 3: Evolution of complex multifactorial traits. This mode of evolution depends on genetic variability and natural selection. Mutation plays a role only as the ultimate source of variations, and the mutation rate is not coupled to the evolution rate.

Mode 4: Evolution of novel gene combinations. This mode, emphasized by Sewall Wright, depends on population structure and random drift. Again, there is no coupling of mutation and evolution rate. Although the other three modes are well established, the importance of this mode remains uncertain.

REFERENCES

Beale, G. (1993). The discovery of mustard gas mutagenesis by Auerbach and Robson in 1941. Genetics 134, 393–399.

Crow, J. F. (1993). How much do we know about spontaneous human mutation rates? Envir. & Molec. Mutagenesis 21, 122–129.

Crow, J. F. & Kimura, M. (1970). An Introduction to Population Genetics Theory, pp. 426, 430. Harper and Row, New York. Reprinted by Burgess Publishing, Minneapolis.

Crow, J. F. & Kimura, M. (1979). The efficiency of truncation selection. Proc. Natl. Acad. Sci. USA 76, 396–399.

Drake, J. W. (1991). A constant rate of spontaneous mutation in DNA-based microbes. Proc. Natl. Acad. Sci. USA 88, 7160–7164.

Gould, S. J. & Eldredge, N. (1977). Punctuated equilibria: the tempo and mode of evolution reconsidered. Paleobiology 3, 115–151.

Haldane, J. B. S. (1937). The effect of variation on fitness. Amer. Natur. 71, 337–349.

Kondrashov, A. S. & Crow, J. F. (1988). King's formula for the mutation load with epistasis. Genetics 120, 853–856.

Mukai, T., Chigusa, S. I., Mettler, L. E., & Crow, J. F. (1972). Mutation rate and dominance of genes affecting viability in *Drosophila melanogaster*. Genetics 72, 335–355.

Wright, S. (1988). Surfaces of selective value revisited. Amer. Natur. 131, 115–123.

RECOMMENDED READINGS

Alberts, B., Bray, D., Lewis, J., Raff, M., Roberts, K., & Watson, J. D. (1989). Molecular Biology of the Cell. Garland Publishing, New York.

Bittar, E. E. (Ed.) (1991). Evolutionary Biology. Fundamentals of Medical Cell Biology, Vol. 1. JAI Press, Greenwich, CT.

Crow, J. F. (1986). Basic Concepts in Population, Quantitative, and Evolutionary Genetics. W. H. Freeman, New York.

Futuyma, D. J. (1979). Evolutionary Biology. Sinauer Associates, Sunderland, MA.

Haldane, J. B. S. (1930). The Causes of Evolution. Harper and Brothers, New York. Reprinted 1990 with Introduction and Afterword by E. G. Leigh. Princeton University Press, Princeton, NJ.

Kimura, M. (1983). The Neutral Theory of Molecular Evolution, pp. 98ff. Cambridge University Press, Cambridge.

Strickberger, M. (1985). Genetics. Macmillan Publishing, New York.

Sturtevant, A. H. (1965). A History of Genetics. Harper and Row, New York.

Chapter 7

Adaptive Evolution at the Molecular Level

AUSTIN L. HUGHES

INTRODUCTION

Molecular data have provided strong support for the hypotheses that most polymorphisms at the DNA level are selectively neutral, and that evolutionary change occurs mainly through the chance fixation of selectively neutral variants rather than as a result of positive Darwinian selection (Kimura, 1983; Nei, 1987). These hypotheses—the key ideas of the neutral theory of molecular evolution—were initially rather strongly resisted by most evolutionary biologists, giving rise to the

Principles of Medical Biology, Volume 1B
Evolutionary Biology, pages 183–203

ISBN: 1-55938-802-1

"selectionist-neutralist" controversies of the 1960s and 1970s. Being accustomed to interpret even subtle phenotypic differences between species as resulting from adaptive evolution, these biologists were reluctant to consider the possibility that at the molecular level nonadaptive evolution might predominate.

The advent of techniques for rapid DNA sequencing in the late 1970s has changed the climate in evolutionary biology dramatically, for DNA sequence data have provided strong support for the neutral theory. For example, molecular analysis has revealed that a surprisingly high proportion of the DNA is noncoding, both in intergenic regions and in intervening sequences (introns) within genes. It is hard to imagine that many evolutionary changes in such noncoding regions are of adaptive significance to the organism.

In coding regions of genes, molecular sequence data do indeed provide ubiquitous evidence of natural selection. But rather than positive Darwinian selection favoring new adaptively favorable variants, the type of selection that is revealed is what is known as "purifying selection." Purifying selection is an essentially conservative force, which in the case of protein-coding genes acts to eliminate the majority of mutations which cause a change in the amino acid sequence. In the case of many proteins, it seems that relatively early in the history of life the basic structure required for the protein's function was hit upon. Amino acid replacements that disrupt that structure are harmful and are thus eliminated from the population by purifying selection. The amino acid replacements that do occur tend to be those that have little effect on the protein's structure and function; in other words, those that are selectively neutral.

The predominance of neutral evolution at the molecular level need not imply that there are *no* cases of adaptive evolution at the DNA level. Although their "selectionist" opponents did not always appreciate the fact, the neutralists were careful to include a role for adaptive evolution on the Darwinian model. Thus Kimura (1969, p. 1188) wrote: "What we have postulated is that, surrounding the adaptive changes that occur by selection, there is a great deal of random noise from near-neutral random changes." A challenge in recent years, as molecular sequence data have accumulated rapidly, has been to devise ways of screening out the "noise" of neutral evolution in order to study the process of adaptive evolution at the DNA level.

Here I review some of the cases in which it has been possible to obtain convincing evidence at the molecular level that adaptive evolution has occurred as a result of positive Darwinian selection favoring certain nucleotide substitutions. Evidence of this sort has been obtained by statistical analysis of DNA sequence data. As I hope will become clear from the following, the development of the neutral theory has made the task of conducting such analyses easier than it would otherwise have been. By making clear-cut predictions about what we expect to observe in the case of neutral evolution, the neutral theory provides a null hypothesis against which to test the alternative hypothesis of adaptive evolution in a particular case.

BALANCING SELECTION

By the term "balancing selection" is meant any type of natural selection that acts to maintain a polymorphism in a population, and such a selectively maintained polymorphism is referred to as a balanced polymorphism. The type of balancing selection most familiar to biologists is overdominant selection or heterozygote advantage. This type of selection was initially modeled by theoretical population geneticists before any evidence was available that such a mechanism ever acted in any natural population. The sickle-cell gene in humans provided the first and, until recently, the only well-established example of overdominant selection. A single amino acid difference in the β chain of hemoglobin causes serious illness in the homozygous state, but in the heterozygous condition confers resistance to *Plasmodium falciparum* malaria. As a consequence, the sickle-cell allele is maintained in the human population in areas where *P. falciparum* is endemic.

Frequency-dependent selection is another type of balancing selection that so far remains hypothetical. There have been a number of theoretical models of frequency-dependent selection. These have in common the property that the fitness of an allele is a function of its population frequency. Actually, certain models of frequency-dependent selection make predictions very similar to models of overdominant selection (Takahata and Nei, 1990). Therefore, in an empirical case it may be very difficult to determine whether the type of selection that is acting is overdominant or frequency-dependent.

In recent years, statistical analysis of DNA sequence data has provided a new way of testing for balancing selection. This approach has proved particularly fruitful in studies of the genes of the major histocompatibility complex (MHC) of vertebrates, and in studies of certain genes of parasites that are recognized by the MHC molecules and other proteins of the vertebrate immune system. In the following two sections I review the methods used in these studies and their major conclusions.

The Major Histocompatibility Complex

The MHC is a multigene family found in all vertebrates; its products are cell-surface glycoproteins that serve to present intracellularly processed foreign peptides to T cells (Klein, 1986). There are two major subfamilies of MHC molecules, which are designated class I and class II. Class I molecules, which are expressed on almost all types of nucleated cells, present peptides to cytotoxic T cells. Class II molecules, which are expressed on antigen-presenting cells of the immune system, present peptides to helper T cells. Both class I and class II molecules are heterodimers with four extracellular domains, but they achieve a similar molecular structure in different ways (Figure 1). The class I α chain consists of three extracellular domains (α_1, α_2, α_3). The α_3 domain associates noncovalently with β2-microglobulin, a molecule encoded outside the MHC but having a distant

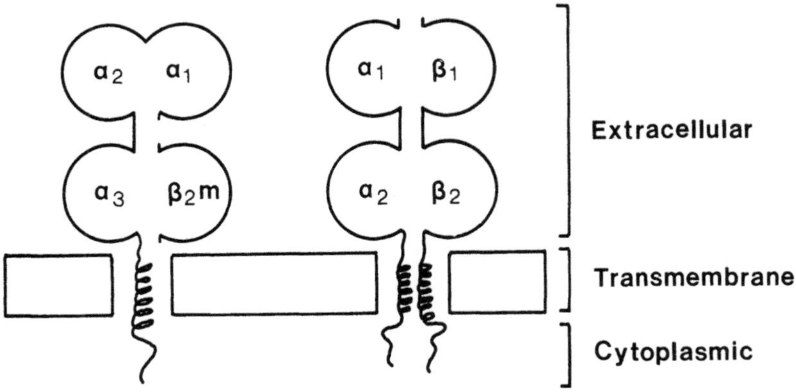

Figure 1. Schematic diagrams of the structure of class I and class II MHC molecules. The class I heavy chain has three extracellular domains: α_1, α_2, and α_3. The α_1 and α_2 domains contain the antigen recognition site, and β_2-microglobulin (β_2-m) associates noncovalently with α_3. The class II molecule consists of α and β chains, each with two extracellular domains. The antigen recognition site is made up of portions of the α_1 and β_1 domains.

evolutionary relationship to MHC molecules. The class II heterodimer consists of an α chain with two extracellular domains (α_1 and α_2) and a β chain with two extracellular domains (β_1 and β_2).

The MHCs of human, mouse, and a number of other species have been known for some time to contain several highly polymorphic loci. In a typical mammal, it appears that there are 1 to 3 polymorphic class I α chain loci, as well as a number of nonpolymorphic class I loci. These latter, nonpolymorphic class I loci, which are called "nonclassical" class I loci, have reduced levels of expression compared to the polymorphic or "classical" class I loci, and their function is not known. In humans, the classical class I loci are designated *A*, *B*, and *C*.

The class II MHC genes of mammals are arranged in a number of separate regions, each of which typically contains one α chain gene and one or more β chain genes. In humans, for example, there are three regions including polymorphic class II loci, designated *DR*, *DQ*, and *DP*. Each of these regions contains a polymorphic β chain locus; in the *DR* region, there are between one and three β chain loci, depending on the haplotype. Only in the *DQ* region is the α chain gene polymorphic as well.

The fact that MHC loci are polymorphic was known for a considerable length of time before the function of MHC molecules was clearly understood. Thus, it was difficult to imagine an evolutionary mechanism that would explain MHC polymorphism. The polymorphism of class I MHC loci was discovered through their role in mediating transplant rejection, yet it seemed very unlikely that this role could be directly related to the selective pressures acting on the MHC in nature. One valiant attempt to relate MHC polymorphism in nature to transplant rejection was the hypothesis that MHC polymorphism is advantageous because it prevents cancer from being a contagious disease by minimizing the chance of the spread of tumors from one individual to another.

Later the function of the class I MHC in immune recognition was worked out. When a cell is infected by a virus or other intracellular parasite, peptides derived from the parasite's proteins are presented on the surface of the infected cell by class I MHC molecules. When a cytotoxic T cell recognizes the complex of foreign peptide and self class I MHC, it kills the infected cell. Class II MHC molecules play a similar role in stimulating helper T cells. The complex of self class II MHC molecule and foreign peptide on an antigen-presenting cell is recognized by the helper T cell; the latter releases lymphokines that stimulate B cells (which produce antibodies) and macrophages to respond to the infection.

The MHC's role in antigen presentation suggested to Doherty and Zinkernagel (1975) a possible mechanism whereby overdominant selection might operate at MHC loci. They had evidence that products of different alleles at MHC loci differ in their ability to bind and present specific foreign peptides. Given such differences, they argued that in a population that is exposed to a variety of pathogens, an individual heterozygous at all or most MHC loci would have an advantage over a homozygote because the former's MHC molecules would be able to present more different kinds of foreign peptides, and thus would confer resistance to a wider array of pathogens.

This argument was initially not very popular with immunologists because it was difficult to obtain evidence of a correlation between MHC heterozygosity and survival. Some evidence suggested that MHC polymorphism might be explainable without having to invoke any form of natural selection. For example, some early studies with mice had suggested that the MHC might have an unusually high mutation rate. Theoretically, a very high mutation rate might lead to a high level of polymorphism without any kind of balancing selection. Later, when DNA sequences began to be available for MHC loci, researchers sometimes found stretches of sequence similarity between alleles at different MHC loci. This led to the hypothesis that MHC polymorphism might be caused by extensive nonreciprocal interlocus recombination ("gene conversion"). Given a gene family whose members were fairly divergent from one another, such a process might cause polymorphism at individual loci. However, this hypothesis did not explain how the MHC loci have come to diverge from each other.

Furthermore, a comparison of population frequencies of alleles at human MHC loci with the expectation of the neutral theory showed that the MHC allelic frequencies are very different from the neutral expectation (Hedrick and Thomson, 1983). In the neutral case, we generally expect to find one common allele and one or a few rare alleles. At polymorphic MHC loci in humans, by contrast, there are a large number of alleles of intermediate frequency.

Hughes and Nei (1988, 1989) approached the question of selection at MHC loci by a comparison of rates of nucleotide substitution. In protein-coding regions of DNA, point mutations (mutations changing a single base pair) can be classified as synonymous or nonsynonymous. A synonymous mutation is one that, because of the redundancy built into the genetic code, does not change the amino acid sequence of the polypeptide encoded. A nonsynonymous mutation is one that changes the amino acid sequence. It turns out that all mutations at the second position of codons are nonsynonymous, as are most at first codon positions, whereas most mutations at third codon positions are synonymous. Mutation is equally likely to occur at each codon position, but not all mutations will result in nucleotide substitutions that are observable in populations. Some mutations will be selectively disadvantageous and will be eliminated by purifying selection. Others will be lost by chance (genetic drift) or fixed by chance. Still others may be fixed by positive selection.

In order to understand the evolutionary forces that have acted on two related genes since they diverged from a common ancestor, it is useful to estimate the number of synonymous nucleotide substitutions per synonymous site (d_S) and the number of nonsynonymous nucleotide substitutions per nonsynonymous site (d_N) (Nei and Gojobori, 1986). In estimating these quantities, we first count the number of synonymous sites and the number of nonsynonymous sites in the two sequences. A synonymous site is one (like the third codon position of glycine codons) at which every possible mutation would be synonymous. A nonsynonymous site is one (like all second codon positions) at which every possible mutation would be nonsynonymous. At certain sites some mutations are synonymous and some nonsynonymous. Such a site is counted as some fraction synonymous and some fraction nonsynonymous. For example, at the third position of the lysine codon AAA, one possible mutation (to AAG) is synonymous, whereas the other two (to AAC and AAT) are nonsynonymous. This site would thus be counted as one third of a synonymous site and two thirds of a nonsynonymous site.

In the case of functional genes, we expect to find that d_S will exceed d_N. This is a consequence of purifying selection. Most amino acid changes will disrupt protein structure in one way or another and thus are harmful to the organism. Therefore, most nonsynonymous mutations will be eliminated by selection. On the other hand, most synonymous nucleotide substitutions are selectively neutral or nearly so, and thus will be fixed or lost by chance. The neutral theory predicts that the rate of fixation of neutral mutations will be equal to the neutral mutation rate; so the rate of substitution at synonymous sites should reflect the mutation rate in the gene in question. Because a much lower proportion of mutations will be neutral at nonsynonymous sites than at synonymous sites, the rate of substitution at nonsynonymous sites will be lower than that at synonymous sites. By contrast, in the case of a nonfunctional pseudogene, we expect d_S and d_N to be about equal, since in this case mutations at all codon positions are equally neutral.

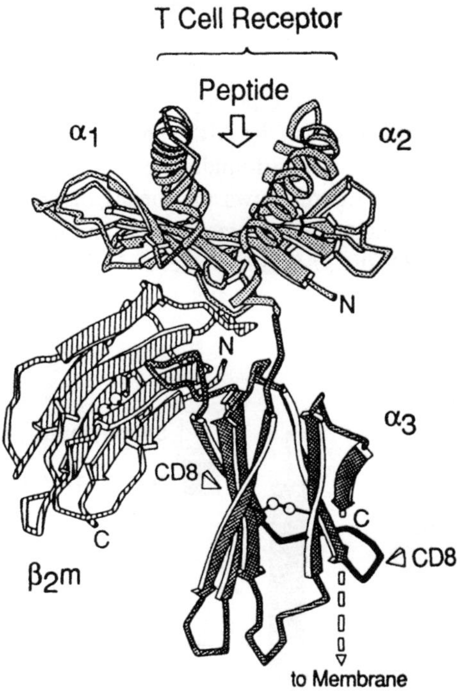

Figure 2. The structure of the α_1, α_2, and α_3 domains and associated β_2-microglobulin of the class I MHC molecule. The β strands are shown as arrows in the amino to carboxyl direction, and the α helices as coiled ribbons. The groove at the top of the molecule is where peptides are bound and recognized by T cells. (Reprinted with permission from Lawlor et al., 1990, © 1990 by Annual Reviews, Inc.).

However, let us suppose that natural selection is acting to favor changes at the amino acid level in a particular protein or in some region of a protein. This is the case of positive Darwinian selection, and in this case we would predict that d_N will exceed d_S. "Selectionists" might have expected to find that this pattern would predominate in comparisons among DNA sequences. In fact, when sequence data became available, comparisons of related sequences overwhelmingly showed $d_S > d_N$. Thus, they showed that purifying selection is nearly universal in protein coding genes and provided powerful support for the neutral theory.

In 1987, Bjorkman and colleagues determined the crystal structure of a human class I MHC molecule, the product of an allele at the *A* locus. The structure showed that portions of the α_1 and α_2 domains together form a groove-like structure at the top of the molecule (Figure 2). This groove is bounded by two α helices, and a β pleated sheet forms its base. Because the groove is at the top of the molecule and thus able to interact with the T cell receptor, it was deduced that this is the site in which foreign peptides are bound and presented to T cells. This site was called the antigen recognition site (ARS) by Bjorkman et al. (1987a,b).

Hughes and Nei (1988) reasoned that, on Doherty and Zinkernagel's (1975) hypothesis that MHC polymorphisms are maintained by overdominant selection relating to peptide binding and thus to pathogen resistance, such selection would act to favor amino acid differences in the ARS. Therefore, d_N should exceed d_S in the ARS. This is precisely what they found. The type of results they found are illustrated in Figure 3. This figure shows the means of d_S and d_N for pairwise comparisons among alleles at the human class I loci *A* and *B*. These quantities were computed separately for: (1) the 57 codons that encode the residues in the ARS; (2)

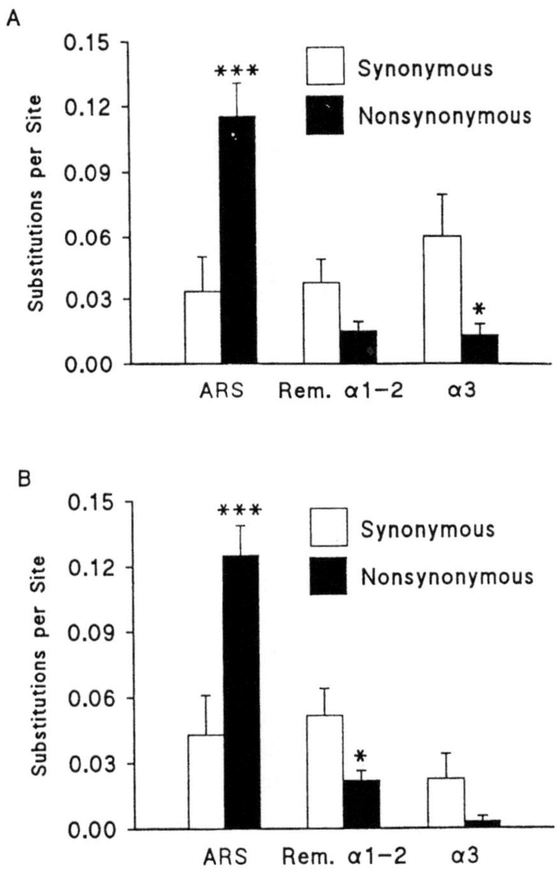

Figure 3. Mean numbers of synonymous (d_S) and nonsynonymous (d_N) nucleotide substitutions per site in pairwise comparisons among (**A**) 21 alleles at the human class I MHC *A* locus; and (**B**) 39 alleles at the human class I MHC *B* locus. d_S and d_N are computed separately for the antigen recognition site (ARS), for the remainder of the α_1 and α_2 domains, and for the α_3 domain. Vertical bars indicate the standard errors of the estimates.

the remainder of the α_1 and α_2 domains; and (3) the α_3 domain. In the ARS, mean d_N is about three times as high as mean d_S. By contrast, in the remainder of the gene $d_S > d_N$, as is true of most genes. Mean d_N is highest in the ARS and lowest in the α_3 domain, which is highly conserved at the amino acid level. Similar results have been found for the class I MHC alleles from the mouse, cat, bovine, and a number of other mammalian species.

In the case of the class II MHC, the α_1 and β_1 domains together form the ARS, with each contributing one of the two α helices and half of the β pleated sheet. Here again, $d_N > d_S$ was found in the codons encoding the ARS (Hughes and Nei, 1989).

These results ruled out the hypothesis that MHC polymorphism is due to a high mutation rate. A high mutation rate should affect all sites equally, both synonymous and nonsynonymous, and would presumably not just be confined to the ARS codons. Furthermore, comparison of human and mouse class I sequences showed that d_S—which should reflect the mutation rate—is actually rather low at MHC loci in comparison with other mammalian loci (Hughes and Nei, 1988, 1989). "Gene conversion" between MHC loci cannot account for MHC polymorphism because this process also should affect gene regions at random, rather than being concentrated on nonsynonymous sites in the ARS. This is not to deny that interlocus recombination occasionally has occurred in the history of the MHC; DNA sequence analysis has revealed evidence of some such events. But such a process cannot be the main cause of MHC polymorphism.

Clearly, MHC polymorphism is maintained by some form of balancing selection that favors diversity in the ARS. This selection presumably relates to the advantage such diversity confers in enhancing the individual's ability to bind a wide array of foreign peptides, and thus to resist a wide array of pathogens. The discovery of a human class I MHC allele (*B53*) and a class II haplotype associated with resistance to severe *Plasmodium falciparum* malaria (Hill et al., 1991) provides further support for the hypothesis that selection on MHC loci is driven by pathogens. Whether the selection acting on MHC loci is overdominant or frequency-dependent is difficult to determine at present. However, certain models of frequency-dependent selection can be ruled out, since they make predictions incompatible with what we know of MHC polymorphism (Takahata and Nei, 1990). And Doherty and Zinkernagel's model of overdominant selection at MHC loci has a strong biological appeal. This model is consistent with all available data, and its biological basis—disease resistance—is easily understood.

Analysis of MHC DNA sequences has revealed still another interesting fact. Certain MHC alleles—or, more accurately, allelic lineages—are very ancient, having persisted since well prior to the speciation events that have given rise to extant animal species. For example, at the human *A* locus, phylogenetic analysis of DNA sequences shows that these can be grouped into a number of families of related alleles (Figure 4). At the corresponding locus of the chimpanzee, all alleles that have so far been discovered belong to just one of these families (Figure 4). This means that the families of *A* locus alleles must have diverged from each other prior

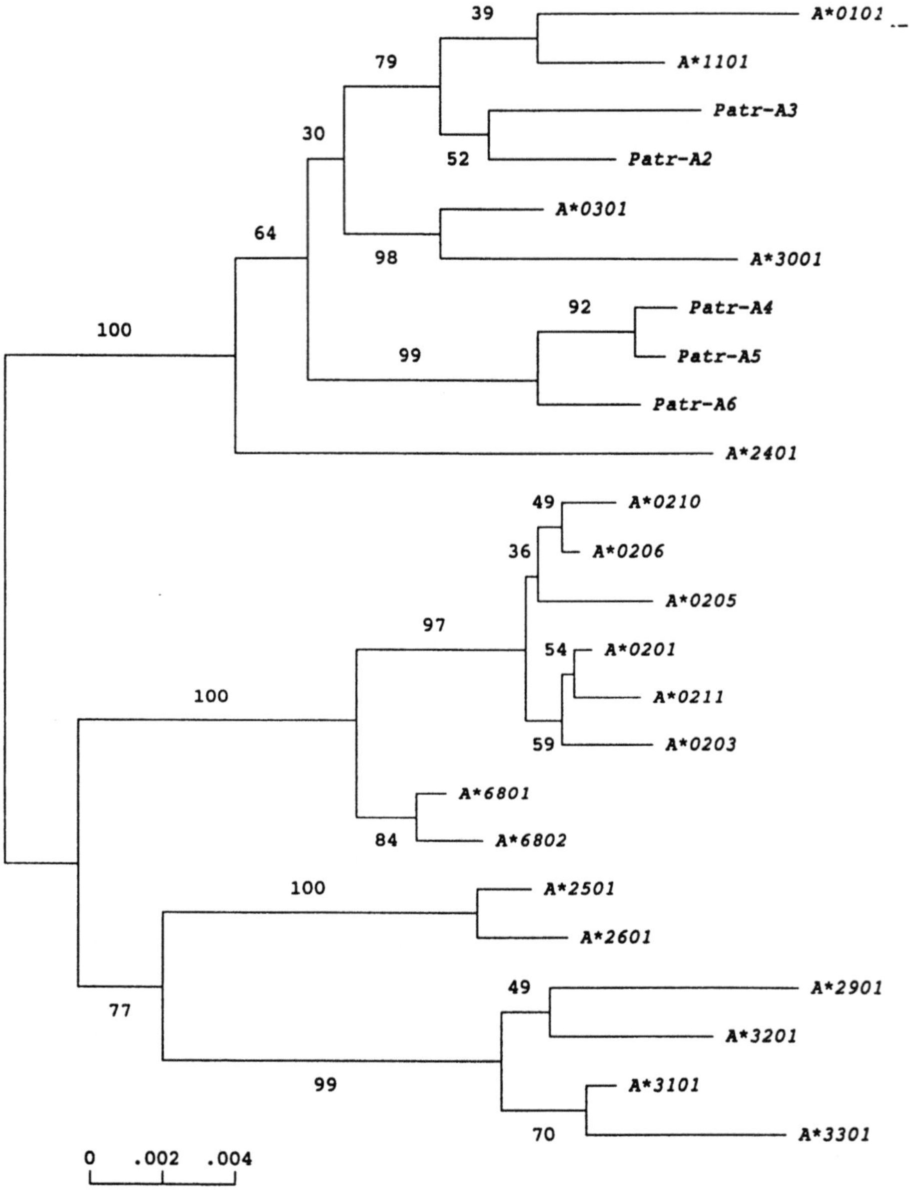

Figure 4. A phylogenetic tree, constructed by the neighbor-joining method (Saitou and Nei, 1987), of human and chimpanzee *A* locus alleles. The tree is based on the number of nucleotide substitutions per site (d). The tree was rooted by using class I MHC sequences from bovine and pig as outgroups. Numbers on the branches are boot strapping percentages. Bootstrapping is a technique whereby one samples from sequence data with replacement and constructs a phylogenetic tree on the basis of the sample data set. This process is repeated numerous times (in this case 1000 times); the bootstrapping percentage is the percentage of samples that supported a particular cluster within the tree.

to the divergence of human and chimpanzee 5 to 7 million years ago. The common ancestor of human and chimpanzee must have had representatives of each of these families. For some reason, all but one of them has been lost in the chimpanzee, but the others have persisted in humans. Certain class II MHC alleles may have persisted for even longer than the *A* locus alleles.

The long-term persistence of MHC alleles has been called "trans-species polymorphism" (Mayer et al., 1988). Actually, such long-term persistence of alleles is further evidence that some form of balancing selection is operating at MHC loci. Neutral polymorphisms cannot last anywhere near as long as polymorphisms maintained by overdominant selection or other forms of balancing selection (Takahata and Nei, 1990).

Immunogenic Proteins of Parasites

The immune system of vertebrates is a complex molecular and cellular system adapted for the recognition and elimination of parasites. (In this case, the term "parasite" refers to any organism living within the body of and at the expense of another, including viruses, bacteria, protists, fungi, and animals.) Given the bewildering variety of parasitic organisms and the strong selective premium placed on resistance to them, it is not surprising to find evidence of selection favoring the ability to recognize and eliminate parasites, as in the case of the MHC. It would also be reasonable to expect that the MHC and other components of the vertebrate immune system exert strong selective pressure on parasites that will favor any mutation that enables the parasite to evade immune recognition or to otherwise counteract the host's defensive apparatus.

One way that parasites have evolved to evade immune recognition is to change their surface proteins over the course of an infection. The most dramatic example of this occurs in the protozoan parasites of the genus *Trypanosoma*, which sequentially express a series of highly divergent surface glycoproteins over the course of an infection (Cross, 1990). Other protozoan parasites which lack this complex system of deception nonetheless pass through a number of life stages which are characterized by different sets of surface proteins.

The malaria parasites of the genus *Plasmodium* provide an example of such a life cycle. The infective stage is called the sporozoite, which enters the host's bloodstream through a mosquito bite and migrates to the liver. One surface protein of the sporozoite, known as the circumsporozoite protein (CSP), has been studied in some detail because of interest in using this protein or some portion of it in an anti-malaria vaccine. In a liver cell, the sporozoite divides mitotically to produce merozoites, which are released into the bloodstream and enter red blood cells. Merozoites have their own set of distinctive surface antigens, including merozoite surface antigen-1 (MSA-1) and merozoite surface antigen-2 (MSA-2).

In the case of the malaria parasites, there appears to be a whole set of additional molecular strategies for evading the immune system. One distinctive characteristic of the surface proteins of these parasites is their tendency to contain simple amino acid motifs repeated many times. For example, in *Plasmodium falciparum* the CSP has a central repeat region containing numerous repeats of the motif asparagine-alanine-asparagine-proline (NANP in the single letter code). The CSP locus is polymorphic, and the number of NANP repeats varies from 40 to 46 depending on the allele. It is known from experimental evidence that large polymeric molecules stimulate a nonspecific and relatively ineffective T cell independent antibody response. This observation has led to the hypothesis that the function of the repeating structure of the CSP and other parasite antigens is precisely to elicit such a response (Kemp et al., 1987; Enea and Arnot, 1988). On this hypothesis, proteins like the CSP are "smoke-screens" whose function is to protect the parasite from a more effective immune response.

Nonetheless, the CSP and other surface proteins of *Plasmodium* include nonrepeat regions, and some of these are known to be presented by the MHC to T cells. Such T cell epitopes (TCE) have been identified in the case of the CSP of *Plasmodium falciparum*. But the TCE turn out to be polymorphic at the population level (Lockyer et al., 1989). Furthermore, estimation of the rates of nucleotide substitution in these and other regions of the CSP gene suggests that amino acid diversity in the TCE is the result of positive selection (Hughes, 1991). CSP alleles from *P. falciparum* have the remarkable property that no synonymous differences are found among them. This seems to be the result of an extremely low G+C content, especially at third codon positions, a factor known to decrease the observed rate of synonymous nucleotide substitution. Nonsynonymous differences are concentrated in the TCE, and d_N in this region is significantly greater than that in other regions of the gene (Hughes, 1991). Positive selection on these particular epitopes is a feature of the CSP of *P. falciparum*, but not necessarily of other species. For example, in the case of CSP genes from *P. cynomolgi* (a parasite of monkeys), the regions corresponding to the TCE of *P. falciparum* do not show evidence of positive selection. However, it does appear that there is positive selection on another portion of the molecule in the case of *P. cynomolgi* (Hughes, 1991).

Thus, it appears that the CSP of *P. falciparum* incorporates two layers of defense: (1) the "smoke-screen" of NANP repeats elicits an ineffective T cell independent response; (2) the T cell epitopes are polymorphic so that an MHC allele able to bind and present the TCE from one CSP allele may not be able to bind that of another allele. Thus, the CSP polymorphism is apparently maintained by some form of balancing selection relating to the host immune system. Interestingly, it was recently found that T cells restricted to the B53 allele, which is associated with resistance to severe malaria, recognized a peptide from a protein known as liver-stage-specific antigen-1 (LSA-1) (Hill et al., 1992), but did not recognize peptides from the repeat or TCE regions of the CSP. This is consistent with the hypothesis that polymorphism at the CSP locus has evolved as a way of evading immune recognition, since the human host is only able to achieve resistance by binding a peptide from another, apparently nonpolymorphic protein.

MSA-1 is another surface protein of *P. falciparum* that is encoded by a polymorphic locus, and shows evidence of balancing selection. In this case, there are two families of alleles which are very ancient, perhaps as old as 35 million years (Hughes, 1992). There have been numerous recombination events between these allelic lineages, so that each modern allele is a patchwork of segments with different evolutionary histories. There are three regions of the MSA-1 gene which show evidence of positive selection favoring diversification at the amino acid level; that is, where d_N is significantly higher than d_S (Hughes, 1992). Here also it seems likely that the host immune system is the selective agent, although no strong evidence of this is available at present.

ADAPTIVE DIVERSIFICATION OF MULTIGENE FAMILIES

Recent study of the genomes of eukaryotes at the molecular level has revealed that many protein-coding genes in these species belong to families of genes that can be shown, on the basis of sequence similarity, to share a common ancestry. Often the members of such gene families encode genes having distinct but related functions. In some cases, several such families can be grouped together into an extensive gene superfamily. The immunoglobulin gene superfamily, for example, encodes a wide variety of cell surface molecules involved in immune recognition in vertebrates, including the immunoglobulins, T cell receptors, and MHC molecules. It also encodes ligands involved in cell–cell adhesion processes. Clearly, the duplication of genes within multigene families and their subsequent functional divergence from one another plays an essential role in the evolution of new protein function, and thus in adaptive evolution.

Two main hypotheses have been proposed to explain the process by which a duplicate gene assumes a new function. According to the first of these hypotheses, the origin of new protein function in a duplicate gene is a fortuitous consequence of mutations that are fixed by chance in the total absence of constraint on one gene copy. This hypothesis was expressed by Ohno (1973, p. 261) as follows:

> The mechanism of gene duplication provides a temporary escape from the relentless pressure of natural selection to a duplicated copy of a functional gene locus. While being ignored by natural selection, a duplicated and thus redundant copy is free to accumulate all manner of randomly sustained mutations. As a result, it may become a degenerate, nonsense DNA base sequence. Occasionally, however, it may acquire a new active site sequence, therefore a new function and emerge triumphant as a new gene locus (Ohno, 1973).

A second hypothesis was implied by Goodman and colleagues (1975) in their study of the hemoglobins of vertebrates. These authors argued that, soon after gene duplication, there is often a period of rapid evolution at the amino acid level, as directional positive selection acts to fix amino acid replacements which adapt one or both of the duplicated genes to separate functions. They also imply that, prior to the gene duplication, these separate functions may have been filled by the product of a single ancestral gene. Gene duplication makes it possible for the two daughter genes to encode specialist proteins that replace an ancestral generalist protein.

The hemoglobin molecule is a tetramer consisting of two α chains and two β chains. Since the α and β chains show clear evidence of an evolutionary relationship, they must have arisen by duplication of a single ancestral gene. Thus, Goodman and colleagues argued that the hemoglobin molecule was originally a homotetramer; that is, the four chains were identical, all being encoded by the same gene. After gene duplication gave rise to separate α and β chain genes, the molecule became a heterotetramer, and the α and β chains were able to specialize to complement each other.

The Goodman et al. (1975) methods of analysis have been criticized (Kimura, 1983), and it has been difficult to decide between these two hypotheses. Most cases of gene duplication of which we have evidence took place in the rather distant evolutionary past. Thus, if gene duplication has been followed by a burst of selectively favored amino acid replacements, it is no longer very easy to obtain evidence of this fact because they are by now overlain by numerous neutral changes that have occurred in the intervening time. For it is to be expected that once the two duplicate genes have adapted to their new functions, they will no longer be subject to positive selection, but will evolve under purifying selection as predicted by the neutral theory.

What would be ideal for studying this question would be an example of a gene family whose members have duplicated rather recently and yet have diverged functionally from each other. Ngai and colleagues (1993) have recently found a very convincing example of this sort, which involves olfactory receptor genes in the channel catfish.

The olfactory receptors are members of a superfamily of transmembrane receptors which are called G protein-coupled receptors. The name derives from the fact that each of these receptors receives an extracellular signal and then transduces it to an intracellular signal by means of a protein called a G protein. Members of this superfamily have an extracellular amino-terminal domain and an intracellular carboxy-terminal domain, and pass through the cell membrane seven times. There are thus four extracellular domains, four intracellular domains, and seven helical trans-membrane domains (designated TM1–TM7). These receptors have also been dubbed "heptahelicals" because of their seven helical membrane-spanning domains. Various heptahelicals act as receptors for light, hormones, and neurotransmitters, as well as for odorants. In different members of the superfamily, different extracellular or transmembrane domains have been suggested as functionally important for the molecule's role as a receptor.

The olfactory receptor gene family was originally described in the rat, where it is estimated that this family may contain 1000 genes (Buck and Axel, 1991). The extraordinary diversity of these receptors is believed to account for the ability of mammals like the rat to discriminate among a wide variety of odors. The channel catfish has a much more modest repertoire of about 100 olfactory receptor genes. This more limited repertoire made it easier for Ngai et al. (1993) to find a subfamily of these genes that had duplicated relatively recently.

Ngai et al. concentrated their analysis on TM3 and TM4, which they found to be the most variable domains at the amino acid level among members of the receptor subfamily. Of course, high variability at the amino acid level need not be evidence that a protein region is under selection favoring diversity. In fact, in most cases, such variability is evidence that the region in question is relatively unimportant functionally, and thus relatively few nonsynonymous mutations will be eliminated by purifying selection. Only a comparison of d_S and d_N can show the reason for a high level of variability at the amino acid level.

When d_S and d_N were estimated for all pairwise comparisons among the olfactory receptor genes in the subfamily, Ngai et al. found that d_N in TM3–TM4 was about twice d_S in the same region, and the difference between the two was statistically significant. By contrast, in the remainder of the gene, d_S was over twice as high as d_N. These results support the hypothesis that the amino acid variability in TM3–TM4 is due to positive Darwinian selection acting to diversify the products of different loci in these two transmembrane domains. Since the basis of such selection is most likely the advantage the catfish derives from having a variety of receptors to cope with a diversity of odorants, these results imply that TM3 and TM4 must be important for binding of odorant molecules by members of this subfamily.

Unlike the MHC, this is not a case of balancing selection acting to maintain polymorphism within a locus, but rather of directional selection promoting diversity between loci. There have been some other cases reported in the literature of apparent directional selection at the molecular level. For example, there is evidence of an enhanced nonsynonymous rate in the active center of serine protease inhibitors (or serpins) (Hill and Hastie, 1987), and in the antigen-binding regions of immunoglobulins (Tanaka and Nei, 1989). In both of these cases, although d_N exceeded d_S in some comparisons, the reverse occurred in certain others. Probably the reason for this is that the directional selection occurred in the past, and now the genes have diverged sufficiently from each other and are evolving neutrally.

These studies have indicated that there are cases in which functionally divergent members of multigene families have diverged from one another as a result of positive selection which occurs shortly after duplication. Of course, molecular analyses have revealed many cases in which a duplicate gene has indeed been redundant and has eventually been silenced by mutations that eliminate transcription, or that produce a nonfunctional protein product. In fact, becoming such a silenced nonfunctional pseudogene is almost certainly the fate of the vast majority of duplicated genes. But in no case has it ever been proved that such a redundant gene has later "emerged triumphant as a new gene locus" (to quote Ohno, 1973) after having by chance acquired a new function.

Kimura and Ohta (1974) included in their list of "principles governing molecular evolution" the following statement: "Gene duplication must always precede the emergence of a new gene having a new function." Actually, it appears that at least in some cases a single gene can have two very different functions simultaneously. One way that this can happen in eukaryotes is that the pre-mRNA transcript from a gene can be alternatively processed to produce two or more different mRNAs, which encode related but different proteins. Another way that a single gene can have two functions is a phenomenon called "gene sharing." This phenomenon, which was discovered through comparative study of the lens crystallins of animals, is especially relevant to the question of how new proteins evolve (Piatigorsky and Wistow, 1991).

In certain cases, a single gene encodes a protein that in the lens of the eye serves as a crystallin, but in other tissues serves some totally different function. For example, δ-crystallin in the duck is encoded by the same gene that encodes the enzyme argininosuccinate lyase. Furthermore, there is evidence that gene sharing has occurred repeatedly in the history of the animals. For example, the α-crystallins of vertebrates are related to small heat-shock proteins, while other vertebrate crystallins are related to enzymes such as enolase and lactate dehydrogenase. The cephalopods—octopus and squid—have eyes that are convergent in structure with the eyes of vertebrates. In cephalopods also the crystallins have been derived from borrowed enzymes, glutathione S-transferases. In all of these cases, it is plausible to hypothesize that originally gene sharing occurred, with a single gene encoding a crystallin and a protein having a very different function. Later, gene duplication occurred, giving rise to a separate crystallin gene.

Gene sharing may seem to be a very isolated case which is unique to the crystallins. However, it provides a counterexample to the idea that gene duplication must precede the evolution of new function. Furthermore, it may be just one example of an alternative pathway by which functionally novel proteins can evolve—a pathway that may, in fact, be the usual way that this process occurs.

If the product of a single gene performs two distinct functions, after gene duplication occurs the products of the two daughter genes may specialize so that each performs one of the two functions. In some cases, this specialization might require changes in the regulatory regions of one or both genes that lead to their expression at different times in development or in different tissues. In other cases, mutations in one or both genes that better adapt the protein to its specific function may be selectively favored—leading to a burst of positively selected amino acid replacements, as is seen in the case of the catfish olfactory receptors. In this model, the ancestor of two closely related olfactory receptor genes would have encoded a receptor that was in a sense a generalist, being able to bind an array of odorant molecules. When the gene duplicated, it was possible for the two daughter genes to encode more specific receptors, and nonsynonymous mutations making this possible were selectively favored.

Jensen's (1976) studies of the evolution of biochemical pathways led to a similar hypothesis. This author concluded that two enzymes that are specialized to catalyze two separate reactions may evolve, after gene duplication, from an ancestral enzyme capable of catalyzing both reactions. And a similar model was implied in the Goodman et al. hypothesis that the hemoglobin molecule was a homotetramer before gene duplication made it a heterotetramer.

CONCLUSION

The hypothesis that the variety of living things on Earth has arisen as a result of a process of "descent with modification," as Darwin expressed it, has been controversial ever since it was first proposed. Not only has the very idea of evolutionary change been assailed by those who feel that it is inconsistent with their religious beliefs, but among biologists themselves there has been a constant debate regarding the mechanisms responsible for evolution. To the antiscientific opponents of evolution, such debate is proof of the weakness and inconsistency of the evolutionary hypothesis. Biologists are likely to take just the opposite view. Fields of scientific knowledge where there is little controversy are generally fields in which most major problems have been solved and little new research is going on; for example, the comparative study of the anatomy of vertebrates. The situation is very different in a field in which new discoveries are being made at a rapid rate. Such a field is typically characterized by lively debate, as researchers actively extend and modify existing theoretical frameworks in order to accommodate new data.

Evolutionary biology has been in such a state of active ferment in the past decade or so, as surprising new findings have extended our views of the evolutionary process at the molecular level. The discovery of introns, the finding that so many genes are members of multigene families, the identification of conserved families of regulatory genes controlling pattern formation in multicellular organisms, and the evidence that most change at the molecular level is due to the chance fixation of selectively neutral variants have all expanded the horizons of evolutionary biologists in ways not previously imagined. In a time of such intellectual activity, it is not surprising that some biologists have begun to wonder whether the mechanism of adaptive evolution by natural selection has any role to play in the new biology. I hope it is obvious from the foregoing that there is by now powerful evidence at the molecular level of adaptive evolution. Indeed, the comparative study of DNA sequences makes it possible to observe the effects of positive Darwinian selection in a much more quantitative and convincing manner than was ever possible before.

In the early decades of molecular biology, from the 1940s to the mid-1960s, evolutionary biology seemed to many biologists a backwater of science. As the structure of DNA was worked out, followed by the mechanisms of transcription and translation and the genetic code, evolutionary biologists seemed to continue a

Nineteenth Century tradition of field- and museum-oriented biology. Of course, molecular biologists in this period paid lip-service to Dobzhansky's famous statement that "evolution is the central organizing idea in biology"; but in practice, evolutionary concepts played little role in their research.

This situation has changed dramatically with the advent of rapid methods of DNA sequencing. Now evolutionary ideas routinely play a role in the newest discoveries in molecular biology. Every time a new gene is sequenced and the researchers search the data base of published DNA sequences for any similar sequence, they are relying on an evolutionary logic. Such searches only make sense under the hypothesis that the new gene and its previously sequenced homologue share an evolutionary relationship; that is, a relationship by descent with modification from a common ancestral gene. Molecular biologists are increasingly interested in understanding the phylogenetic relationships among genes and the evolutionary forces that have acted to make these genes the way they are today.

Nor is this interest merely speculative. It has a strong practical component. It has become increasingly obvious that the only way we will ever understand the operation of complex biological processes, such as the development of a multicellular animal from a fertilized egg or the operation of the vertebrate immune system, is in an evolutionary framework. For the properties of biological systems cannot be predicted from any set of first principles or even by considerations of optimal design. Because any biological system is the product of a historical process, understanding that history is crucial to understanding the system as it is today.

Nowhere is this clearer than in our effort to understand and therefore to combat the organisms responsible for infectious disease. The struggle to develop vaccines against such ancient scourges of our species as malaria and schistosomiasis, as well as newer scourges such as AIDS, will depend on an understanding of the mechanisms that pathogens have evolved to evade recognition and elimination by the immune system. As indicated by the results presented above, in the case of *Plasmodium falciparum*, an understanding of the role of past adaptive evolution on the part of both parasite and host will be essential to finding a vaccine strategy that works. The same is certain to be true in the case of the HIV virus, which evolves rapidly within a single patient. Far from being an obsolete Victorian parlor game, the study of adaptive evolution will play a key role in the biomedical science of the future.

SUMMARY

Although neutral evolution predominates at the molecular level, nucleotide substitutions fixed by positive Darwinian selection are believed to provide the basis for adaptive evolution. With the advent of DNA sequencing, it has been possible to study adaptive evolution at the molecular level by estimating rates of synonymous and nonsynonymous (i.e., amino acid-altering) nucleotide substitution per site. In a typical protein-coding gene, the rate of synonymous substitution exceeds the rate of nonsynonymous substitution; this occurs because most nonsynonymous nucleotide substitutions are harmful to protein function and thus are eliminated by purifying selection. If positive selection is acting to favor diversity at the amino acid level, this pattern is expected to be reversed. In recent years a number of such cases have been studied. The best documented case involves the highly polymorphic major histocompatibility complex (MHC) loci of vertebrates. MHC molecules bind foreign peptides and present them to T cells. In the region of the gene encoding the peptide-binding groove of the MHC molecule, the rate of nonsynonymous substitution exceeds the rate of synonymous substitution, indicating that these loci are subject to positive diversity-enhancing selection relating to peptide binding and thus, presumably, to disease resistance. A similar type of analysis has indicated that immunogenic regions of certain cell surface proteins of malaria parasites are likewise subject to a form of diversity-enhancing selection. New proteins having new functions evolve when genes are duplicated, but the mechanism by which proteins evolve new functions is not understood. In a number of cases in which functionally distinct proteins have evolved after recent gene duplication, gene duplication has been followed by a period of accelerated nonsynonymous nucleotide substitution. This finding contradicts the widely held theory that a new gene with a new function can only emerge after a period in which a nonfunctional duplicated gene accumulates random mutations. Rather, the most plausible theory of the evolution of new gene function is that after duplication the daughter genes are able to specialize for functions previously performed by a single ancestral gene.

ACKNOWLEDGMENT

The author's research on MHC and parasite evolution is supported by a grant from the National Institutes of Health.

REFERENCES

Bjorkman, P. J., Saper, M. A., Samraoui, B., Bennett, W. S., Strominger, J. L., & Wiley, D. C. (1987a). Structure of the human class I histocompatibility antigen, HLA-A2. Nature 329, 506–512.

Bjorkman, P. J., Saper, M. A., Samraoui, B., Bennett, W. S., Strominger, J. L., & Wiley, D. C. (1987b). The foreign antigen binding site and T cell recognition regions of class I histocompatibility antigens. Nature 329, 512–518.

Buck, L. & Axel, R. (1991). A novel multigene family may encode odorant receptors: a molecular basis for odor recognition. Cell 65, 175–187.

Cross, G. A. M. (1990). Cellular and genetic aspects of antigenic variation in trypanosomes. Ann. Rev. Immunol. 8, 83–110.

Doherty, P. C. & Zinkernagel, R. M. (1975). Enhanced immunological surveillance in mice heterozygous at the H-2 gene complex. Nature 256, 50–52.

Enea, V. & Arnot, D. (1988). The circumsporozoite gene in Plasmodia. In: Molecular Genetics of Parasitic Protozoa (Turner, M. J. & Arnot, D., Eds.), pp. 5–11. Cold Spring Harbor Laboratory, Cold Spring Harbor, NY.

Goodman, M., Moore, G. W., & Matsuda, G. (1975). Darwinian evolution in the genealogy of haemoglobin. Nature 253, 603–608.

Hedrick, P. W. & Thomson, G. (1983). Evidence for balancing selection at HLA. Genetics 104, 449–456.

Hill, A. V. S., Allsopp, C. E. M., Kwiatowski, D., Austey, N. M., Twumasi, P., Rowe, P. A., Bennett, S., Brewster, D., McMichael, A. J., & Greenwood, B. M. (1991). Common West African HLA antigens are associated with protection from severe malaria. Nature 352, 595–600.

Hill, A. V. S., Elvin, J., Williams, A. C., Aidoo, M., Allsopp, C. E. M., Gotch, F. M., Gao, X. M., Takiguchi, M., Greenwood, B. M., Townsend, A. R. M., McMichael, A. J., & Whittle, H. C. (1992). Molecular analysis of the association of HLA-B53 and resistance to severe malaria. Nature 360, 434–439.

Hill, R. E. & Hastie, N. D. (1987). Accelerated evolution in the reactive centre regions of serine protease inhibitors. Nature 326, 96–99.

Hughes, A. L. (1992). Circumsporozoite protein genes of malaria parasites (*Plasmodium spp.*): evidence for positive selection on immunogenic regions. Genetics 12, 345–353.

Hughes, A. L. (1992). Positive selection and interallelic recombination at the merozoite surface antigen-1 (MSA-1) locus of *Plasmodium falciparum*. Mol. Biol. Evol. 9, 381–393.

Hughes, A. L. & Nei, M. (1988). Pattern of nucleotide substitution at major histocompatibility complex loci reveals overdominant selection. Nature 335, 167–170.

Hughes, A. L. & Nei, M. (1989). Nucleotide substitution at major histocompatibility complex class II loci: evidence for overdominant selection Proc. Natl. Acad. Sci. USA 86, 958–962.

Jensen, R. A. (1976). Enzyme recruitment in the evolution of new function. Annu. Rev. Microbiol. 30, 409–425.

Kemp, D. J., Coppel, R. L., & Anders, R. F. (1987). Repetitive proteins and genes of malaria. Annu. Rev. Microbiol. 41, 181–208.

Kimura, M. (1969). The rate of molecular evolution considered from the standpoint of population genetics. Proc. Natl. Acad. Sci. USA 63, 1181–1188.

Kimura, M. (1983). The Neutral Theory of Molecular Evolution. Cambridge University Press, Cambridge.

Kimura, M. & Ohta, T. (1974). On some principles governing molecular evolution. Proc. Natl. Acad. Sci. USA 71, 2848–2852.

Klein, J. (1986). Natural History of the Major Histocompatibility Complex. Wiley, New York.

Lawlor, D. A., Zemmour, J., Ennis, P. D., & Parham, P. (1990). Evolution of class I MHC genes and proteins: from natural selection to thymic selection. Annu. Rev. Immunol. 8, 23–64.

Lockyer, M. J., Marsh, K., & Newbold, C. I. (1989). Wild isolates of *Plasmodium falciparum* show extensive polymorphism in T cell epitopes of the circumsporozoite protein. Mol. Biochem. Parasitol. 37, 275–280.

Mayer, W. E., Jonker, M., Klein, D., Ivanyi, P., van Seventer, G., & Klein, J. (1988). Nucleotide sequences of chimpanzee MHC class I alleles: evidence for trans-species mode of evolution. EMBO J. 7, 2765–2774.

Nei, M. (1987). Molecular Evolutionary Genetics. Columbia University Press, New York.

Nei, M. & Gojobori, T. (1986). Simple methods for estimating the numbers of synonymous and nonsynonymous nucleotide substitutions. Mol. Biol. Evol. 3, 418–426.

Ngai, J., Dowling, M. M., Buck, L., Axel, R., & Chess, A. (1993). The family of genes encoding odorant receptors in the channel catfish. Cell 72, 657–666.

Ohno, S. (1973). Ancient linkage groups and frozen accidents. Nature 244, 259–262.

Saitou, N. & Nei, M. (1987). The neighbor-joining method: A new method for reconstructing phylogenetic trees. Mol. Biol. Evol. 4, 406–425.

Piatagorsky, J. & Wistow, G. (1991). The recruitment of crystallins: new functions precede gene duplication. Science 252, 1078–1079.

Takahata, N. & Nei, N. (1990). Allelic genealogy under overdominant and frequency-dependent selection and polymorphism of major histocompatibility complex loci. Genetics 124, 967–978.

Tanaka, T. & Nei, M. (1989). Positive Darwinian selection observed at the variable-region genes of immunoglobulins. Mol. Biol. Evol. 6, 447–459.

RECOMMENDED READINGS

Hughes, A. L. (1993). Coevolution of immunogenic proteins of *Plasmodium falciparum* and the host's immune system. In: Mechanisms of Molecular Evolution: Introduction to Molecular Paleopopulation Biology (Takahata, N. & Clark, A. G., Eds.), pp. 109–128. Sinauer, Sunderland, MA.

Hughes, A. L., Hughes, M. K., & Watkins, D. I. (1993). Contrasting roles of interallelic recombination at the *HLA-A* and *HLA-B* loci. Genetics 133, 669–680.

Klein, J. & Takahata, N. (1990). The major histocompatibility complex and the quest for origins. Immunol. Rev. 113, 5–25.

Lee, Y.-H. & Vacquier, V. D. (1992). The divergence of species-specific abalone sperm lysins is promoted by positive Darwinian selection. Biol. Bull. 182, 97–104.

Nei, M. & Hughes, A. L. (1991). Polymorphism and evolution of the major histocompatibility complex loci in mammals. In: Evolution at the Molecular Level (Selander, R. K., Clark, A. G. & Whittam, T. S., Eds.), pp. 222–247. Sinauer, Sunderland, MA.

Watkins, D. I., McAdam, S. N., Liu, X., Strang, C.R., Milford, E. L., Levine, C. G., Garber, T. L., Dogon, A. L., Lord, C. I., Ghim, S. H., Troup, G. M., Hughes, A. L., & Letvin, N. L. (1992). New recombinant *HLA-B* alleles in a tribe of South American Amerindians indicate rapid evolution of MHC class I loci. Nature 357, 329–333.

Chapter 8

The Neutral Theory of Molecular Evolution

NAOYUKI TAKAHATA and MOTOO KIMURA

Principles of Medical Biology, Volume 1B
Evolutionary Biology, pages 205–234

ISBN: 1-55938-802-1

INTRODUCTION

There are a variety of molecular mechanisms that are responsible for changing the genome. However, molecular evolution has not been driven solely by these mechanisms. Mechanisms acting at the individual and population levels have played an equally important role in determining the process of molecular evolution. Of particular importance are natural selection and random genetic drift. The neutral mutation and random genetic drift theory of molecular evolution, hereafter called the neutral theory for brevity, claims that the overwhelming majority of evolutionary changes at the molecular level are caused by random fixation of selectively neutral (i.e., selectively equivalent) mutants. The theory also asserts that most of the protein or DNA variation within species is maintained by the balance between mutational input and random extinction. This chapter overviews molecular evolutionary mechanisms, and then describes the neutral theory and its predictions in the light of recent observations.

MOLECULAR EVOLUTIONARY MECHANISMS

The genome of all organisms undergoes a variety of molecular changes associated with DNA or RNA replication and homologous chromosome pairings in both mitosis and meiosis, and some such changes may also be mediated through transposable elements. These molecular mechanisms generate raw genetic materials based on which the genome has evolved. The nature of these changes and their effects on organisms are discussed in Chapter 6. In this section, we briefly overview these molecular evolutionary mechanisms, emphasizing those of importance for molecular population genetics.

Point Mutation

A mutation that produces a change at a single nucleotide or amino acid position is called a point mutation. Deoxyribonucleic acid (DNA), of which the genome consists, has two components: purine and pyrimidine. Since there are two kinds of each, there are four different nucleotides. Purines are adenine (A) and guanine (G), while pyrimidines are thymine (T) and cytosine (C). There have been few direct observations about the pattern and frequency of single nucleotide mutations, mainly because their occurrence is too low for reliable estimates. However, comparative studies of homologous DNA sequences between different species have revealed that the rate of these nucleotide changes is rather constant in limited taxonomic groups (molecular evolutionary clock) and in primates it is as low as about 2×10^{-9} per site per year. Since the human genome, for example, is composed of some billions of nucleotides, the total number of point mutations per genome becomes 10 or more every year.

Comparative studies of homologous DNA sequences have also revealed that changes among the four nucleotides are not equally likely. If they were, transitions between A and G and between T and C would be half as frequent as transversions between purines and pyrimidines. The opposite is the case. Transitions are more frequent than transversions, and the most extreme case is found in primate mitochondrial DNA (mtDNA) in which transitions are about 17 times as frequent as transversions.

It is important to realize that the pattern and rate of mutations thus estimated may not reflect the true pattern and rate of DNA changes. This is because DNA sequence evolution is not driven solely by mutation pressure. Mechanisms at the population level must also have played important roles in shaping DNA sequence evolution. For instance, if we compare the first and second codon positions between aligned protein coding sequences, the requirement that this protein functions properly evokes the action of natural selection and obscures the true pattern and rate of mutation. For this reason, the term "mutation" is carefully distinguished from the term "substitution" in the literature of population genetics and molecular evolution. Substitutions mean evolutionarily accepted mutations, or mutations that have become shared by the majority of individuals in a species.

However, the substitution rate becomes synonymous with the mutation rate if natural selection has not been involved and random genetic drift has been the sole evolutionary mechanism at the population level. This is one important conclusion from the neutral theory, as we shall see later. Genomic regions in which natural selection has had the least effect, if any at all, are possibly noncoding regions, pseudogenes (dead genes), as well as introns and synonymous sites in protein coding regions. The synonymous sites are defined for each codon and they are those sites at which at least one type of nucleotide change does not alter the existing amino acid composition. There are no such sites in the second codon positions in the universal code and most are located at the third codon position. Most transitions at the third positions are synonymous. The number of synonymous sites at a third codon position is defined as one in fourfold degenerate codon families. However, there are some ambiguities for twofold degenerate codon families. In this case, the number depends on the relative frequency of transitions and transversions, since transversions are nonsynonymous. Only when transitions and transversions are equally frequent can the number of synonymous sites at the third position in twofold degenerate codon families be counted as one-third. For primate mtDNAs, this number must be close to one because of the extreme transition bias. A recent estimate of synonymous substitution rate is about 1.2×10^{-9} per site per year for primate nuclear DNAs, and 2.4×10^{-8} per site per year for primate mtDNAs, the latter being approximately 20 times as high. The synonymous substitution rate of nuclear DNA is believed to be much higher in other vertebrates, insects, and bacteria with short generation times.

Recombination

Recombination merely reshuffles existing sequences in a population of individuals, but the new combinations of words and sentences are often as important as the printing errors. In the human genome, one centimorgan (cM = 1% recombination frequency per meiosis) is approximately 10^6 nucleotides long, although the value differs in different organisms. There is growing evidence, however, that recombination does not occur at a uniform rate along the chromosome. It is a complex event involving a number of enzymes and other proteins. In addition, for recombination to be initiated, a particular nucleotide sequence must be recognized. Empirically, both centromeric and telomeric regions are reduced in recombination. The absence of recombination leads to linkage disequilibrium, particular combinations of genetic markers remaining together through many generations.

Recombination may occur not only between different genes on a chromosome, but even within a gene (intragenic recombination). In the latter, new allelic forms can be produced.

Recombination provides a means of gathering together mutations that occurred independently in different individuals, thereby accelerating the evolutionary process by advantageous mutations, or alternatively eliminating several deleterious mutations at once. Another evolutionary feature of recombination is called the "hitch-hiking effect." When an advantageous mutation arises, it will rapidly spread or be fixed in a population (i.e., all individuals come to share this mutation). The same thing happens at other loci tightly linked to the first, even though they do not have advantageous mutations. Fixation of an advantageous mutation sweeps out otherwise existing genetic variation at linked loci. Therefore, the hitch-hiking effect provides one explanation for heterogeneity in the extent of genetic variation along the genome, and it is more likely to manifest in genomic regions in which recombination frequency is reduced.

Unequal Crossing-Over

The number of homologous genes often differs from species to species. For example, the human C_H region (constant regions of the immunoglobulin heavy chain) contains 11 genes, of which nine are functional and two are nonfunctional, whereas the mouse counterpart contains eight functional genes. Similar changes may involve DNA segments smaller than an entire gene. The number of V_H segments (variable domains in the immunoglobulin heavy chain) is about 100 in the human but it is more than 300 in the mouse genome.

The mechanism most responsible for these changes in the number of homologous genes and DNA segments is unequal crossing-over. This may increase or decrease the copy number. If a duplicate gene is produced, there are two genes with a same function in the genome, at least at the initial stage of duplication. If one gene product is sufficient for survival, the other will be free to accumulate changes. The most likely fate of such a duplicate gene is to become nonfunctional, thus creating a pseudogene. However, it is also conceivable that a duplicate gene may accumulate some useful changes, permitting it to acquire a new function. Thus, the copy gene diversifies functionally, and further changes that affect transcription may lead to the developmental and tissue-specific expression. There are a large number of such examples.

Unequal crossing-over is, however, a double-edged sword. Because it translocates a gene in one chromosome into the other during pairing, the gene is deleted in the donor chromosome. Individuals that carry such a deficient chromosome are likely to die, leaving no descendents: The chromosome lacking an indispensable gene will disappear from the population forever.

Gene Conversion

Gene conversion is a result of DNA repair following recombination. The term was originally introduced to explain abnormal gene segregation in fungi. When DNA heteroduplexes are produced between chromatids and the exchange takes place, the noncomplementary nucleotides between two alleles are unpaired. The unpaired region is then recognized by proofreading enzymes which excise one of the nucleotides and replace it by a nucleotide complementary to the one on the other chromosome (Figure 1). Gene conversion may be either reciprocal or nonrecipro-

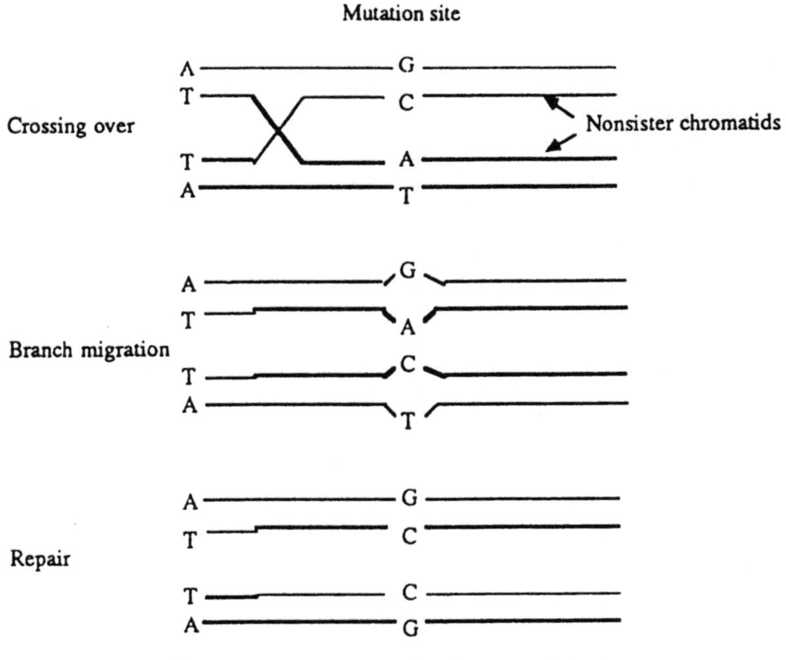

Figure 1. Gene conversion by recombination.

cal, though there is little information in mammals. The size of the gene conversion tract varies greatly, ranging from a few nucleotides to a number of genes. If gene conversion includes an entire gene, it acts so as to homogenize the variation at this locus in a population. On the other hand, if conversion occurs within a gene, it may create a new allelic form. Molecular biologists often designate a change as gene conversion if stretches of DNA sequences are shared between nonallelic genes, or if DNA sequences of some alleles generated by multiple point mutations are found in other alleles. The essential feature of gene conversion for these cases that were observed in mammals is the unidirectional transfer of DNA segments. This is somewhat different from the phenomenon observed in fungi where, in about a half of the cases, conversion is accompanied by the reciprocal recombination of flanking makers.

Other Molecular Mechanisms

There are many other changes or mechanisms that are responsible for genomic evolution. Some of them involve drastic changes. These include gene shuffling, duplicated transposition of genes, chromosomal rearrangements (inversion, translocation, fission, fusion, etc), and polyploidization. Gene shuffling is a mechanism which brings together different genes scattered in chromosomes into a single combined one, making a larger gene. It is thought to have played an important role in the early evolutionary life, generating various genes by a combinatorial means.

At present, it is not entirely clear to what extent natural selection has favored or disfavored these drastic changes at the genomic level, although it is probable that the more drastic effect on the phenotype the more harmful. However, it is certain that all of genomic changes could not have been free from the effect of random genetic drift. Random genetic drift occurs ceaselessly in any population of organisms. It is this force that the neutral theory regards as the main driving force in the molecular evolutionary process. It is therefore essential to understand what happens with new forms of genetic materials when they are subject to random genetic drift. The following section introduces the concept in an elementary way.

EVOLUTIONARY MECHANISMS AT THE POPULATION LEVEL

We first consider a simple situation: (1) Individuals are diploid and monoecious; (2) they mate at random, permitting also self-fertilization; (3) generations are discrete and nonoverlapping; (4) the number of breeding individuals in the population is finite and constant throughout time; and (5) natural selection is absent. There are N_e breeding individuals (or more precisely N_e is the effective number of individuals defined in a later section), and therefore $2N_e$ genes at a particular autosomal locus in the population.

Selective neutrality means that these $2N_e$ genes are equal in their effect on the fitness of individuals, regardless of the form of the gene (allele). Whether or not particular genes are transmitted from generation to generation is simply determined by chance.

Random Genetic Drift

Because the number of breeding individuals in the population remains finite, if some genes are transmitted more than once to the next generation, others are not at all. Which genes are multiplied follows a probability law. Suppose that $N_e = 3$ so that there are six genes in the population. These six genes at a certain generation (generation 0 in Figure 2) are labeled by integers $\{1, 2, 3, 4, 5, 6\}$. The frequency of each labeled gene in the population is 1/6. We throw a die to select which gene has a descendant. If the face is 1, we regard gene 1 as having a descendant, and record it. In the second trial, one of the six genes, including gene 1 which produced a descendant in the first trial, is again chosen at random. This process is repeated six times to obtain six descendants in the next generation (generation 1). There are a variety of possible outcomes. Some genes may not appear at all, others may appear singly, and still others may have two or more copies.

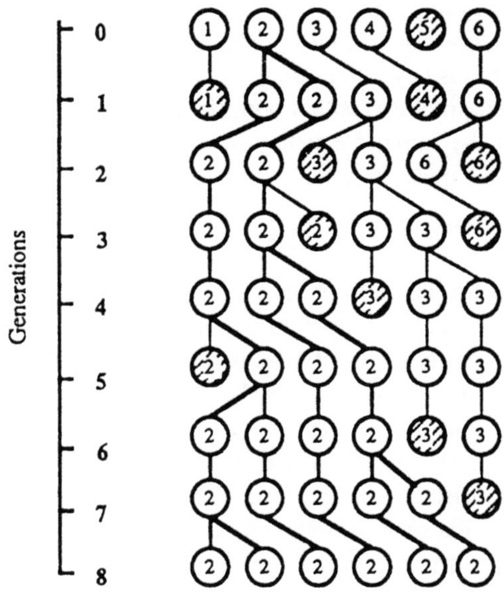

Figure 2. A model of random genetic drift and gene genealogy. Different numbers indicate different genes (circles) and solid lines between generations indicate the family relationships among genes. Shaded circles represent no offspring or extinction of lines of descent.

The occurrence of a particular outcome has a certain probability that we can specify. This probability is known to be described by the multinomial distribution. If some genes are not sampled for generation 1 (gene 5 in Figure 2), they have disappeared from the population. There is no way for these to reappear, so it is said that they have gone to extinction. The cause of extinction is not due to their inferiority in survival but simply to their bad luck. If there are such unlucky genes, there must be lucky genes which were multiplied more than once (gene 2). The population in generation 1 is then written as {1, 2, 2, 3, 4, 6}. To generate the population of genes in generation 2, we follow a procedure similar to that for generation 1. However, the die that we now use is different from the usual one. Face 5 of the previous die is replaced by face 2, so that there are two faces with gene 2. The frequency of gene 2 is 2/6 = 1/3. If we throw this die six times, the outcome may be something like {2, 2, 3, 3, 6, 6} as in Figure 2.

If we continue to play this game, we will find a certain generation at which all six genes have a same label for the first time; they are descended from a single ancestral gene at generation 0. The frequency of such a gene in the population is of course 100%. This event is called fixation because nothing will change after this generation until a *mutation* changes one of the labels. The elapsed time until fixation is called the fixation time.

Both fixation probability (u) and fixation time (t) of a particular form of a gene (allele) have been long-standing subjects in population genetics. A mathematical tool commonly used to study these quantities is diffusion processes or approximations which can describe the stochastic behavior of allele frequency change. The detail of such analyses is beyond the scope of this chapter. One important feature revealed by diffusion processes is that both fixation probability and fixation time depend strongly on the initial frequency of an allele. Under the assumption of neutrality, we have the mean fixation probability:

$$E\{u\} = p \tag{1}$$

in which $E\{\ \}$ stands for the mean of the argument and p is the initial frequency of a particular allele. In a small region of the genome, such as a protein coding region, mutation is a rare event and almost always produces a new allele (the so-called infinitely-many-allele model proposed by M. Kimura and J. F. Crow), so that it is customarily assumed that $p = 1/(2N_e)$ or $E\{u\} = 1/(2N_e)$.

This result can be derived from a simple argument. Neutrality implies, by definition, that all ancestral alleles are equivalent in their survival and fate. It immediately follows that each gene has an equal probability of being fixed in the population, or $E\{u\} = 1/(2N_e)$ without any sophisticated computation. It may be noted that the mean probability of loss of an allele, which is given by $1 - E\{u\}$, is very high in a large population. Even in our example of $N_e = 3$, the mean probability of loss is 5/6.

The analysis of fixation time by diffusion processes is more difficult. It is necessary to determine separately the time until fixation and that until loss. We merely present the result of the mean number of generations until fixation (excluding the cases of loss). It is given by:

$$E\{t\} = -4N_e(\frac{1}{p} - 1)\log(1-p).$$

The variance of t, $E\{t^2\} - E\{t\}^2$, is more complicated, but the distribution of fixation time has been worked out by M. Kimura.

Of particular interest is $E\{t\}$ when p is small. The approximate formula of $E\{t\}$ for $p = 1/(2N_e)$ is given by:

$$E\{t\} = 4N_e, \tag{2}$$

or it takes $4N_e$ generations on average for a new neutral allele to become fixed in a population of size N_e. R. A. Fisher noted long ago that the mean fixation time does not greatly exceed N_e generations. We shall derive the mean and standard deviation of t from a genealogical consideration.

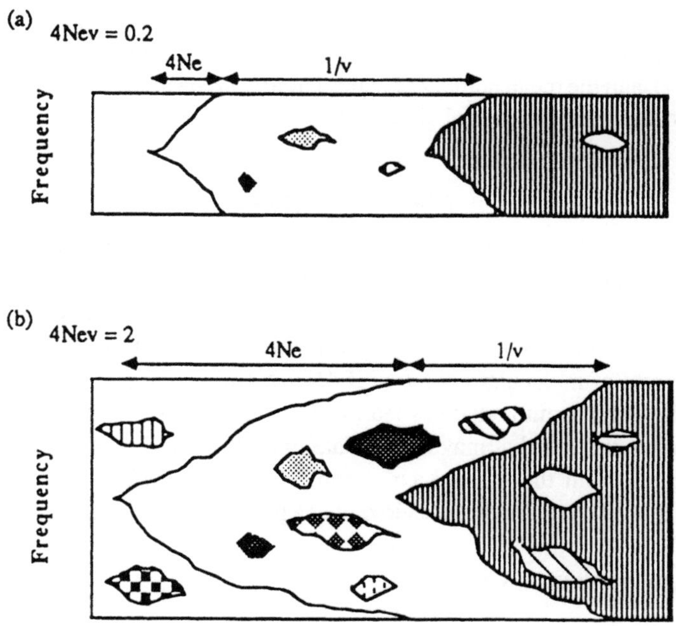

Figure 3. Relationships between mean fixation time $4N_e$ and mean rate of gene substitutions $1/v$. In a small population (**a**), fixation of alleles can occur and the population is monomorphic for most of the time. In a large population (**b**), complete fixation of alleles is prevented by frequent introduction of new alleles and therefore the population is almost always polymorphic. However, in both cases, the mean time until the majority of individuals come to share a new mutation is the same, $1/v$.

During the period from the first appearance of an allele to fixation, there are at least two different alleles in the population, generating the state of polymorphism. In a finite population and without a further supply of new mutations, the polymorphism is never immune to extinction owing to fixation of one of the alleles which contributed to that polymorphism. It is therefore said that polymorphism is a transient phase of molecular evolution. Fixation of a new allele is called gene substitution.

Whether or not further mutations appear at a polymorphic locus depends on the magnitude of the neutral mutation rate per locus per generation (v) relative to $1/(4N_e)$. The mean waiting time for a new mutation to be incorporated, or go to fixation, is the reciprocal of v (see below). Hence, if $1/v > 4N_e$ or $4N_ev < 1$, this locus is likely to be monomorphic for some time after the fixation of one allele took place (Figure 3a). On the other hand, if $4N_ev \geq 1$, it is likely that further mutations are incorporated into the population before any fixation can take place. Under this situation, the population is polymorphic most of the time (Figure 3b) and allelic turnover occurs continuously without complete fixation.

The mean rate of molecular evolution, $E\{k\}$, is defined as the mean number of mutations that arise per unit time and are destined for fixation. There are $2N_e$ genes at a locus in a population, and if the neutral mutation rate is v, there will be $2N_ev$ such mutations per generation in the whole population. Each of these mutations has $p = 1/(2N_e)$, and the mean number of neutral mutations destined for fixation is given by $2N_ev$ times the fixation probability $E\{u\} = 1/(2N_e)$. It follows that:

$$E\{k\} = 2N_evE\{u\} = v. \tag{3}$$

This relationship is independent of N_e. The larger N_e the more mutations. However, $E\{u\}$ becomes correspondingly smaller, which cancels exactly the increased number of mutations. Therefore, the neutral theory predicts that molecular evolution occurs at the same speed as the neutral mutation rate. We may ask how to define the mean rate of molecular evolution if $4N_ev \geq 1$ in which case fixation may not occur. The same formula that has led to $E\{k\} = 2N_evE\{u\} = v$ is still valid, if we ignore any mutations that may accumulate in all lines of descent since they have been derived from the most recent common ancestor (MRCA). It is implicitly assumed that $E\{k\}$ should be measured over a period of time much longer than $4N_e$.

Genealogy of Neutral Genes

A more visual interpretation of the molecular evolutionary rate is provided by genealogical consideration of genes. We look at Figure 2 in a slightly different way. Since we have recorded the family relationships among genes, we know which genes at generation 8 were descended from which ones at generation 7, and so on. We ignore all the gene lineages that became extinct by generation 8. Figure 4 depicts the family relationships among six extant genes up to the MRCA. Beyond the

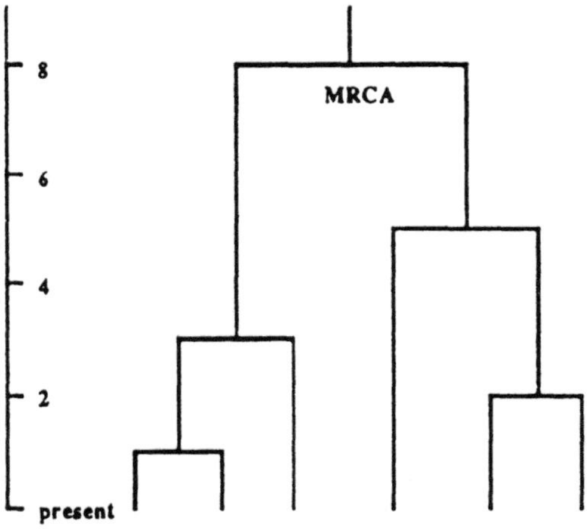

Figure 4. Family relationship among the extant genes in Figure 2. Generations are measured back in time. The MRCA indicates the most recent common ancestor for those genes. Each multiplication of a gene is called gene divergence or coalescence when the process is looked at backward in time.

MRCA, there is only one lineage which goes back in time until this gene itself was produced *de novo*.

Substitution Rate

Assume that during the persistence of such an ancestral gene lineage, the population was split into two and these two sister populations have not interbred for a number of generations (Figure 5): They are now different species. Assume also that we know the splitting time of these two species from paleontological data. Usually, the rate of molecular evolution is measured by the number of amino acid or nucleotide substitutions (d) between a gene taken at random from one species and another gene taken from the second species. A common assumption is that the divergence time of these genes is the same as the species divergence time (t_S). The rate of molecular evolution is then estimated by $d/(2t_S)$. Some caution is necessary when applying this formula since d must be inferred from the differences between sequences; an identical site does not necessarily mean no substitutions since a change may have been reversed, and a different site does not necessarily mean one substitution since there may have been more than one.

Furthermore, if $4N_e v \geq 1$, the lineage of a sampled gene is likely to have accumulated some mutations since it was descended from the MRCA which existed $4N_e$ generations ago on average. The same thing happens in other gene lineages taken from the same population, so that the population is polymorphic and different gene lineages may have accumulated different numbers of mutations since the

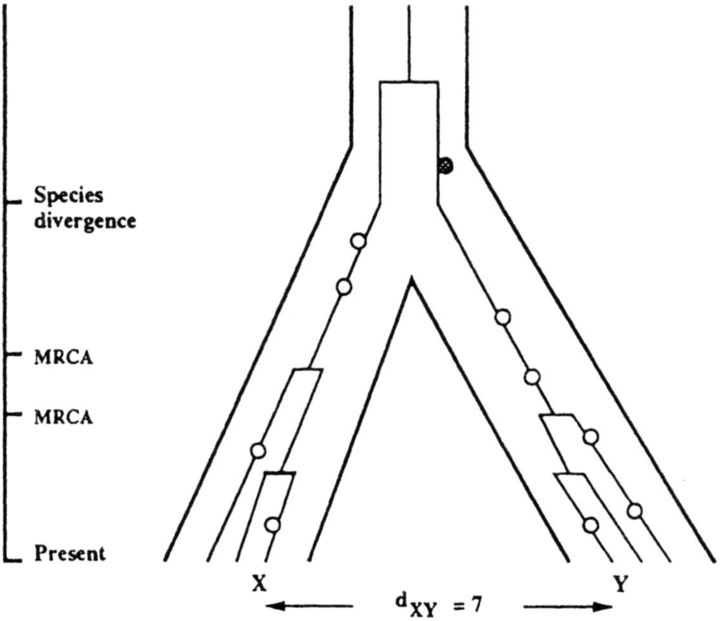

Figure 5. Genealogy of genes sampled from different species. Each circle along lines indicates a mutation; open circles stand for mutations that had occurred after species splitting and a single shaded circle stands for a mutation before species divergence. The number of mutations between genes X and Y is denoted by d_{XY}. The value differs for different pairs of orthologous genes. In this example, $d_{XY} = 7$, one of which occurred between two ancestral lineages when they were distinct in the ancestral population.

MRCA. The rate of molecular evolution, therefore, differs among sampled genes. However, if the species divergence time t_s is much longer than the mean fixation time $4N_e$ most of mutations that contribute to d must have occurred before the MRCA. If this is the case, the existing polymorphism will not affect the d value in any substantial way. This argues that the extent of polymorphism in an extant population is immaterial for estimating the mean rate of molecular evolution, if t_s *is* in fact much longer than $4N_e$. Rather, what matters is t_s and the neutral mutation rate, v. If we measure t_s in units of generations, the expected value of d is simply the product of $2t_s$ (because two genes in different species have been separated that long) and v (in units of generations). Hence, the mean rate of molecular evolution is well approximated by $E\{k\} = E\{d\}/(2t_s) = v$. The qualifier "mean" emphasizes the fact that mutations do not occur deterministically. They may occur in a manner similar to the process of a radioactive decay. For this reason, d is a random variable and it is often approximated by a Poisson process with mean $2t_sv$. Since the variance of a Poisson variable is the same as the mean, our estimate of d is always subject to a fairly large fluctuation.

Polymorphism

The genealogical approach has demonstrated its power in studying various problems in population genetics and molecular evolution. We derive two formulas for quantities that measure the extent of polymorphism at a locus. Suppose that we randomly sample two genes from the population and ask the probability that they are different in state. The difference is, of course, due to the occurrence of mutations since these two genes diverged from the MRCA. This probability of differences between two randomly chosen genes is called the mean heterozygosity, $E\{H\}$. To derive $E\{H\}$, it is crucial to determine the rule concerning the divergence time (x_2) between two genes randomly sampled from the population. Like the fixation time t, x_2 is a random variable, depending on effective size N_e. Without giving the proof, we note that $x_2 = x$ (≥ 0) follows an exponential distribution $p(x) = 1/(2N_e)\, e^{-x/(2N_e)}$. The mean of x_2 is $2N_e$. For a given value of $x_2 = x$, two genes are expected to have accumulated mutations following a Poisson distribution with mean $2xv$. The probability of no mutation is therefore given by e^{-2xv}. Recalling that x_2 is a random variable, we integrate e^{-2xv} over the possible range $(0, \infty)$ of x, each value having a weight $p(x)$. Thus, the probability of identity between two genes (mean homozygosity denoted by $E\{F\}$) is $1/(1 + 4N_e v)$. If two genes are not identical in state, they differ by at least one mutation, so that we have:

$$E\{H\} = 1 - E\{F\} = \frac{4N_e v}{1 + 4N_e v}. \tag{4}$$

Figure 6 shows $E\{H\}$ for a certain range of $4N_e v$.

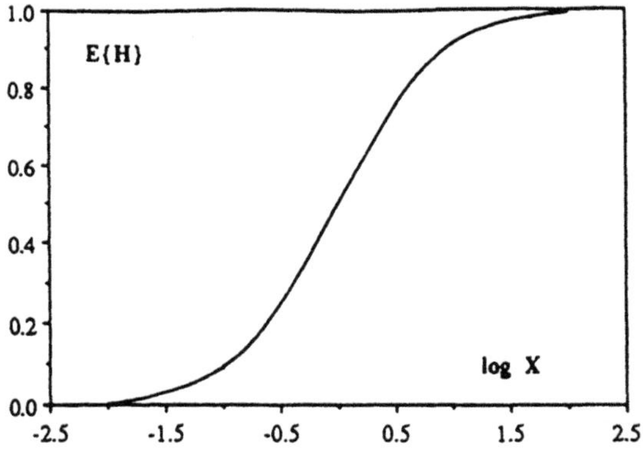

Figure 6. Expected heterozygosity $E\{H\} = X/(1 + X)$ where $X = 4N_e v_T f$. The abscissa is measured in $\log_{10}X$. When $v = v_T f$ is constant, the figure represents the N_e dependence of $E\{H\}$. For a given value of $4N_e v_T$, it also represents the f dependence ($0 \leq f \leq 1$). Note, however, that when $4N_e v_T$ is small, $E\{H\}$ is also small even for $f = 1$.

There is another way to obtain formula (4). As discussed in Chapter 6, we know that the probability of nonidentity, $E\{H\}$, decreases by a factor $1/(2N_e)$ per generation. Therefore, under the infinitely-many-allele model, and ignoring the higher order terms of v, we have:

$$E\{H'\} = E\{H\}(1 - \frac{1}{2N_e}) + 2vE\{F\}$$

in which the left stands for the quantity one generation later. The first expression on the right includes no mutation term since a mutation in a heterozygote always changes it into another heterozygote and reproduction takes place after mutation. The second expression indicates that either of two genes in a homozygote can mutate so as to produce a new heterozygote. At equilibrium, $E\{H'\} = E\{H\}$ so formula (4) holds.

Another quantity often used in analyzing DNA sequence data is called the nucleotide diversity, π. This is defined as the number of nucleotide differences per site between a pair of sequences. We can derive the formula of the mean nucleotide diversity, $E\{\pi\}$. First we note that the mean number of mutations that occurred between two sequences with divergence time $x_2 = x$ is given by $2xv$. However, since the probability density of $x_2 = x$ follows the exponential function of $p(x)$, we take the average of $2xv$. Because the mean of x_2 is $2N_e$, the mean number of mutations between the two sequences is $4N_ev$ per gene. Letting v be $m\mu$ where m is the number of nucleotides compared, and μ is the mutation rate per site, we obtain:

$$E\{\pi\} = 4N_ev/m = 4N_e\mu. \quad (5)$$

We have used the formula for coalescence time x_2, the time since two randomly chosen genes were descended from the MRCA. We can extend this formula to a sample of an arbitrary number of genes. Suppose that we sample i of such genes ($i = 6$ in Figure 4), and consider the time until these i genes were descended from $i - 1$ distinct ancestral genes for the first time. This waiting time, denoted by x_i, depends on i as well as N_e. The time x_i is again an exponentially distributed random variable, and the probability density of $x_i = x$ is given by:

$$\frac{i(i-1)}{4N_e} \exp\{-\frac{i(i-1)x}{4N_e}\}$$

with mean $E\{x_i\} = 4N_e/\{i(i-1)\}$ and variance $[4N_e/\{i(i-1)\}]^2$. If $i = 2$, $E\{x_2\} = 2N_e$, as already derived, and the variance, $E\{x_2^2\} - E\{x_2\}^2$, is given by $4N_e^2$. It is important to note that when two of the i genes coalesced to a common ancestor there were $i - 1$ distinct ancestral genes for a certain period of time in the past, and that these $i - 1$ genes again follow an exponential law for the next coalescence among them. (Here and subsequently, the term *coalescence* indicates the event that one of ancestral gene lineages of a sample diverged into two distinct lineages.)

Which two genes coalesced among i genes is determined by choosing a pair of genes at random from all $i(i-1)/2$ possible combinations.

For $i-1$ ancestral genes, the time x_{i-1} is independent of the previous coalescence and we further trace back the ancestral lineages of $i-1$ genes. In this way, we can continue to follow this process until the MRCA for a sample of i genes is found. Since the mean time for the j-th coalescence ($j = 2, 3, ..., i$) is $E\{x_j\} = 4N_e/\{j(j-1)\}$, the mean time until the MRCA for i genes is found is given by:

$$\sum_{j=2}^{i} E\{x_j\} = 4N_e[(1-\frac{1}{2})+(\frac{1}{2}-\frac{1}{3})+\cdots+(\frac{1}{i-1}-\frac{1}{i})] = 4N_e(1-\frac{1}{i}) .$$

Hence, when i is large or when we sample all extant genes in the population, the above formula reduces to formula (2) for the mean fixation time derived by the diffusion process. The standard deviation of $\Sigma_{j=2}^{i} x_j$ is given by:

$$4N_e\sqrt{\sum_{j=2}^{i}\frac{1}{\{j(j-1)\}^2}}$$

which indicates a large stochastic variation in the coalescence process.

Functional Constraints and Negative Selection

Although the neutral theory claims that the main driving forces of molecular evolution are mutation and random genetic drift, it does not deny the importance of natural selection. In fact, without incorporating one common type of natural selection (negative selection; see later for the definition), the theory would have failed to explain most observations and have already been rejected. The occurrence of deleterious mutations and the operation of negative selection due to selective constraint are both very important parts of the neutral theory in explaining some prominent features of molecular evolution.

In 1974, five principles which govern molecular evolution were enumerated by M. Kimura and T. Ohta. One of the principles states that functionally less important molecules or parts of a molecule evolve (in terms of mutant substitutions) faster than more important ones. Those mutant substitutions that are less disruptive to the existing structure and function of a molecule occur more frequently in evolution than more disruptive ones. The more conservative the mutational change, the more likely it is to turn out to be selectively neutral. When this principle was first proposed, much opposition was voiced, but it has become a part of common sense among molecular biologists. It is now a routine practice to search for various signals by comparing a relevant region of homologous DNA sequences of diverse organisms and to pick out a constant (consensus) pattern, but disregard variable ones as unimportant.

To emphasize that the mean rate of molecular evolution, $E\{k\}$, is determined by random accumulation of *neutral* mutations, we rewrite formula (3) as:

$$E\{k\} = v = v_T f \qquad (6)$$

where v_T is the total mutation rate and f is the fraction of neutral mutations. Advantageous mutations may occur, but the neutral theory assumes that they are so rare as to be negligible. In this formulation, $1 - f$ represents the fraction of definitely deleterious mutations which are eliminated from the population by negative selection and which do not contribute to molecular evolution. The value of f is small for functionally important molecules or parts of a molecule, whereas it is large for unimportant ones. The conservative mode of molecular evolution is one feature that distinguishes it from the opportunistic mode of phenotypic evolution.

Population Structure, Mating System and Mode of Inheritance

The extent of polymorphism due to neutral mutations depends strongly on effective size (N_e), as mentioned earlier. It is critical to examine factors which determine N_e. We consider three such factors: population subdivision, mating system, and mode of inheritance.

Since any species is geographically subdivided to some extent, we first consider the effect based on the finite island model in which a population is subdivided into L subpopulations of size N and migration occurs from one subpopulation to another with probability $m/(L-1)$ per generation. S. Wright derived an equation for N_e, which is given by $N_e = NL/(1 - F_{ST})$ where F_{ST} is the fixation index, representing the relative proportion of genetic variability attributed to the differences between subpopulations to the total. Since F_{ST} is given by $F_{ST} = 1(1 + 4Nm\alpha)$ in which $\alpha = \{(L/(L-1)\}^2$, close to 1 for large L, N_e becomes:

$$N_e = NL\left(1 + \frac{1}{4Nm\alpha}\right) . \qquad (7)$$

Formula (7) can be derived from genealogical considerations, and various important quantities in molecular population genetics can be predicted fairly well from the theory for randomly mating populations if the effective population size for the finite island model is used. This formula indicates that N_e can be substantially greater than the actual number of individuals (NL) in the entire population when the migration rate among subpopulations is small. The reason for large N_e in a subdivided population is shown diagrammatically in Figure 7. With limited amounts of migration, it takes a long time to trace back to the common ancestors of a sample of genes taken from different subpopulations. This is because two genes sampled from different subpopulations must have experienced at least one migration event since they diverged from the common ancestor, and this event is assumed

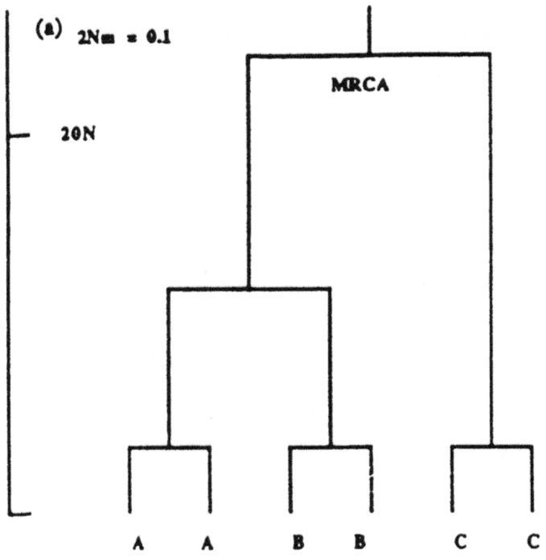

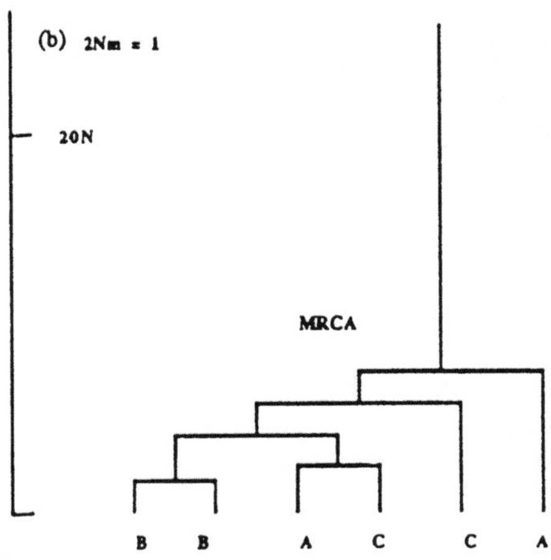

Figure 7. Genealogy of six neutral genes sampled from three subpopulations A, B and C. These subpopulations, each being of size N, are assumed to have diverged simultaneously $20N$ generations ago and to have exchanged genes with the frequency of $2Nm$. When $2Nm$ is small (**a**), two genes from each subpopulation coalesce rapidly, but those from different subpopulations coalesce slowly. When $2Nm$ is large (**b**), all genes coalesce rapidly, irrespective of their sampled locations. The effective population size is large, nearly $20N$ in (**a**), but it is small, about $3N$ in (**b**).

to be rare during a given time period. The value of $4Nm$ for $N_e >> NL$ is clearly $<< 1$, or much smaller than one migrant every other generation, as shown by S. Wright.

However, formula (7) depends on the assumption that the population structure remains the same for a long evolutionary time. In many natural populations this assumption may not hold. In some species such as *Escherichia coli* and *Drosophila*, subpopulations of a species are often subject to extinction and rapid multiplication. Under this circumstance, the effective population size can be much smaller than the actual size (NL). Nevertheless, there are cases in which population structure remains more or less the same for a long period of time, and in which genetic variability between subpopulations is much higher than that within a subpopulation.

In reality, such an ideal situation as random mating, even within a subpopulation, may not be realized. The effective population size is a concept with which complicated situations are reduced to the equivalent ideal case which we understand and for which we have formulas. One way of defining the effective population size is to consider the probability (P) that two randomly chosen gametes come from the same parent, and N_e is equated to $1/P$. N_e will be different, depending in particular on progeny distribution as well as on whether self-fertilization is permitted (monoecious) or not (separate sexes).

In separate sexes, it is important whether or not both sexes contribute their gametes with equal frequency. Genes on autosomes in either sex are equally transmitted to later generations, but genes on sex chromosomes are not. The effective size of genes on the X chromosome is $9N_fN_m/(4N_m + 2N_f)$ and that on the Y chromosome is N_m, where N_f and N_m are the numbers of breeding females and males in a randomly mating population. On the other hand, the effective size of genes on autosomes is given by $2N_e$ where:

$$N_e = \frac{4N_fN_m}{N_f + N_m} \cdot \tag{8}$$

If the sex ratio is 1, $N_e = 2N_m = 2N_f$, but if the ratio is biased, then N_e is mainly determined by the smaller value of N_f and N_m.

Corresponding to $N_e = N_m$ for the Y chromosome, it is interesting to evaluate N_e for cytoplasmic genomes such as mtDNA and chloroplast DNA in plants. Since these cytoplasmic genomes are usually, if not completely, transmitted maternally, the relevant effective size must be approximated by N_f. This holds true even though there are a large number of copy genomes within a germ cell. The reason is that these copy genomes tend to be genetically identical within a cell because the transmission occurs maternally, and because there is random assortment of genomes following the replication and cell division. Therefore, there is an approximately fourfold difference in the effective numbers of autosomal and cytoplasmic genes in a population.

Natural Selection

Natural selection is the most important evolutionary mechanism acting at the individual and population level. Natural selection is important for explaining not only phenotypic but also molecular evolution. It is therefore essential to have a correct understanding of its role in changing and maintaining the genetic architecture of natural populations.

Natural selection is defined as the differential multiplication of variant or mutant types. It occurs when there are differences in fitness between genotypes; the term fitness represents the organism's ability to leave offspring to the next generation and, for each genotype, fitness is measured by the number of offspring per individual. The number of offspring must be counted at the same stage of development as the parents. Natural selection acts through survival and reproduction of individuals based on phenotypes rather than on genotypes. The relationship between phenotypes and genotypes is often not straightforward, particularly when a phenotypic character is determined by many genes for which environmental effects also play important roles. There are a variety of types of natural selection, although not all will be discussed in this chapter.

Two distinct types in its simplest classification are called positive and negative selection. Positive selection operates if a mutant type arises in a population and if it enhances the average fitness of individuals carrying the mutation. It is the type of natural selection that Darwin assumed as the main cause of adaptive evolution in the biological world. For this reason, it is often called Darwinian selection. On the other hand, if a mutant reduces the fitness of individuals due to its deleterious effects on viability or fertility, then it tends to be eliminated from the population. This type of selection is called negative selection. Examples of negative selection are abundant, whereas those of positive selection are rare at the molecular level.

The third type is called balancing selection (from Th. Dobzhansky). This leads to genetic heterogeneity or polymorphism and includes overdominance (heterozygote advantage) and frequency-dependent selection with minority advantage. An example of balancing selection is demonstrated in genes involved critically in the vertebrate immune system. They are called *Mhc* (major histocompatibility complex) genes, whose products present self and nonself processed peptides to T cell receptors and play a crucial role in the cellular immune response.

In this context, it is important to distinguish the cause and mechanism of natural selection. For instance, minority advantage is an efficient mechanism for generating polymorphism, but this term itself does not specify the cause. J. B. S. Haldane once mentioned that in every species at least one of the factors which kills or lowers its fertility must increase in efficiency as the species becomes denser. He speculated that a very small biochemical change gives a host species a substantial degree of resistance to a highly adapted microorganism. Therefore, it is an advantage to individuals to possess a rare biochemical phenotype. Such a rare biochemical phenotype is the basis of the mechanism of minority advantage. The cause is disease resistance of particular genotypes, and the agent is a microorganism that is highly adapted to the majority of host individuals. Further studies of natural selection are required to address such a cause and agent explicitly (see Chapter 6).

PREDICTIONS AND TESTS

Predictions

The neutral theory assumes that a certain fraction of mutations are deleterious and the rest are selectively neutral. Favorable mutations no doubt occur, but the theory assumes that these are too rare to be of significance in explaining the major features of molecular evolution and polymorphism.

Prediction 1. The fraction (f) of neutral mutations depends on the functional constraint on a molecule: the stronger the constraint, the larger the fraction of deleterious mutations, and therefore the smaller the chance of a mutation being selectively neutral. Stated another way, the weaker the functional constraint, the larger the fraction of neutral mutations.

Prediction 2. There is a maximum value of the substitution (evolutionary) rate which is set by the total mutation rate. That is to say, $E\{k\} = v_T f \leq v_T$ since $0 \leq f \leq 1$. By contrast, there is no such an upper bound if molecular evolution has been driven by positive selection.

Prediction 3. If v_T is constant throughout the evolutionary course and if f remains also constant, the evolutionary rate must be constant and equal to the neutral mutation rate ($v = v_T f$). Conversely, if the evolutionary rate were constant despite changing values of v, the theory must be wrong. The neutral theory predicts, however, that the evolutionary rate must be constant *per generation* rather than *per year*.

Prediction 4. The rate of molecular evolution is independent of environmental factors. Molecular changes occur as frequently in species that undergo rapid changes in morphology as in those conservative species, including so-called living fossils such as coelacanths, horseshoe crabs, and *Lingula*.

Prediction 5. Polymorphism is simply a transient phase of molecular evolution and should be controlled by the same causes as govern molecular evolution. More specifically, the level of polymorphism is strongly influenced by the degree of functional constraints, but almost independent of environmental conditions in which the species is placed. Loci coding for proteins that are subject to stronger constraints tend to show lower genetic variabilities. A great deal of polymorphism should be found in spacer regions, introns and synonymous sites even for genes coding for highly conserved proteins such as histones.

Prediction 6. There must be a significant correlation between the extent of polymorphism (e.g., measured by H) and the effective population size (N_e). For example, the polar bear whose effective size is known to be quite small must show a very low level of polymorphism.

Prediction 7. The faster the evolutionary rate of molecules, the higher their level of polymorphism. If the mutation rate per nucleotide site is more or less the same over the genome, the difference in both evolutionary rate and polymorphism among genes can be attributed to the difference in the degree of functional constraints.

Statistical Tests

Because of these testable predictions, many statistical tests have been proposed. There are two categories of such tests, one being concerned with the molecular evolutionary rate and the other with polymorphism.

Relative Rate Test

A commonly used method for testing the rate constancy is called the relative rate test. It can avoid the problem of inaccurate dating of species divergence times and uses only three species. Consider three species A, B, and C in Figure 8, and assume that A and B are known to be more closely related to each other than to C. Suppose that we have estimated *distance* d_{AB}, d_{AC}, and d_{BC} among the three species. There are various kinds of distances. When V. M. Sarich and A. C. Wilson first proposed the relative rate test, they measured the intensity of immunological reactions between antigens and antisera prepared from different species and used the immunological distance. For allele frequency data from different populations

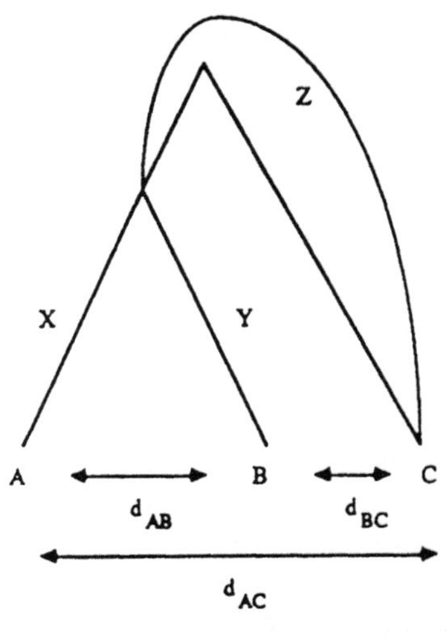

Figure 8. Relative rate test. Three orthologous genes are taken from species A, B and C.

or relatively closely related species, the standard genetic distance by M. Nei has been most widely used. Where amino acid or DNA sequences are available, the distance can be measured by the number of amino acid changes, the number of nucleotide changes, the number of synonymous and nonsynonymous changes, or even the number of deletion/insertion events. A problem is how to infer the actual number of changes (distance) from the number of differences in a specified category. Such work was initiated by M. O. Dayhoff and her coworkers for amino acid sequences, and by T. H. Jukes and C. R. Cantor for nucleotide sequences. We assume here that the distances among the three species are already at hand.

In the case of three species, we can compute three branch lengths X, Y, and Z in Figure 8 as:

$$X = \frac{1}{2}(d_{AB} + d_{AC} - d_{BC}), \quad Y = \frac{1}{2}(d_{AB} - d_{AC} + d_{BC}),$$

$$Z = \frac{1}{2}(- d_{AB} + d_{AC} + d_{BC}) .$$

If the substitution rate is constant, the expectations of X and Y are equal to each other. The hypothesis to be tested here is whether $X = Y$ (or $d_{AC} = d_{BC}$) can be rejected, considering the standard errors of the estimated distances.

Star Phylogeny Test

Another method for testing rate constancy is based on use of species which diverged at the same time. Mammals which diverged simultaneously at the end of the Cretaceous are a good example of a star phylogeny. Suppose that n ($n \geq 4$) species are studied, and let X_j be the distance that occurred in the j-th species after separation from the common ancestor (Figure 9). Such distances cannot be directly inferred and therefore need be estimated from pairwise distances between extant species. The method proposed by M. Kimura is based on the mean and variance of X_j among $n(n-1)/2$ different pairs of species. Then this variance is compared with the expected variance under a specified substitution process.

Maximum Rate Test

If the neutral theory is valid, there is a maximum evolutionary rate that is set by the total mutation rate (*P*2). Genes used in these tests must be orthologous. The term "orthologous" means that those genes in different species started to diverge after the species splitting (X and Y in Figure 5) and have accumulated mutations independently in each species. Therefore, orthologous genes mean that the divergence time between these genes can be equated to the species divergence time (t_s), and the rate is computed as $d/(2t_s)$, as mentioned earlier. At present, we can examine many genes or various parts of the genome. Since different genes are composed of

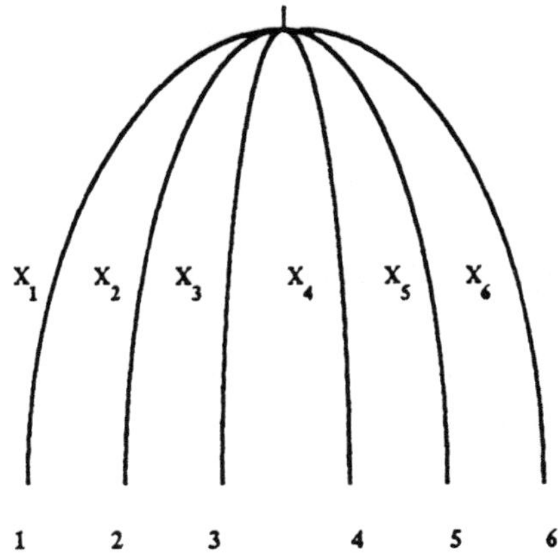

Figure 9. Star phylogeny test. Each branch length X_j is inferred from the pairwise distances and the mean and variance among those *X*s are used to test the constancy of evolutionary rate. For instance, the ratio (*R*) of the mean to the variance is 0.7 for the nonsynonymous substitutions and 6.0 for the synonymous substitutions of hemoglobin α among human, goat, rabbit, and mouse. The former does not significantly deviate from the variation expected under a Poisson process, but the latter does. Similarly, if we compare hemoglobin β among human, bovine, rabbit and mouse, $R = 6.2$ for the nonsynonymous substitutions but $R = 1.4$ for the synonymous substitutions. The former indicates that the evolutionary rates in some lineages are different.

different numbers of nucleotides, the evolutionary rate is conveniently calculated per amino acid or nucleotide site for comparisons. To compute the synonymous substitution rate from orthologous genes, we must count the number of synonymous differences and that of synonymous sites (taking into account substitution biases). The former is divided by the latter to obtain the per-site difference. This per-site difference is then converted into the distance by correcting for multiple-hit substitutions.

In contrast to orthologous, the term "paralogous" means that those genes have evolved independently by gene duplication. Since there is no information about the divergence of paralogous genes other than their sequence divergences themselves, dating of duplication of paralogous genes is a matter of inquiry. The essential assumption is also the constancy of molecular evolution. If we know the rate and the distance of paralogous genes, then we can conversely date the duplication event. In some instances, the rate may change due to changes in functional constraint fostered by duplication. In an extreme case, one paralogous gene might have accumulated disabling mutations and become a pseudogene. In this case, it is also interesting to estimate the evolutionary rate of pseudogenes.

The neutral theory predicts that the evolutionary rate of pseudogenes must be highest because of lack of any functional constraint ($f = 1$). Introns and synonymous sites also exhibit rapid accumulations of mutational changes. It is very difficult to imagine that these high evolutionary rates are the result of positive selection for two reasons: (1) there is no obvious biological reason for their functional importance, and (2) the rate in functionally important regions is almost always slower than that of pseudogenes. It is now a common practice to find regions of the genome which show the rate significantly lower or higher than that of the maximum rate. If the rate is lower in some regions, they have been conserved and are playing some important roles in function. If, on the other hand, the rate is significantly faster in other regions, we can conclude that these regions have been subject to positive selection.

Homozygosity Test

Under neutrality, the mean homozygosity is $E\{F\} = 1/(1 + 4N_e v)$, as in formula (4). W. J. Ewens and G. Watterson worked out the number of alleles (K) in a sample of i genes taken from the population. The mean of K is given by:

$$E\{K\} = \sum_{j=1}^{i} \frac{4N_e v}{4N_e v + j - 1} \tag{9}$$

where i is the number of genes sampled. Since F and K are determined by a single parameter $4N_e v$, there must be a correlation between F and K. Roughly speaking, the larger $4N_e v$, the smaller F and the larger K. The statistical test developed requires the probability of F, provided that there are K alleles found in a sample of i genes. This test is based on a single-locus so that the statistical power may not be very high. A successful application may be to *Mhc* loci in which case, for a given K at each locus, the F value is too low to be compatible with neutrality.

Nei's Test

We mention one more test concerning neutral polymorphism. The test first uses formula (4) and observed mean heterozygosity H over many loci. The test estimates $4N_e v$ by $H/(1 - H)$ and uses this value to predict the variance of H. This predicted variance can be compared with the corresponding variance observed at a large number of loci. In this test, it is necessary to set a confidence limit of the expected variance. One problem is that neutral mutation rate v may differ from locus to locus because of the differences in the size of genes. Taking account of this possibility, M. Nei and his coworkers showed that the observed variance of H is compatible with that predicted from the neutral theory.

OBSERVATIONS

Rates of Nucleotide Substitutions

The first truly favorable evidence for the neutral theory was the finding that synonymous substitutions occur almost always at much higher rates than nonsynonymous substitutions. It was also found that substitutions at other silent sites, such as introns, occur at comparably high rates. A more dramatic example was the discovery of very high evolutionary rates of pseudogenes. Importantly, the rates of substitutions were found equally high in all three codon positions. These and other observations suggested that molecular changes which are less likely to be subject to natural selection occur more frequently in evolution, and therefore show higher evolutionary rates (*P*1). From the viewpoint of the neutral theory, such changes are considered to be more likely nondeleterious. On the other hand, these observations do not fit nicely with the neo-Darwinian theory of evolution, which claims that the speed and direction of evolution are almost completely determined by natural selection. If positive selection is the driving force of molecular evolution, we can expect that synonymous and other silent changes should show lower evolutionary rates than nonsynonymous changes, because natural selection acts on the phenotype of the organism, in the determination of which the structure and function of proteins play a decisive role.

Another interesting example is the observation that the evolutionary rate of the eye lens protein, αA-crystallin, has been much enhanced in the blind mole rat, *Spalax ehrenbergi*. This animal is adapted to a burrowing, subterranean way of life, and possibly has been for the last 25 million years according to fossil evidence. It is completely blind, although the crystallins are still expressed in the atrophied lens cells. The αA-crystallin is generally a slowly evolving protein, with the average replacement rate of 0.3×10^{-9} per codon per year in rodents and other vertebrates. However, in the mole rat lineage, this rate has increased several fold, to about 2.5×10^{-9}. Yet, the rate is only one-sixth of that of some rodent pseudogenes (1.5×10^{-8} per codon). This slow rate is quite understandable. In this animal, the αA-crystallin gene is expressed, and even if the eyes are no longer used for visual purposes, vestigial amounts of the protein exist in the body. Therefore, some selective constraint should still remain and the evolutionary rate does not exceed the maximum rate (*P*2).

It is also reported that RNA viruses undergo very rapid evolutionary changes. The rate is enhanced a million times. This high rate is consistent with the finding that RNA viruses have a mutation rate approximately one million times higher than DNA organisms. Remarkably, however, synonymous substitutions still predominate over nonsynonymous substitutions, similar to what has been found in genes of DNA organisms. For instance, sequence analyses of human influenza A virus show that the synonymous substitution rate is about 3.5 times as high as the nonsynonymous substitution rate. If we measure the degree of selective constraint against the nonsynonymous changes by the ratio of the nonsynonymous to synonymous substitution rate, we have $f = 1/3.5 = 0.29$. In other words, the degree of selective constraint, $1 - f$, is about 70% despite the very high substitution rates.

In fruit flies, even if the evolutionary rate of nucleotide substitutions is not as dramatically high as in RNA viruses, it is nevertheless very high as compared with the substitution rates in mammals. The rate of synonymous substitutions in *Drosophila* lineages is roughly 10^{-8} per site per year, which is about twice as high as that of rodents and 5 to 10 times higher than in higher primate nuclear DNAs. In this case too, synonymous substitutions greatly predominate over nonsynonymous substitutions.

Molecular Clock or Constant Evolutionary Rates

The term *molecular evolutionary clock* expresses the clock-like progression of mutant substitutions in the course of evolution at the molecular level. From the standpoint of the neutral theory, this feature of molecular evolution can be explained by assuming that the neutral mutation rate *per year* remains constant among diverse lineages (*P*3). There has been a great deal of controversy on this topic.

Traditional mutation studies on "visible" and "viability" traits (including lethals) have strongly suggested that the spontaneous mutation rate *per generation* is roughly equal among different animals whose generation spans are very different. This observation has been used repeatedly to criticize the neutral theory. For example, it is claimed that the actual data on the rates of molecular evolution do not exhibit the degree of dependence on the generation time. It appears, however, that many of these "visible" and "viability" mutations are caused or controlled by transposons and insertion sequences. On the other hand, it is likely that errors in DNA replication and repair are the main causes of DNA base changes responsible for molecular evolution. Thus, the mutation rate for nucleotide substitutions may depend on the number of cell divisions in the germ lines, particularly in the male line, and this may make the mutation rate for nucleotide substitutions roughly proportional to year.

A universally valid and exact molecular evolutionary clock would exist only if, for a given molecule, the neutral mutation rate *per year* were *exactly equal* among all organisms at all times (which is rather unlikely in nature). Any deviation from the exact equality of neutral mutation rate per year makes the molecular clock less exact. For one thing, mutation rate itself may possibly change over long evolutionary time, because of changes in proofreading mechanisms. For another, if successive mutant substitutions cause a fluctuation of "neutral space" (degree of selective constraint), this also contributes to making the molecular clock less exact. Clearly, from these considerations, the variance of the evolutionary rates among different lineages for a given molecule tends to become larger than expected under an ideal situation, as often noted in actual observations.

Polymorphism

The nucleotide diversity, π, observed at synonymous sites is about 0.076% in human populations. If the synonymous substitution rate is 1.2/2 = 0.6 of that of pseudogenes, the $E\{\pi\}$ value for the latter would be 0.076% × 1.7 = 0.13%. On the other hand, based on rare variant alleles at protein loci, the average fraction of neutral mutations, f, is estimated as 0.14. If we assume that the average coding region of each enzyme locus consists of $L = 1000$ nucleotide sites, we can compute the mean heterozygosity per enzyme locus with respect to electrophoretically detectable alleles. The estimate of $4N_e v$ for the *average* enzyme may be given by $E\{\pi\} \times L \times f \times (3/4) \times (1/3)$. Here, 3/4 is the approximate probability that a random base change within a codon leads to an amino acid change, and 1/3 is a rough estimate for the probability of a random amino acid change leading to a change in electric charge of the protein. Substituting the estimated value of $4N_e v = 0.044$ for formula (4), we have $E\{H\} = 4.2\%$. This is a little smaller than the actual mean H value per locus for electrophoretically detectable alleles in European populations (6.7%), but not very much.

Among various species of fruit flies, *Drosophila melanogaster* has been studied most extensively for the DNA variation. The estimated π per nucleotide site is 0.6% for the alcohol dehydrogenase (*Adh*) gene region, 1.2% for the *white* locus region, 0.7% for the *Notch* locus region, 0.3% for the *rosy* locus region, and 0.2% for the 87A heat-shock protein locus. The average of these estimates turns out to be π = 0.6%, which is eight times as high as the estimate for synonymous sites obtained for humans. If this value is representative for the whole genome, and if we apply the same calculation as we did above, we obtain $E\{H\} = 17\%$ with respect to electrophoretically detectable alleles in *D. melanogaster*. The corresponding observed value of $10.2 \pm 1.4\%$ is somewhat smaller than expected, but not very much.

In *D. melanogaster*, the observed π value for mtDNA is 0.22%, and it is shown that the evolutionary rates (μ) of silent substitutions are roughly equal between the mtDNA and the *Adh* gene. To analyze these observations, we note the formula $E\{\pi_n\}/E\{\pi_m\} = 4N_e\mu/2N_m\mu = 4$ if $N_f = N_m$ or the sex ratio is 1. Here, $E\{\pi_n\}$ and $E\{\pi_m\}$ are the expected values of the nucleotide diversity for nuclear and mitochondrial genes, respectively, and the ratio represents the differences in inheritance and ploidy. On the other hand, the observed ratio is 0.6%/0.22% = 2.7, which does not differ significantly from 4, considering the relatively large standard error involved in the ratio.

If we make a similar comparison for primate mitochondrial and nuclear DNAs, we should note that the synonymous substitution rate of mtDNA is approximately 20 times as great as that of single-copy nuclear DNA. Using the observed value of $\pi_n = 0.076\%$, we expect that $E\{\pi_m\}$ for the mtDNA is $20 \times \pi_n \times 0.25 = 0.38\%$. This value agrees well with the observed values of π_m for various mitochondrial genes in humans (0.3%–0.5%).

There has been much interest in the predicted positive correlation between genetic distance and heterozygosity (*P7*). Recently, a very strong correlation between these measures was found from electrophoretic data for 42 proteins from over 200 invertebrate and 300 vertebrate species. We note here two possible causes which obscure this correlation. Neither of them affects the rate greatly, but they have strong influences on the extent of polymorphism. One is temporal changes in effective population size N_e. Generally, when N_e changes, it takes a long time for polymorphism to reach a new equilibrium value. Because of the stochastic nature of random genetic drift and mutation, the time actually required to reach a new equilibrium can be substantially different from gene to gene. The other is the hitch-hiking effect. Occasional advantageous mutations may occur at some loci tightly linked to genes in question and sweep out the existing neutral polymorphism. The effect depends on recombination rates, as well as the frequency with which advantageous mutations occur. This makes it possible that the extent of neutral polymorphism is low at loci which nevertheless have been evolving fast.

GENERAL REMARKS

In the neutral theory, mutation pressure plays a predominant role in molecular evolution, as well as in phenotypic evolution. If the rate of mutant substitutions is equal to the neutral mutation rate (*P4*), we can estimate the intrinsic mutation rate due to base substitutions per genome per generation. We regard the observed evolutionary rate of pseudogenes (2×10^{-9} per site per year) as the intrinsic mutation rate. The human genome consists of about 3 billion (3×10^9) nucleotide sites. Assuming that the average generation span is 20 years for humans, we obtain 120 as the average number of mutations that occur in the human genome within a single generation. The corresponding number per zygote is 240. This enormous number of mutations newly arisen within a single generation strongly suggests that most of them should not have serious deleterious effects on the fitness of individuals. In fact, this was the very reasoning for neutrality when the theory was first proposed in 1968.

If the neutral theory applies not only to evolution at silent sites but also to that at the amino acid altering sites, this must mean that, in general, the neutral space of a protein is reasonably wide. In other words, a replacement of amino acid within a protein often preserves the activity of that protein unaltered. In fact, extensive studies by many investigators have revealed that proteins are surprisingly tolerant of amino acid substitutions as long as the proteins can keep essentially the same structure and function. M. F. Perutz concluded that adaptations leading to response to new chemical stimuli have evolved by only a few (one to five) amino acid substitutions in key positions, while most of the amino acid replacements between species are functionally equivalent.

As stressed, mutational changes among the four nucleotides are often strongly biased. In this context, of interest is the deviant coding system discovered by S. Osawa and his coworkers in *Mycoplasma capricolum*. In this bacterial species, TGA, a stop codon in the universal code, encodes for tryptophan instead of TGG, the ordinary codon for tryptophan. This organism is characterized by a very high A + T content (75%) in its genomic DNA. A/T directed mutation pressure, that is, a much higher mutation rate from G/C to A/T than the reverse direction can explain the evolution of this codon. The A/T directed mutation pressure will lead to the replacement of TGA (as a stop codon) by TAA (another stop codon), followed by a change in the anticodon of one of the duplicated copies of the tRNA gene for tryptophan from CCA to TCA. What is important is that duplicated copies then existed simultaneously in the genome. After these changes, there is nothing to hinder the gradual replacement of TGG by TGA as the major codon for tryptophan under A/T directed mutation pressure. Such a "capture of a stop codon" must have been brought about by mutation pressure. However, directional mutation pressure can be a cause of molecular evolution only when the changes involved are selectively neutral.

It is important to ask how we can understand evolution at two levels—molecular and phenotypic—in a unified way. It is generally believed that evolutionary changes at the phenotypic level are almost exclusively adaptive and caused by Darwinian selection. However, even at the phenotypic level, there must be many changes that are so nearly neutral that random genetic drift plays a significant role, particularly with respect to "quantitative characters." If a large number of segregating nucleotide sites each with a small effect are involved in a quantitative character and if the character is subject to the most prevalent type of selection (called stabilizing selection), the average selection coefficient per mutant becomes exceedingly small. Under such a condition, extensive neutral evolution can occur through random genetic drift.

As regards the problem of progressive evolution, we should consider the possibility that gene duplication and subsequent accumulation of new mutants in the duplicated genes must have played a very important role. As a kind of mutation, gene duplication is constantly fed into the population, and many of them may have so little deleterious effect that they become fixed in the population through neutral evolution. Because of substantial reduction of selective constraint after gene duplication, some of the mutants that are definitely deleterious and that would have been rejected before duplication can now reach a high frequency by random genetic drift. Such a store of new mutants may contain a few variants that will turn out to be useful for adaptation to a new environment.

Paleontological studies have revealed that great evolutionary changes usually result from exploitation of a new ecological niche by a species. In this connection, we emphasize the importance of a condition that may be called "liberation from selective constraint" as a main cause of macroevolution. If a species is confined for a long time to a constant ecological niche with all other available niches being occupied by other species, further adaptive shifts become impossible. Only when new vacant niches are presented, will the possibility be open for macroevolution to occur. This is clearly shown by spectacular evolution at the early Cambrian (often called "Cambrian explosion"), where large scale evolutionary experiments were performed in the then new way of life as multicellular organisms. Relaxation of natural selection is the prerequisite for new evolutionary progress. In other words, "liberation from selective constraint" enables extensive neutral evolution to occur, creating new variants, some of which turn out to be useful in a new environment.

SUMMARY

The neutral theory claims that the great majority of evolutionary changes at the molecular level are controlled by random genetic drift under continued input of mutations, and that most of the genetic variation within species is maintained by the same mechanisms. The theory leads to a general view that since the origin of life on Earth, neutral evolutionary changes have played a most important role in evolution and predominated over Darwinian evolutionary changes, at least in number, throughout the whole history of life.

RECOMMENDED READINGS

Crow, J. F. & Kimura, M. (1970). An Introduction to Population Genetics Theory. Harper & Row, New York.

Hartl, D. L. & Clark, A. G. Principles of Population Genetics. Sinauer Associates, Sunderland, MA.

Kimura, M. (1968). Evolutionary rate at the molecular level. Nature 217, 624–626.

Kimura, M. (1983). The Neutral Theory of Molecular Evolution. Cambridge University Press, Cambridge.

Kimura, M. (1990). The present status of the neutral theory. In: Population Biology of Genes and Molecules (Takahata, N. & Crow, J. F., Eds.), pp. 1–16. Baifukan, Tokyo.

Nei, M. (1987). Molecular Evolutionary Genetics. Columbia University Press, New York.

Chapter 9

A Darwinian View of Cytochromes and Channels

PETER NICHOLLS

INTRODUCTION: MOLECULAR DARWINISM

The evolution of any molecular entity, and especially of electron transfer enzymes and transmembrane proteins, presents special problems for an evolutionist. According to Charles Darwin (1859, cf. also the 6th edition, 1872): "If it could be demonstrated that any complex organ existed, which could not possibly have been formed by numerous successive slight modifications, my theory would absolutely break down. But I can find no such case."

Principles of Medical Biology, Volume 1B
Evolutionary Biology, pages 235–251

ISBN: 1-55938-802-1

In the theory of evolution, as developed by Darwin and Wallace, change occurs gradually, with adaptations at every stage suited to the lifestyle of the organism involved. In the nineteenth century, these limitations presented difficulties in understanding the evolution of macroscopic organs like eyes and wings. This evolutionary model was therefore challenged from time to time, yet it remains part of the present day neo-Darwinian consensus (see, e.g., Dawkins, 1986). There are no trends in adaptation; one cannot tell, by looking at a present organic feature, what its evolutionary future may be. The alternative idea, of "orthogenesis," is no longer respectable.

But limitations of other kinds may also restrict the possible pathways of evolutionary change. Gould and Lewontin (1979) point out that: "Architectural constraints . . . arise from . . . restrictions that never were adaptations but the necessary consequences of materials and designs selected to build basic *Baupläne*."

The problems which were evident at the organismic level recur at the molecular level. Darwin knew little about molecules. In his time even the atomic theory was as yet not well established (see, e.g., Nye, 1972). And all the developments of molecular biology have taken place within living memory (Judson, 1979). We now know that each evolutionary step is represented by the spread through a given population of a change at the level of DNA in its primary structure (base sequence), and at the protein level by a similar change in the primary structure (amino acid sequence). What are the molecular limitations to molecular evolution? The genetic material is molecular in nature—an empirical discovery and not a tautology (Schrödinger, 1944; Judson, 1979)—and so such limitations have consequences at the level of the whole organism.

First, we have a kind of "molecular quantization" of the possible mutations. If a change takes place at the active site of an enzyme, this will involve total substitution of one amino acid for another. The technique of site-directed mutagenesis, for which Mike Smith of Canada received a 1993 Nobel Prize, now permits the experimenter to do this routinely in the laboratory. Even the smallest such change typically has a dramatic effect upon enzyme activity, contrary to the Darwinian principle of gradualness. The enzyme binding pockets that we see today have a specificity that can be almost "perfect" (Knowles, 1992). And evolutionary analysis suggests that the amino acid sequences involved have been subjected to stabilizing ("negative") selection, but not positive selection. Sickle cell hemoglobin remains one of the few examples of an apparent recent evolutionary change driven by positive selection (against the malarial parasite). As the homozygous bearer of this gene is markedly disadvantaged when compared with either the heterozygote or the "wild-type," it is not an example that is likely to persuade a creationist (Perutz, 1992). Yet, functional molecules must have evolved in gradual stages, each of which must have had adaptive value at the time. This may well be one reason why proteins are so large. Large molecules contain distant residues from their active sites which can be effective in modulating the chemistry of the sites in an appropriately graduated way.

Second, the changes that can take place in the function of a protein molecule are restricted by the chemical nature of its environment. If the site is in an aqueous phase, a hydrophobic substitution will lead to molecular instability, and vice versa for a site in a membrane phase subjected to a hydrophilic substitution. Yet the cellular environment cannot usually be modified. It is part of the *Baupläne* (Gould and Lewontin, 1979). The "architectural" limitations of medium physics and chemistry may apply more forcefully at the molecular than at the macroscopic level.

Third, enzymes and channels that are membrane-bound commonly catalyze the transmembranous movement of ions, metabolites, and charges (Stein, 1990; Hille, 1992). Such movement, to be effective, must extend over the full membrane thickness of 2 or 3 nm, a substantial part of the cross section of the protein molecule involved. How then is it possible for the complex molecules required for membrane enzyme activity and for ion transfer to have evolved gradually?

CLASSIFICATION OF MEMBRANE PROTEINS AND CHANNELS

According to Bangham (1983), biological membranes may have evolved before specific proteins. At any rate, the two have evolved at least concurrently with the development of mutual specificity. But there may be a need for the occurrence of "sudden" changes of a dramatic kind. If protein pumps derive from smaller soluble entities, they must have changed into their present forms over a relatively short period. There seems to be no use in having a pump that only partially penetrates a bilayer. Yet the Darwinian model requires a gradual change of something into a pump. What might that something have been?

Channels and membrane enzymes comprise four types, or family groups, listed in Table 1 and illustrated schematically in Figure 1.

Table 1. Ion-Translocating Membrane Complexes

Family Group	*Functions*	*Scalar Chemistry*
Type 1. Channels and carrier systems	Ion or solute movement (exergonic)	None
Type 2. Enzymes (aqueous phase)	Ion or solute movement (endergonic)	ATP hydrolysis Decarboxylation Transhydrogenation
Type 3. Electron transfer complexes	Charge movement; redox coupling	QH_2-*c* reduction H_2O-Q photoreduction *c*-Q photoreduction
Type 4. Photo/redox systems	Ion movement; redox coupling	NADH-Q reduction Cytochrome *c* oxidation None (bacteriorhodopsin)

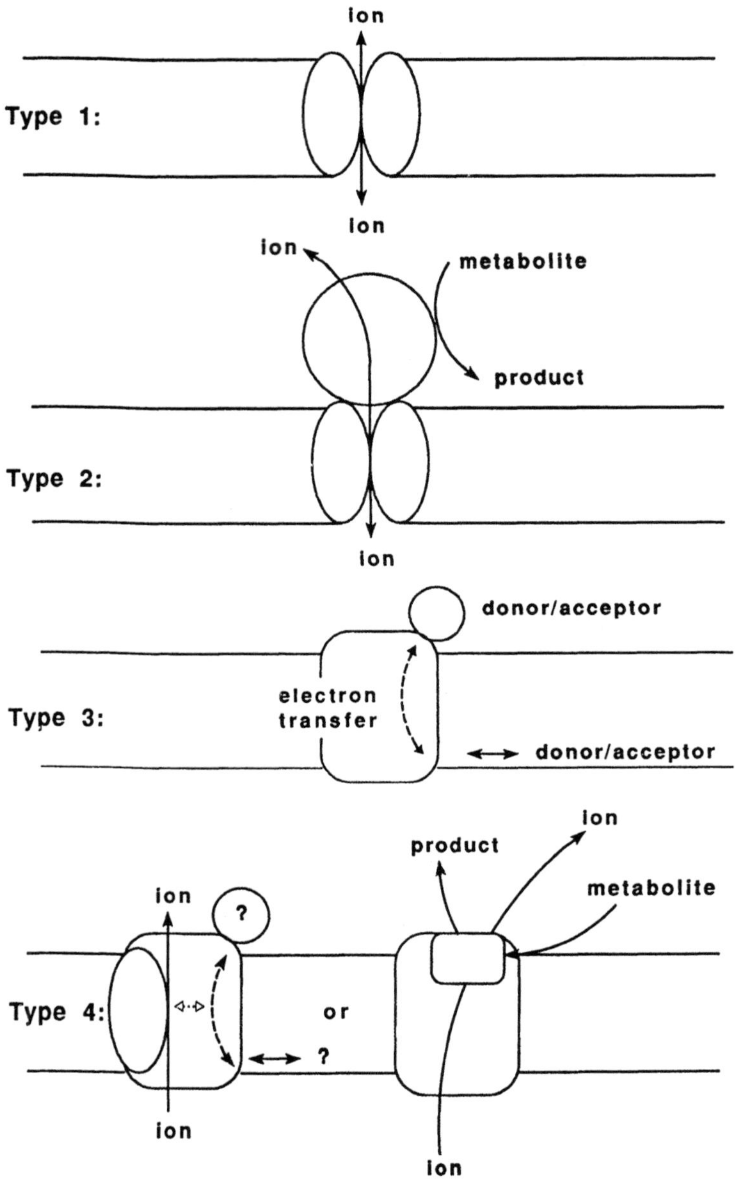

Figure 1. The four families of membrane protein complexes. **Type 1.** Carriers and channels: ion or metabolite transfer without any linked scalar processes. **Type 2.** Pumps and ATPases: ion or metabolite transfer with linked scalar processes such as ATP hydrolysis; possible hybrid systems. **Type 3.** Pure cytochrome complexes: transmembrane electron transfer only. **Type 4.** Combined pump-channels: electron transfer or other scalar chemistry is linked to the movement of unrelated ions; possible hybrid systems (*right-hand scheme*) or unknown mechanisms (*left-hand scheme*).

Type 1: Carriers and Channels

All plasma membranes, and most organellar membranes, contain proteins the sole role of which is to transfer specific metabolites and ions across the membrane from the medium on one side to that on the other. These proteins include the carriers and the channels. The turnovers of carriers are similar to those of enzymes; their "catalytic" movement of ions or metabolites may be coupled to movement of other species by metabolically independent processes of "symport" or "antiport." Channels permit specific passage of particular ions at high rates, and may be "gated," but do not couple such movement to that of other ions. Neither channels nor carriers catalyze any scalar biochemical reactions (Figure 1, Type 1).

Hille (1992) describes the specificity and mechanisms of transmembrane channels, and points out that such channels (Type 1) probably evolved from a small number of precursors. What might those precursors have been like? Channel specificity, like that of protein binding pockets, cannot be varied continuously. Only a small number of ions engage in physiologically significant distributions. Either a monocation channel must be specific for protons, for potassium, or for sodium, or it must remain rather nonspecific (like the acetylcholine-dependent channel). The channel dimensions and the nature of the channel-lining groups will determine this "quantization of cation specificity." Each pattern of size and chemistry will be characterized by a specificity set defined by an appropriate "Eisenman" series (Hille, 1992).

Gene duplication has led to multi-transmembrane peptide complexes with evolutionarily later specificities for specific ions. The original ion channel was apparently of lower specificity with fewer transmembrane segments. It could have derived from an ion-binding soluble protein that turned itself inside out as it interacted with the membrane.

Similar problems arise in considering the evolution of carriers such as the bacterial sugar carriers. Henderson (1990) has summarized the families of sugar carriers present in prokaryotes which are also evolutionarily related to those occurring in eukaryotes. Substantial evolutionary homologies can be detected that extend over the eukaryotic and prokaryotic superkingdoms. Most of these carriers are derived from a very small number of ancient precursors, but the prior functioning and possible extramembranous origins of those precursors themselves remain unknown.

Type 2: Pumps and ATPases

In addition to carriers and channels, most biomembranes include enzymes physically linked to the hydrophobic membrane regions by means of transmembrane segments. These enzymes also catalyze reactions whose scalar chemistry is aqueous, coupled to a vectorial event within or across the membrane. A number of

ATPases are of this type, and some bacterial decarboxylases and the mitochondrial and prokaryotic transhydrogenases may be included (Figure 1, Type 2).

The ion or metabolite pumping takes place against a thermodynamic gradient (chemical or electrochemical) using the free energy of the scalar metabolic reaction. Some of the pumps are partially soluble enzymes bound to transmembrane channels or carrier complexes. The functions may be clearly separated into membrane-located and aqueous components, as with the F_1F_0 ATPases which couple ATP synthesis to electrophoretic H^+ movement in: mitochondria, chloroplasts, and prokaryotic membranes; the evolutionarily related V-ATPases of vacuoles (Nelson and Taiz, 1989; Nelson, 1992); and the sarcoplasmic and other Ca^{2+} ATPases. In the H^+ ATPases it is possible physically to separate the two components. Isolated F_1 ATPase is water-soluble, while F_0 is a hydrophobic proton channel (Pedersen and Amzel, 1993). Sarcoplasmic Ca^{2+} ATPase is a single polypeptide with two clear domains: one aqueous, associated with ATP hydrolysis, and the other membrane-located, associated with calcium transfer (cf. Stein, 1990; Laüger, 1992).

One striking experiment indicating the probable hybrid origin of the F_1F_0 ATPases was carried out by Laubinger et al. (1990). They were able to show that a hybrid membrane ATPase complex composed of the F_1 ATPase of *E. coli* (part of that bacterium's H^+-ATPase) and the F_0 of *P. modestum* (part of a bacterial Na^+-ATPase) could be created, and that it pumped sodium ions. The specific channel component (F_0) thus determines the specificity of these systems and the ATPase is a specificity-independent source of scalar energy.

Other clues to the evolution of type 2 enzymes are given by the behavior of some present-day soluble cytosolic enzymes towards mitochondria. Creatine kinase and hexokinase are proteins which can associate with mitochondrial outer membranes (McEnery, 1992). Mitochondrial association of yeast hexokinase was originally postulated (Chance and Hess, 1959) on kinetic grounds. It binds to outer membrane (VDAC) channels at sites which make contact with ATP/ADP translocators (carriers) in the inner membrane. This gives hexokinase preferred access to mitochondrial ATP. Creatine kinase has a similar privileged access to such ATP via the same "contact sites" between inner and outer membrane (Laüger, 1991).

Although this reminds us of F_1F_0 ATPase, the kinase/VDAC interactions are metabolically most effective in their present reversible forms. The kinases are not *en route* to being fully membranous species, because association and dissociation are functionally important. Similarly, the periplasmic proteins of bacterial transport systems when bound to the appropriate sugars associate with the transmembrane pumps, but remain fully soluble in their absence (Quiocho, 1990). *E. coli* proline dehydrogenase associates and dissociates reversibly from inside the plasma membrane (Wood, 1987). In no case is the reversible complex evolving "towards" another more closely associated state orthogenetically. They are what they are. But these associations may nonetheless provide clues to the evolution of other transmembrane proteins, such as those of Type 2.

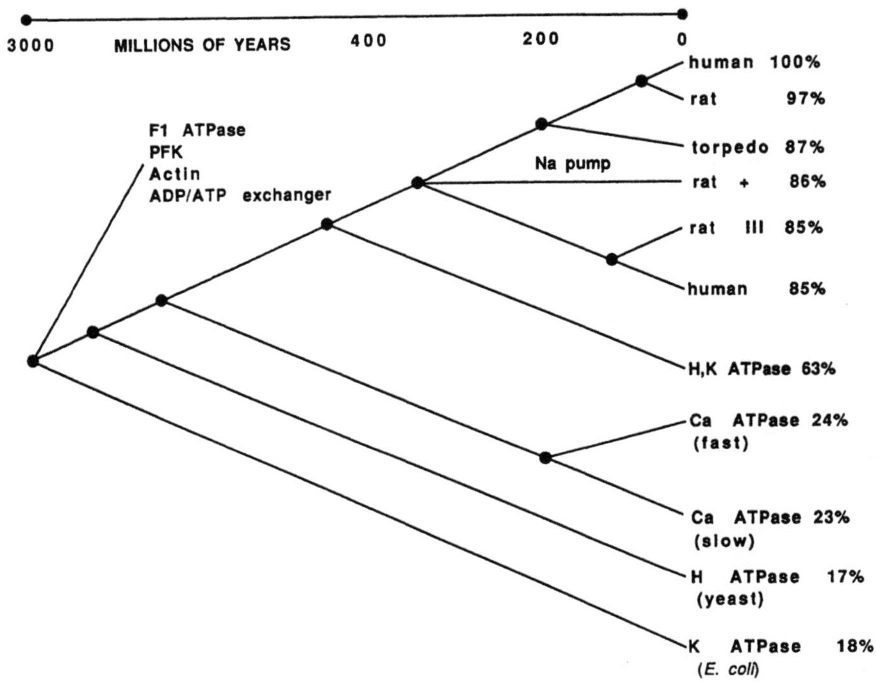

Figure 2. A possible evolutionary tree for the α-subunit of the Na^+-K^+-ATPase and related ATPases. A comparison of DNA sequences for Na^+-K^+-ATPases [α subunits, compared to the α subunit of human Na^+-K^+-ATPase (*top corner*)], with those for gastric K^+-H^+ ATPase, Ca^{2+}-ATPases, the membrane H^+ ATPase of yeast and the K^+-ATPase of *E. coli* (*bottom corner*). These are all more closely related than any are to soluble kinases, ATP-binding proteins or the F_1 ATPase of eukaryotes and prokaryotes. The percentages (*right-hand side*) indicate sequence homologies compared to the human Na^+-K^+-ATPase α-subunit. Horizontal axis in approximate millions of years. [Redrawn from Stein (1990) after Jorgensen (1988)].

Sufficient examples of the Na^+-K^+-ATPase have been sequenced for a provisional evolutionary tree to be proposed (cf. Figure 2). All the vertebrate α subunits diverged from a common ancestor at the point of divergence of the major vertebrate classes 300 to 400 million years ago. Some 10^8 years prior the gastric H^+-K^+-ATPase had diverged from the Na^+-K^+-ATPase (Jorgensen, 1988; Stein, 1990). A more remote homology is seen with the Ca^{2+}-ATPases, plasma membrane H^+-ATPase, and prokaryotic cation ATPases. And this set of enzymes may or may not be evolutionarily homologous with the other groups of ATP-binding proteins, including the F_1 ATPase, phosphofructokinase, actin, and the mitochondrial ATP/ADP exchanger.

Type 3: Pure Cytochrome Complexes

The cytochromes were discovered by Keilin in 1925 (see Keilin, 1966). Keilin was not a conventional biochemist, and he was almost alone among his contemporaries in being able to deal conceptually with both soluble proteins and membrane-bound ("particulate") forms. The influence of Darwin on French science was less direct than elsewhere. Lamarckian ideas remained influential for many years. Keilin's mentors were Maurice Caullery and the earlier Alfred Giard, who were interested in parasite evolution. In his early years Keilin too was a parasitologist. Ambiguity in the idea of adaptational perfection may have stood him in good stead when studying what seemed less than "clean" enzyme systems. He did not hesitate to describe bacterial and eukaryotic cytochrome systems that appeared incomplete or defective in comparison with the eukaryotic prototype. Some of his cytochromes form parts of complexes which are clear examples of pure electron transfer in membranes. Prokaryotic, mitochondrial, and chloroplast membranes contain such systems in which both chemistry and vectorial events involve only the electron. Bacterial and chloroplast photosynthetic centers, as well as the cytochrome bc_1 complex found in all three sorts of membrane, are of this kind (Figure 1, Type 3).

Electron transfer complexes, which move only electrons across natural membranes, are arranged with their electron-carrying prosthetic groups appropriately arranged across the membrane from one side to the other, as shown by X-ray analysis for the bacterial photosynthetic centers (Deisenhofer et al., 1985) and by indirect deduction for the cytochrome bc_1 complex. The electron-transferring (heme) groups are located internally with the transmembranous helices packed around them. Electron donors and acceptors occur in the aqueous regions on the appropriate sides of the membrane (Trumpower, 1991). The rate of electron transfer across the membrane is then controlled by the relative distances between and the redox potentials of the centers involved.

Type 4: Combined Pump-Channels

The most puzzling family is Type 4 which includes cytochrome oxidase, some NADH-Q reductases, and bacteriorhodopsins (bR). Here either a photoreaction (bR) or a redox reaction (oxidases and reductases) is linked to the transmembrane movement of unrelated ions, usually protons but sometimes another ion (Rich, 1992). The origins of these proteins are quite unclear, although the pyridine nucleotide-Q reductases may be hybrids of type 1 and type 3 systems (left-hand cartoon, Figure 1, Type 4). The others are "black boxes" in which a membrane-bound biochemical process drives an unrelated ion movement (right-hand cartoon, Figure 1, Type 4). These latter systems show few functional or evolutionary signs of hybrid origin.

If the history of the biological membrane is coextensive with the history of life, not all membrane proteins may have evolved from hydrophilic ancestors. But quite simple proteins—water-soluble and functional in water—may exist in alternative forms that are soluble and even functional in organic solvents (Zaks and Klibanov, 1988). Enforced transition from an aqueous to an organic phase can trap water molecules inside the newly interior hydrophilic region as the molecules "evert." Such molecules persist in their catalytic activity. They could in theory be placed in the hydrophobic part of a bilayer. Were insertion of such a "hopeful monster" to leave hydrophilic regions on either or both sides of the membrane, functional continuity and "emergent" properties would be possible. Only partial conformation change, involving a few peptide subunits or domains, might occur. Any one of the four membrane protein types could derive from such an ancestor.

A TEST CASE: EVOLUTION OF CYTOCHROME AND QUINOL OXIDASES

Cytochrome *c* oxidase is in the inner mitochondrial membrane and in the plasma membrane of many bacteria and cyanobacteria (Cooper et al., 1991). It transfers electrons from cytochrome *c* to molecular oxygen, and also catalyzes transmembranous proton movement (Wilson and Bickar, 1991). If electron and proton transfer were distinct and parallel events, the oxidase would be a hybrid system, sharing some properties of Type 3 transmembrane electron transfer systems and some properties of Type 2 enzymes with ion-pumping capability.

This is not the case; the oxidase is a member of a unique class of membrane proteins. It is a more complex type of transmembrane molecular device whose evolution needs an explanation. The most important respiratory enzyme for most aerobes, in vertebrates and most other complex eukaryotes oxidizes cytochrome *c*, the penultimate electron carrier, and reduces molecular oxygen to water. It was initially regarded by Peter Mitchell as a simple type 3 system, with electron donation occurring on one side of the membrane and oxygen reduction on the other. Such electron transfer involved either a single transmembrane complex carrying two electron transfer groups (like cytochrome *b*), or interaction of subunits with metal centers on two different sides of the membrane (cf. left- and right-hand cartoons, Figure 3A). Subsequent debate forced Mitchell to accept a direct ion-pumping role for the eukaryotic oxidase; he modelled this as an hydroxyl ion movement involving the so-called "invisible" copper atom (Figure 3B).

Most terminal oxidases contain at least three metal centers (Figure 4A). One center accepts electrons and then transmits them to the center(s) reacting with oxygen. The oxygen-reacting center contains two metals: one which binds oxygen, and the other being an accessory metal, probably trapping some of the intermediate oxygen reduction products. This configuration may lend itself to a direct mechanism for ion transfer using the chemistry of the accessory metal atom (Figure 4B). In the case of cytochrome aa_3, this is Cu_B (Figure 5).

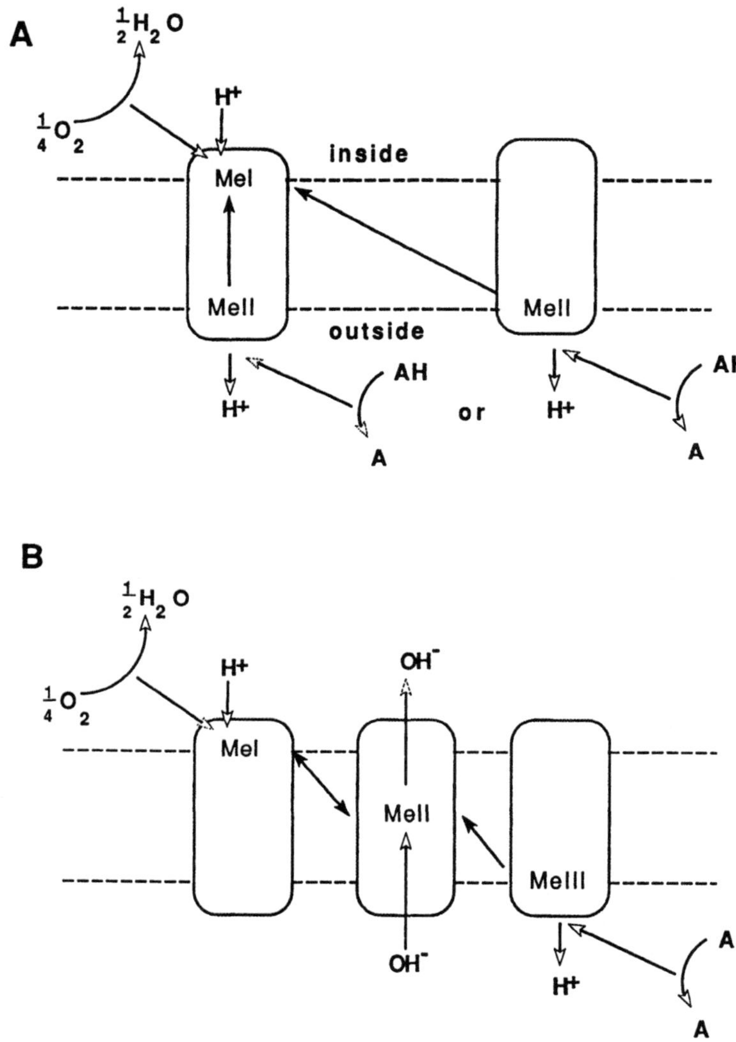

Figure 3. Two Mitchell models for cytochrome *c* oxidase action. **A.** The original nonproton translocating model: cytochrome oxidase as a transmembrane "wire." **B.** The final ion translocating model: cytochrome oxidase as an electron transfer system coupled to intermediate hydroxyl ion movement (cf. Mitchell, 1976, 1988).

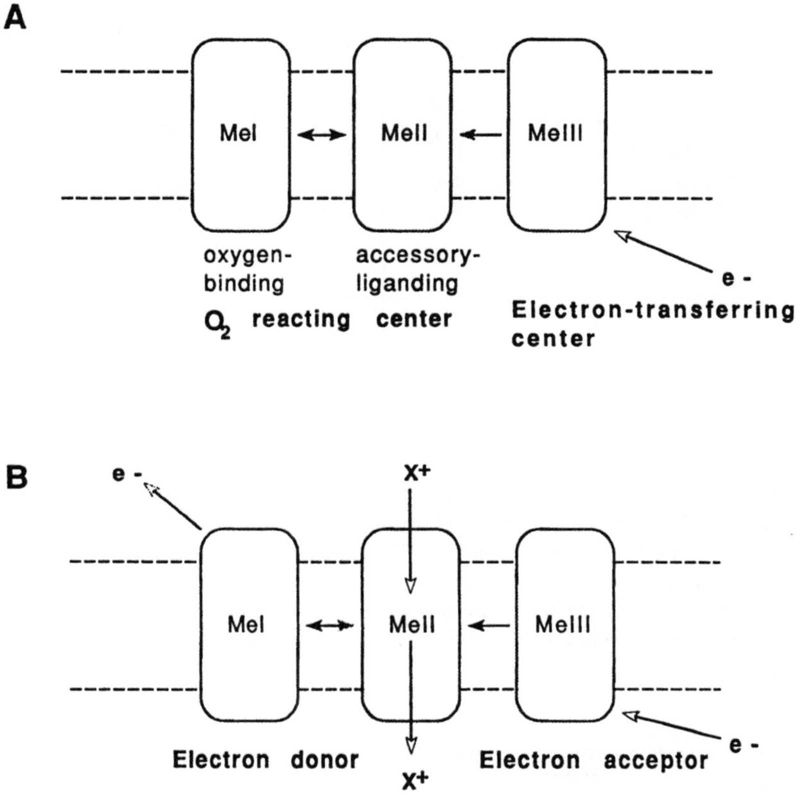

Figure 4. Generic features of a terminal oxidase (a Type 4 membrane protein). **A.** The centers involved in the reduction of oxygen to water. **B.** The centers involved in moving ions and electrons.

The relative contributions of transmembrane electron transfer and proton transfer to the generation of a membrane potential remained uncertain. The close location of cytochrome *a* and the binuclear cytochrome a_3Cu_B center, probably at or towards the "C" (cytochrome *c*) side of the membrane (Brown et al., 1993)—as in Figure 5—indicates that nearly all charge transfer involves ion (proton or cation) movement. Electron transfer is parallel to the membrane plane.

Cytochrome oxidase is a difficult case. The eukaryotic oxidase is one among a very large family of oxygen-reacting enzymes, whose genes are related and whose mechanisms are similar (Saraste et al., 1991; Gennis, 1992). This family of oxidases contains members which use either a *c*-cytochrome or a quinol as electron donor. The metal centers and prosthetic groups involved can be very different (Table 2). But the key difference between the cytochrome *c* oxidases (Table 2 and Figure 5) and the quinol oxidases (Table 2; Figures 6 and 7) is that the latter do not contain

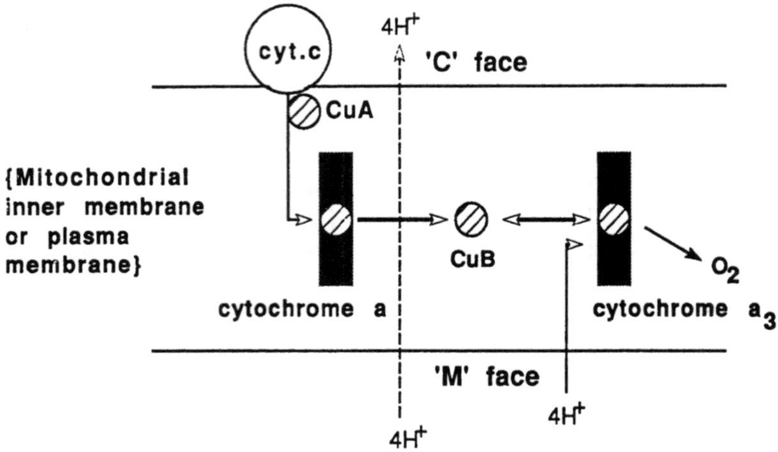

Figure 5. A cartoon of the membrane location, electron and proton-translocating centers in a eukaryotic or prokaryotic (*R. spheroides*) cytochrome *c* oxidase (cytochrome aa_3). Cytochrome *c* (at the external or "C" face) is oxidized by molecular O_2. Electron transfer is essentially parallel to the plane of the membrane. Protons are pumped from the "M" (internal) to the "C" (external) face. Four metals are involved: Cu_A as electron acceptor: the heme iron center cytochrome *a* as electron transfer system; the Cu_B plus cytochrome a_3 as binuclear Fe–Cu center reacting with oxygen. (See Table 2 for the relationship of this enzyme with other members of the oxidase family).

Table 2. The Evolutionary Variety of Prokaryotic Cytochrome/Quinol Oxidases

Oxidase Type	*Metal Centers*	*Number of Peptides*	*Electron Donor*	*Ions Pumped*	*Oxygen Affinity*	*O_2 Reactive Center*	*Bacterium*
Cyt. aa_3	2 hemes a, Cu_B, Cu_A	3–4?	cyt. *c*	H^+	high	a_3-Cu_B	*B. subtilis, R. spheroides, P. denitrificans*
Cyt. aa_3'	2 hemes a, 1 Cu_B	3?	QH_2	?	?	a_3-Cu_B	*B. subtilis*
Cyt. caa_3	2 hemes a, 1 heme c, Cu_B, Cu_A	3–4?	cyt. *c*	H^+	high	a_3-Cu_B	*Thermus thermophilus; PS3*
Cyt. *cao*	hemes a, o & c, Cu_B, Cu_A	3–4?	cyt. *c*	?	?	o_3-Cu_B	*Acetobacter spp.*
Cyt. ba_3	hemes a & b, Cu_B, Cu_A	1–2?	?	?	?	a_3-Cu_B	*T. thermophilus*
Cyt. bo_3	hemes o & b, 1 Cu_B	3	QH_2	H^+	medium	o_3-Cu_B	*E. coli*
Cyt. oo_3	2 hemes o, 1 Cu_B	3	QH_2	H^+	medium ?	o_3-Cu_B	*E. coli*
Cyt. *bd*	2 hemes b, 1 heme d	2	QH_2	Na^+(?)	high (*E. coli*)	*d*-b_{595}(?)	*E. coli* (low O_2 tension)

N.B. The eukaryotic oxidase is closely similar to cytochrome aa_3 of *Paracoccus* (1st row).

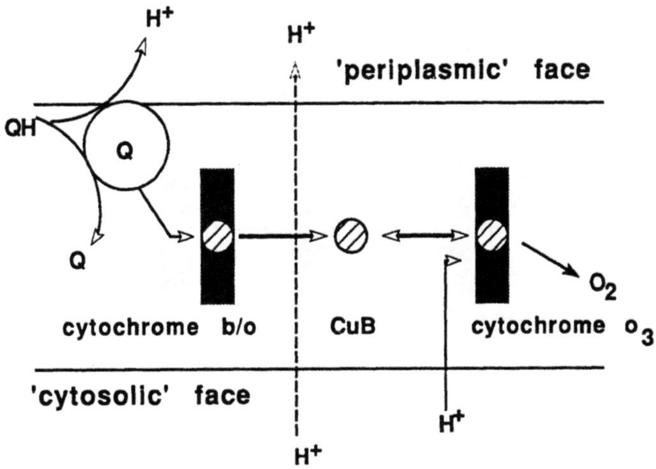

Figure 6. A cartoon of the membrane location, electron and proton-translocating centers in a prokaryotic quinol oxidase (cytochrome *b/o* of *E. coli*). Coenzyme Q (at the "periplasmic" face) is oxidized by molecular O_2. Electron transfer is essentially parallel to the plane of the membrane. Protons are pumped from the cytosolic to the periplasmic face. Three metals are involved: the heme iron cytochrome *b* as electron acceptor and transfer system; and Cu_B plus cytochrome o_3 as the binuclear Fe–Cu center reacting with oxygen. (See Table 2 for the relationship of this enzyme with other members of the oxidase family).

the Cu_A that accepts electrons directly from cytochrome. The quinol oxidases do possess a subunit homologous to the one that contains the Cu_A in the cytochrome *c* oxidases. They are capable of pumping either protons, like the cytochrome *c* oxidases, or sodium ions, like the sodium pump and some NADH-Q reductases (Semeykina et al., 1989; Skulachev, 1992). The alternative oxidase of *E. coli* and *Azotobacter* species, cytochrome *bd* (Figure 7) may be a true oxygen-dependent sodium pump.

How is the ion—a proton or a sodium ion—moved by an oxidase, as for example, in a Type 4 system? Unlike Type 2 enzymes, Type 4 systems contain no obvious channels. Yet, water is intimately implicated in the activity of the enzyme. Kornblatt and Bon Hoa (1990) showed that even in detergent-dispersed systems the catalytic process involves obligatory entry and exit of water as the enzyme cycles. The electron transfer reaction is differentially sensitive to hydrostatic pressure and to osmotic pressure. In its intermediate states, the enzyme is also more reactive with other ligands such as cyanide. Analyses with tritiated water show that several water molecules may actually be occluded sufficiently to survive column fractionation (Kornblatt and Kornblatt, 1992).

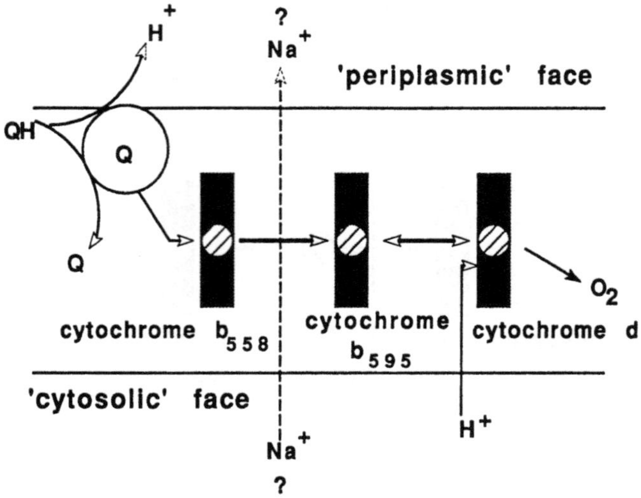

Figure 7. A cartoon of the membrane location, electron and sodium-translocating centers in a prokaryotic quinol oxidase (cytochrome *bd* of *E. coli* or *A. vinelandii*). Coenzyme Q (at the periplasmic face) is oxidized by molecular O_2. Electron transfer is essentially parallel to the plane of the membrane. Sodium ions may be pumped from the cytosolic to the periplasmic face. Three metals may be involved: the heme iron cytochrome b_{558} as electron acceptor and electron transfer system; and cytochrome b_{595} plus cytochrome *d* as binuclear Fe–Fe center reacting with oxygen. (See Table 2 for the relationship of this enzyme with other members of the oxidase family).

RECONCILIATION AND CONCLUSIONS: WATER AS A CEMENT OF THE *BAUPLÄNE*

Many enzymes, both membrane-associated and aqueous, cycle through states involving binding and release of water molecules. Several molecular and membrane processes have been reexamined from the point of view of "water" binding. Theoretical and experimental work of Parsegian and co-workers (Colombo et al., 1992) demonstrates that up to 60 molecules bind hemoglobin *together with* oxygen (as the protein changes from the "T" to the "R" form). Conversely, in yeast hexokinase, nearly 100 H_2O molecules are *removed* from the enzyme surface (or crevice) as glucose binds (Rand et al., 1993). The conformational changes of hexokinase are linked to molecular events that cannot be seen by X-ray crystallography, but which determine the thermodynamics of the enzymatic reaction.

Cytochrome oxidase has no convenient equilibrium behavior, but in the catalytic pathway as many as 10 molecules of osmolyte inaccessible water are bound. Kornblatt and Kornblatt (1992) find that some of these molecules are tightly associated and may be related to the proposed proton channel. Ion channels release over 1000 water molecules as they close (Zimmerberg and Parsegian, 1986).

Because these waters are in regions inaccessible to large osmolytes, the systems show characteristic responses to osmolarity and pressure changes.

Ligand binding by soluble proteins can involve large changes in the amounts of bound water. This could have affected dramatically the permeability of membranes with which the ligand-binding proteins became electrostatically or hydrophobically associated. Ion permeability changes would then spontaneously be linked with scalar catalysis. Water binding must accompany any conformational change that exposes new areas of hydrophilic surface. Such behavior could give rise to channel opening events when the evolutionary descendants of the aqueous proteins became membrane-embedded.

In cytochrome oxidase conformational changes that accompany electron transfer to oxygen demand water binding and release at different temporal points in the catalytic cycle (Kornblatt and Bon Hoa, 1990). These points are energetically separated. When the enzyme is membrane-oriented, energetically separate water binding and release can become channel opening and closing. If the right ionizing species are present in the water pocket, the opening and closing can catalyze ion movement.

Gradual Darwinian change could therefore have been preserved as such an enzyme evolved. Although charge transfer efficiency may have improved over time no "macromutations" involving sudden jumps in enzymatic properties need to be postulated. The architectural limitations of material strength and composition which channeled the "evolution" of church roof supports towards the "spandrels of San Marco" (Gould and Lewontin, 1979) have their counterparts in the evolution of membrane and other proteins. One of these is the chemistry of water and water-binding surfaces.

SUMMARY

Protein evolution presents problems for Darwinians. At the molecular level change is necessarily discontinuous, as one amino acid residue is substituted for another. The discontinuity may be more general. Transmembrane pumps that only partially penetrated a membrane would be useless. Yet, functional entities have to evolve in stages each with adaptive value.

Membrane proteins form four families: (1) channels and carriers which catalyze no scalar reactions; (2) enzymes with both transmembranous and extramembranous regions which couple aqueous chemistry to vectorial events—these may have arisen from soluble enzymes bound to transmembrane channels; (3) electron transfer systems whose chemistry and vectorial processes involve the same intramembranous species, the electron (bacterial photosynthetic centers and some cytochrome complexes are in this class); (4) a fourth class in which intramembranous photo/redox events are linked to the transmembranous movement of an unrelated ion—these include cytochrome oxidases, NADH dehydrogenases, and bacteriorhodopsins, and show few signs of hybrid origin.

Most enzymes cycle through states involving binding and release of many H_2O molecules. Hydrophilic proteins which do this would have dramatically affected the behavior of membranes with which they became reversibly associated. Adaptations of water-binding steps that originally accompanied conformational changes in solution could give rise to permeability changes of ions. Gradual molecular change could thus have given rise to known membrane proteins; evolutionary pathways are however more readily conceivable with some types than others.

REFERENCES

Chance, B. & Hess, B. (1959). Science 129, 700–708.

Chepuri, V., Lemieux, L., Au, D. C.-T., & Gennis, R. B. (1992). Biol. Chem. 265, 11185–11192.

Colombo, M. F., Rau, D. C., & Parsegian, V. A. (1992). Science 256, 655–659.

Cooper, C. E., Nicholls, P., & Freedman, J. A. (1991). Biochem. Cell Biol. 69, 586–607.

Deisenhofer, J., Epp, O., Miki, K., Huber, R., & Michel, H. (1985). Nature 318, 618–624.

Gennis, R. B. (1992). Biochim. Biophys. Acta 1101, 184–187.

Jorgensen, P. L. (1988). In: The Ion Pumps (Stein, W. D., Ed.), pp. 15–24. Alan Liss, New York.

Knowles, J. (1992). To build an enzyme. Proc. Roy. Soc. (Lond.) B 332, 115–121.

Kornblatt, J. A. & Bon Hoa, G. H. (1990). Biochemistry 29, 9370–9376.

Kornblatt, J. A. & Kornblatt, M. J. (1992). Biochim. Biophys. Acta 1099, 182–184.

Laubinger, W., Deckers-Hebestreit, G., Altendorf, K., & Dimroth, P. (1990). Biochemistry 29, 5458–5463.

McEnery, M. (1992). J. Bioenerg. Biomemb. 24, 63–69.

Mitchell, P. (1988). Annals N. Y. Acad. Sci. 550, 185–198.

Nelson, N. (1992). Bioenerg. Biomemb. 24, 407–414.

Nye, M. J. (1972) Molecular Reality. MacDonald, London & American Elsevier.

Pedersen, P. L. & Amzel, L. M. (1993). J. Biol. Chem. 268, 9937–9940.

Perutz, M. (1992). Protein Structure: New Approaches to Disease and Therapy. W. H. Freeman, New York.

Rand, R. P., Fuller, N., Butko, P., Francis, G., & Nicholls, P. (1993). Biochemistry 32, 5925–5929.

Quiocho, F. A. (1990). Phil. Trans. Roy. Soc. B 326, 341–351.

Saraste, M., Holm, L., Lemieux, L., Lübben, M., & van der Oost, J. (1991). Biochem. Soc. Trans. 19, 608–612.

Semeykina, A. L., Skulachev, V. P., Verkhovskaya, M. L., Bulygina, E. S., & Chumakov, K. M. (1989). Eur. J. Biochem. 183, 671–678.

Soper, J. W., Decker, G. L. & Pedersen, P. L. (1979). J. Biol. Chem. 254, 11170–11176.

Stein, W. D. (1988). (Ed.). The Ion Pumps: Structure, Function and Regulation. Alan R. Liss, New York.

Steitz, T. A., Shoham, M., & Bennett, W. S. (1990). Phil. Trans. Roy. Soc. B293, 43–51.

Trumpower, B. L. (1991). J. Bioenerg. Biomemb. 23, 241–255.

Wilson, M. T. & Bickar, D. (1991). J. Bioenerg. Biomemb. 23, 755–771.

Wood, J. M. (1987). Proc. Natl. Acad. Sci. USA 84, 373–377.

Zaks, A. & Klibanov, A. (1988). J. Biol. Chem. 263, 3194–3201.

Zimmerberg, J. & Parsegian, V. A. (1986). Nature 323, 36–39.

RECOMMENDED READINGS

Bangham, A. D. (1983). Liposomes: An historical perspective. In: Liposomes (Ostro, M. J., Ed.), pp. 1–26. Marcel Dekker, New York.

Brown, S., Moody, A. J., Mitchell, R., & Rich, P. R. (1993). Binuclear centre structure of terminal proton-motive oxidases. FEBS Lett. 316, 216–228.

Darwin, C. (1859, 1872). The Origin of Species (1st. & 6th eds.). John Murray, London.

Dawkins, R. (1986). The Blind Watchmaker. W. W. Norton, New York & Penguin Books, London.

Gould, S. J. & Lewontin, R. C. (1979). The spandrels of San Marco and the Panglossian paradigm. Proc. Roy. Soc. B205, 581–598.

Henderson, P. J. F. (1990). Proton-linked sugar transport systems in bacteria. J. Bioenerg. Biomemb. 22, 525–569.

Hille, B. (1992). Ion Channels (2nd. ed.). Sinauer Associates, Sunderland, MA.

Judson, H. F. (1979). The Eighth Day of Creation. Simon & Schuster, New York.

Keilin, D. (1966). The History of Cell Respiration and Cytochrome. Cambridge University Press.

Lauger, P. (1991). Electrogenic Ion Pumps. Sinauer Associates, Sunderland, MA.

Mitchell, P. (1976). Vectorial chemistry and the molecular mechanics of chemiosmotic coupling: Power transmission by proticity. Biochem. Soc. Trans. 4, 399–430.

Nelson, N. & Taiz, L. (1989). The evolution of H^+-ATPases. Trends in Biochemical Sciences 14, 113–116.

Rich, P. (1992). In: Perspectives in Vectorial Metabolism and Osmochemistry (Mitchell, P., & Pasternak, C. A., Eds.). Glynn Research, Bodmin, U.K.

Schrödinger, E. (1944). What is Life? Cambridge University Press, Cambridge, U.K.

Skulachev, V. (1992). The laws of cell energetics. Eur. J. Biochem. 208, 203–209.

Stein, W. D. (1990). Channels, Carriers and Pumps. Academic Press, San Diego, CA.

Chapter 10

Biological Periodicity with Reference to Higher Mammals and Humans

A. LIMA-DE-FARIA

Principles of Medical Biology, Volume 1B
Evolutionary Biology, pages 253–319

ISBN: 1-55938-802-1

INTRODUCTION

Periodicity is a well established fact in chemistry, as evidenced by the existence of the Periodic Table. However, in biology such a concept has not yet been expounded because molecular biology was until recently not sufficiently advanced. We have now reached a stage in which periodicity of biological functions and structures can be understood at the molecular level. Moreover, this periodicity can be directly correlated with periodicity occurring at the level of the chemical elements. In chemistry, periodicity has turned out to be a most fertile concept. At the beginning it was considered a curiosity, and hence most chemists paid hardly any attention to it. Today, by contrast, practically every phenomenon in chemistry is interpreted on the basis of the mechanism of periodicity.

Periodicity in biology is characterized by the recurrence of functions and structures that appear at relatively regular intervals. This is a concept that may sound rather strange to biologists, mainly because it represents a completely novel approach. It actually may well become a most helpful concept in the sorting out of many of the evolutionary problems being encountered at present.

Periodicity provides an explanation of the sudden emergence of functions that had been absent in many species for millions of years of evolution. Organisms belonging to remote groups of invertebrates happen to have the same function found in most advanced mammals, yet during the intervening period, many species lacked the corresponding properties. A prime example is the placenta. This organ appears at first in primitive invertebrates, turns up in fishes, as well as other vertebrates, but is absent or rudimentary in the marsupials, which are the mammals most closely related to the placentals, to which humans belong. Several other functions and structures manifest the same type of periodic occurrence along the evolutionary scale. This chapter is concerned with seven other functions that shed some new light on the problems of the evolution of higher mammals and humans.

CHEMICAL AND MINERAL PERIODICITY AND ITS TRANSFER TO LIVING ORGANISMS

The Discovery of Chemical Periodicity

Order in the organization of the chemical elements was a concept foreign to chemistry until the end of the last century. Although information since 1772 had been accumulating on the regularity of certain properties of the elements, it took 100 years before this phenomenon was established firmly.

Newlands proposed his law of octaves in 1865, showing that the properties of certain elements repeated themselves, and that the simple chemicals could be put together in groups of eight. The publication that he sent to *Nature* was rejected and openly ridiculed. No one could think of any order or any periodicity in the emergence of chemical properties.

It was only in 1871 that the Russian chemist Mendeleev arranged the elements in several large tables. His main contribution is that of showing that there were empty spaces in the charts that should represent elements that had not yet been discovered. He predicted the existence of 10 such elements and described the properties that should characterize them. A few years later his predictions were confirmed. Periodicity became an established concept, but had still a long way to go.

Two basic aspects remained to be elucidated. First, no chemical or physical mechanism was available that could explain such a startling regularity. Second, a series of exceptions existed that are still today part of the rule.

Only when the Danish physicist Niels Bohr discovered the structure of the atom did it become possible to put forward a mechanism that explained the repetition of the properties. He established that it is mainly the number of electrons that occur in the outer shell of an atom that governs its properties. Chemical elements with similar properties have the same number of peripheral electrons. Other evidence that accumulated during the 1920s, 1930s, and 1940s led to the discovery of atomic numbers, which explained the ordered sequence of the elements. Other atomic features were also found that better elucidated exceptional situations.

Chemical periodicity is characterized by the following (Figure 1):

1. Chemical elements exhibit similar properties, irrespective of their degree of complexity. The simplest ones, such as beryllium (Be), can have the same properties as the most complex, such as radium (Ra).
2. It is not just one, but a package of properties that repeat themselves in the other elements. These may be as many as 10.
3. The periodicity is evident both with respect to structure and function; that is, both the electron pattern and the properties of the chemical elements reappear at relatively regular intervals.

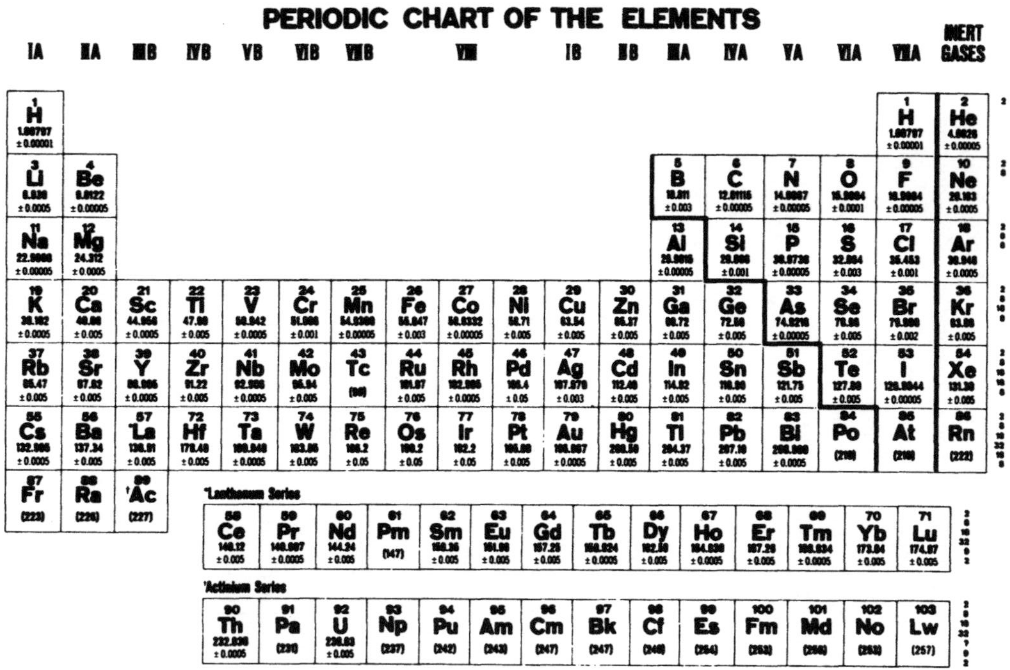

Figure 1. Periodic Chart of the Elements. Based on chart published by Fisher Scientific Co. (From Sanderson, 1967).

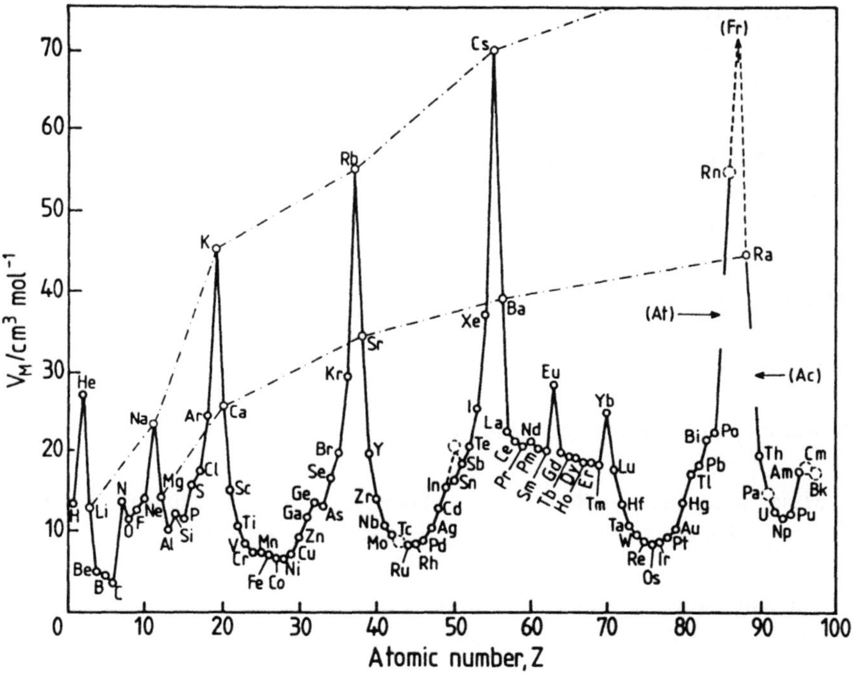

Figure 2. Periodicity in chemical elements. The periodicity found among the little over 100 atoms known is usually displayed as a periodic table of the elements. The periodicity refers to the fact that atoms in the same group have similar chemical properties. The atomic number of an element denotes the number of protons contained in the nucleus of an atom of the element. Graphic representation of the variation in volume per mol, as the atomic number (Z) increases. The periodicity is revealed by the peaks that occur at Na (sodium), K (potassium), Rb (rubidium), Cs (cesium), and Fr (francium). (From Greenwood and Earnshaw, 1989).

4. The interval at which the same package of properties recurs tends to have a certain length. However, the regularity is not total. This is why over 700 different types of periodic tables have been published (Mazurs, 1974) . An example is the location of helium (He) in the table which varies from author to author. There are also elements, such as copper (Cu) and chromium (Cr), that do not fit well into the general pattern of periodicity.

Some examples of periodicity may be mentioned. The most obvious is furnished by the noble or inert gases: helium (He), neon (Ne), argon (Ar), krypton (Kr), xenon (Xe), and radon (Rn). All build a single group and all are chemically inert due to the great stability of their electronic structures. This results from their filled electron orbitals. Another example is the first group of the periodic table (IA) composed of lithium (Li), sodium (Na), potassium (K), rubidium (Rb), and cesium (Cs). They are all alkali metals characterized by being soft, having a silvery-white color and great chemical reactivity. Francium (Fr) also belongs to this group, but has a short half-life (Figure 2).

It is noteworthy that the recognition of periodicity, and the discovery of its electronic mechanism, turned out to be most fruitful for the development of chemistry. At present, nearly every chemical phenomenon is explained on the basis of periodicity (Sanderson, 1967; Greenwood and Earnshaw, 1989).

Periodicity of the Elements Governs That of The Minerals

Minerals consist of nothing else but chemical elements. It is thus not surprising that the periodicity of the elements determines the periodicity of properties at the mineral level.

Three minerals—arsenic (As), antimony (Sb), and bismuth (Bi)—are classified as semimetals. They differ from metals due to their atom packing, which is characterized by the formation of covalent bonds between their atoms. This covalency is related to their location in the periodic table as elements (group VA). As minerals, they have common properties, which are due to this strong bonding. All three have: (1) a layered structure that results in good cleavage; (2) poor conductivity of heat and electricity; and (3) lower symmetry.

Another example is the gold group of minerals formed by gold (Au), silver (Ag), and copper (Cu), which are native metals. As elements they belong to a single group in the periodic table (group IB) due to their similar properties (Figure 1). Minerals of these elements also have properties in common: (1) they are rather soft; (2) they are malleable and ductile; (3) they are good conductors of heat and electricity; (4) they have hackly fracture; (5) they have metallic luster; and (6) they have low melting points.

The chemical elements number 106, but only 18 are the most abundant in minerals. This restriction is also related to their periodicity. The 18 occupy the first four rows of the periodic table (Dickerson and Geis, 1979).

It is the electrons within the atom that govern its chemical behavior, and that decide how atoms become packed into minerals (Jaffe, 1988). It is the original electronic mechanism that has led to the transfer of periodicity of properties to the next level of organization.

Structural Periodicity at the Mineral Level

Structural periodicity becomes most evident in minerals. The little over 3000 mineral species described by mineralogists are grouped into 12 classes based on their chemical similarity. However, all known minerals crystallize in only seven crystallographic systems. This represents a strong structural restriction. One class of minerals is formed by the native metals: gold, silver, copper, platinum, and iron. Although the atoms of these metals differ considerably in the number of their protons and electrons, they all crystallize in the cubic system. Moreover, minerals from the sulfide class, such as pyrite (FeS_2), also form cubic crystals, and the same holds for magnetite (Fe_3O_4) which belongs to the oxide class (Klein and Hurlbut, 1985).

Another example is chalcanthite ($CuSO_4 \cdot 5H_2O$), albite ($NaAlSi_3O_8$), and anorthite ($CaAl_2Si_2O_8$) which belong to the same triclinic crystal system. A copper sulfate so different in chemical composition from an aluminum silicate is able to produce the same type of crystal pattern. The recurrence of form is characterized by similar crystal faces and angles.

Especially significant for the understanding of biological periodicity is the fact that very different types of atoms and molecules can produce exactly the same structural pattern. Identical properties and structures arise from different atomic and molecular configurations.

The Properties of Proteins and Other Macromolecules are Mainly Determined by their Mineral Components

The cell is not able to put together protons, neutrons, and electrons and in this way build a zinc, a cobalt, a magnesium or an iron atom. As Weinberg (1989) points out, none of these atoms can be created or destroyed by the cell. They (as well as other atoms) are obtained directly from the mineral world. What is remarkable is that metal atoms become the key components of the cell's most important macromolecules. The basic properties of chlorophyll, hemoglobin, and many enzymes are not primarily dependent on the long sequences of carbon atoms, or of amino acids, but are located in the atoms coming from minerals that become an integral part of the macromolecule.

Chlorophyll is not a protein, but it is one of the most important cell molecules. It acts as a catalyst in the photosynthesis of carbohydrates in plants. The formula of chlorophyll *a* is $C_{55}H_{72}MgN_4O_5$. Although it is such a large molecule, the atom that decides its key property is magnesium (Mg), inherited from minerals, which is located at the center of the molecule. It is not the long molecular tail of chlorophyll, consisting of carbon, hydrogen, and oxygen atoms, that has the main function of photosynthesis. Instead, this function is performed by the magnesium

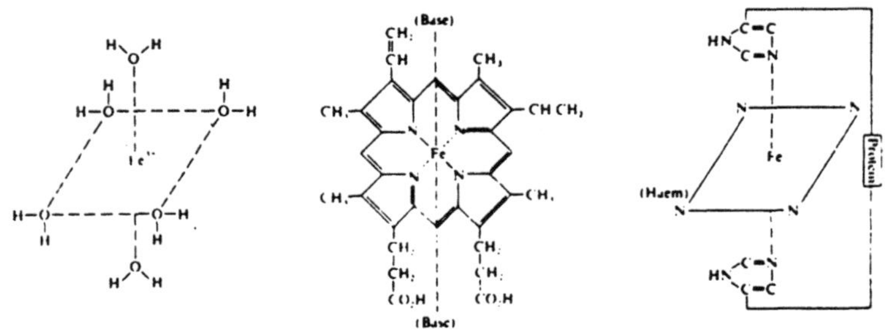

Aqueous ferric ion | Haem | Catalase

10^{-5} | 10^{-2} | 10^{5}

Catalytic activity $ml^{-1}s^{-1}$ at 0° C

Figure 3. Catalytic activity occurred in ferric ions before iron appeared in enzymes. Catalytic activity (ml^{-1} s^{-1} at 0 °C) in an aqueous ferric ion is 10^{-5}; in haem (a combination of the iron atom with a protoporphyrin) it increases to 10^{-2}, and in the enzyme catalase it attains the value of 10^{5} (From Calvin, 1983).

atom and its surrounding nitrogen and carbon atoms which generate the electron transfer needed to produce photosynthesis (Alberts et al., 1983).

This is also the case with the protein hemoglobin in which iron atoms are responsible for the main function of this macromolecule. Iron is situated at the center of a porphyrin ring that combines with the amino acids of globin. Actually, the catalytic activity occurred in ferric ions before it appeared in hemoglobin and in the enzyme catalase (Figure 3).

In other proteins—many of them enzymes—the key atoms are not those that are part of their long amino acid sequences, but the ones that they inherited from minerals. Zinc is one of them, which is mainly responsible for the properties of carboxypeptidase (Branden and Tooze, 1991).

The same situation is found in vitamins. For example, cobalt is the atom that governs the basic properties of vitamin B_{12} (cyanocobalamin).

It is noteworthy that the cell does not contain a genetic code for magnesium, iron, zinc, or cobalt. These atoms were incorporated into the living systems directly from the external environment. It is thus not surprising that the functions and structures of the cell were obliged to follow and perpetuate the periodicity inherent in the chemical elements and the minerals.

The problems related to the molecular organization and evolution of the chromosome were treated earlier (Lima-de-Faria, 1983). The way elementary particles, atoms, and minerals have contributed to shape and to canalize biological evolution was also discussed previously (Lima-de-Faria, 1988).

BIOLOGICAL PERIODICITY OF FUNCTIONS IN LIVING ORGANISMS

A search for periodicity among living organisms discloses that, as in the case of the chemical elements and minerals, not just one property, but a series of biological functions and structures display periodicity. The list is long, but let us concentrate on the functions in which the phenomenon is most evident. These include: (1) the occurrence of the placenta; (2) the formation of the penis; (3) the return to aquatic life in terrestrial animals; (4) the production of "carbon copies" among the marsupial and placental mammals; (5) the emergence of high mental ability; (6) the distribution of vision in invertebrates; (7) bioluminescence; and (8) the capacity to fly.

Since this chapter is mainly directed at medical students, it seems appropriate to start with the functions that are carried out by organs that are present in humans and higher mammals. Other functions may be mentioned only to the extent that they elucidate this problem.

The Occurrence of the Placenta

Humans have a fetal life that requires the existence of a placenta. This is formed during embryonal development in the uterus of the human female. It is an organ consisting of a double vascular spongy structure built by the interlocking of fetal and maternal tissues which allows nutritive, respiratory, and other exchanges between the mother and the fetus. The exchanged materials include: oxygen, food, and waste products.

This organ is so characteristic that it establishes the division between the two groups of living mammals: the eutherians or placental mammals (which have a well developed placenta), and the marsupials or pouched mammals (which lack it or only have a rudimentary one) (Savage and Long, 1986).

One would expect that such a complex and vital organ would be unique to higher animals. This is, however, not the case. It is present in simple invertebrates, lower vertebrates, and even in plants.

Onychophorans are a phylum closely related to the myriapods (which include the centipedes) and are known as velvet worms. A well studied species is *Peripatus*, which is a primitive animal with an advanced system of live birth. The embryo is connected to the oviduct of the mother by a placenta and an umbilical cord. These two organs have the same functions that appear later in vertebrates (Margulis and Schwartz, 1982).

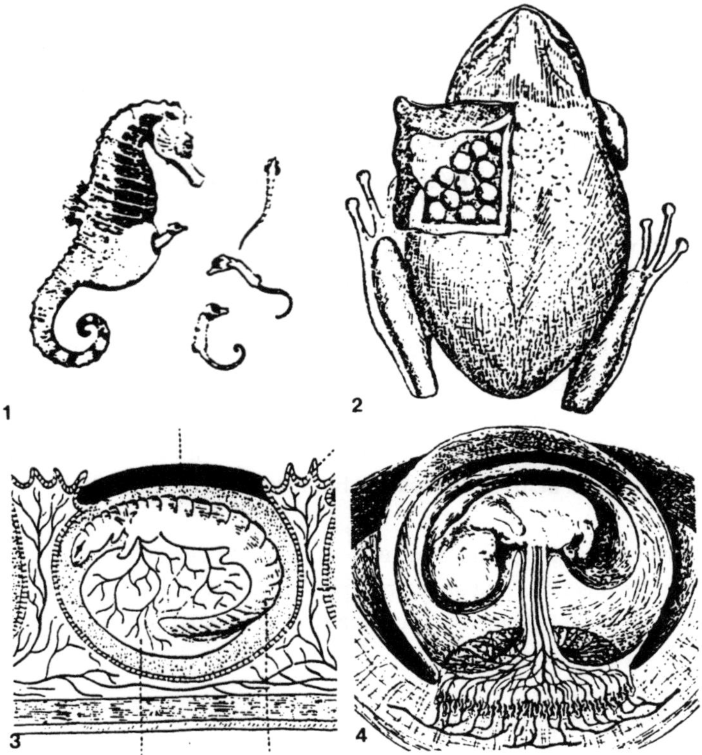

Figure 4. Periodicity of the placenta. (**1**) The male of the sea-horse *Hippocampus* sp. (a fish) with young that are emerging from its pouch. (**2**) The dorsal pouch of the frog *Nototrema marsupiatum* with part of the skin lifted to show the developing eggs. (**3**) Placenta in *Pipa americana* (a frog) showing the embryo with the adjacent tissues and blood vessels. (**4**) Placenta in the human female with an embryo, surrounding tissues and blood vessels. *Source*: (**1**) Beazley, 1980; (**2**) and (**3**) Rabaud, 1934; (**4**) Frisch, 1938.

Teleost fishes, like the sea horse *Hippocampus*, carry their eggs in pouches where they are nourished by means of a placenta. This is irrigated by blood vessels, with oxygen and nutrients being exchanged through it. The young are released into the seawater at a later stage of development. The placenta emerges in these fishes without direct relationship to that of the low invertebrates. Instead of developing in the female and in the uterus, it is formed in the pouch of the male *Hippocampus*. Thus, the same function can be produced by a different sex and in connection with a different organ (Figure 4).

Many species of fishes release their eggs directly into the water where they are fertilized and the embryos develop independently of their parents. In sharks the situation is different. They have developed a placenta that is nearly as complex as that of humans. The embryo has a membranous sac which contains the food supply for its development, and which fuses with the tissues of the oviduct. The blood vessels of the embryo and of the female are in intimate contact (Burton, 1987).

Some amphibians and reptiles also possess this organ. The black salamander that lives in the European Alps is viviparous; that is, the tadpoles develop within the mother instead of in the water, as is the case in related species. The gills of frog tadpoles are respiratory organs, but in the salamander they function as a placenta. Here, again, it is another organ, the gills, which takes over this function. Not only the gills but also the tails of tadpoles may function as a placenta. This last situation occurs in *Pipa pipa* (the Surinam toad). The embryos develop in contact not with the uterus, but with the back of the female, and each embryo absorbs the nutrients from the mother's body by using its tail as a placenta (Figure 4). Sóme lizard and snake species display a placenta.

Flowering plants also possess this organ. The use of the same nomenclature by botanists is not accidental but it actually corresponds to the same functional and physiological solution. The ovules of angiosperms are provided with nutrients through vascular vessels from their point of origin in the placenta (Denffer et al., 1971). The trophoblast in humans is the tissue surrounding the embryo from which the placenta is formed. The suspensor of plants and the trophoblast have comparable anatomical locations in relation to their embryos and are both involved in the transport of nutrients necessary for embryonal development (Nagl, 1973, 1976).

The Periodicity of the Placenta

What leads to the establishment of the concept of periodicity is that the emergence of the placenta displays the following features:

Occurrence in a wide range of organisms. Plants, invertebrates, fishes, amphibians, reptiles and mammals have placentas. This means that it occurs throughout the whole of the evolutionary scale (Figure 5).

Absence in closely related species or groups. Some families of frogs have a placenta (Hylidae), whereas close relatives (Ranidae) do not possess it. Species or groups of fishes and reptiles may have this organ, whereas it is absent in their close relatives. Most evident is the case of the marsupials which are the closest relatives of the placental mammals, but they usually lack a placenta.

The degree of complexity is independent of phylogenetic position. The placenta of amphibians is not more complex than that of simple invertebrates, and the sharks and other fishes, such as *Heterandria formosa*, have a placenta not different from that of mammals (Needham, 1950). The placenta found in plants is similar in many details to that present in humans.

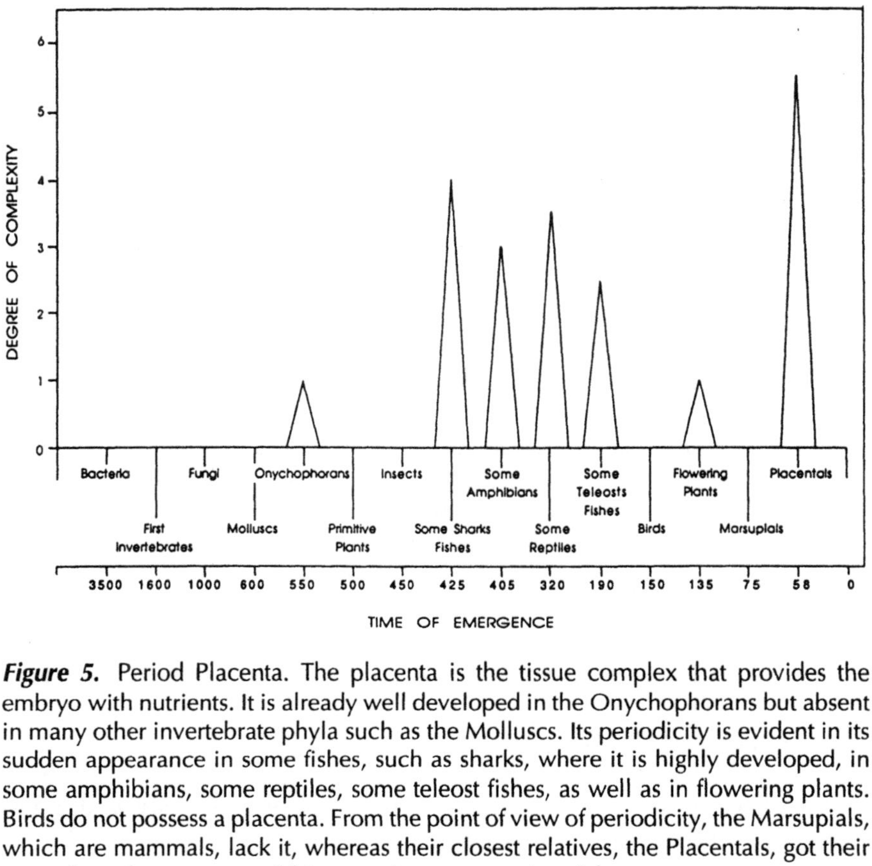

Figure 5. Period Placenta. The placenta is the tissue complex that provides the embryo with nutrients. It is already well developed in the Onychophorans but absent in many other invertebrate phyla such as the Molluscs. Its periodicity is evident in its sudden appearance in some fishes, such as sharks, where it is highly developed, in some amphibians, some reptiles, some teleost fishes, as well as in flowering plants. Birds do not possess a placenta. From the point of view of periodicity, the Marsupials, which are mammals, lack it, whereas their closest relatives, the Placentals, got their name from the presence of this organ in every species of the group.

Occurrence at relatively regular intervals. Periodicity not only demands repetition but also a certain regularity of occurrence. The periodicity cannot be absolute since one is dealing with a biological phenomenon. All living phenomena tend to display order, but within the limits that are imposed by the fact that they are dynamic processes. Moreover, as one reaches the biological level, the number of variables increases and less regularity is to be expected. A total regularity would imply an absence of evolution. Transformation of form and function demands alternatives. These do not need to be disparate but can occur within a frame of order that reveals itself in the repetition of the same solution and in a certain regularity of emergence. The velvet worms appeared 550 million years ago (Margulis and Schwartz, 1982). The placenta emerges again in the sharks which are found as early as 425 million years ago. The teleosts only developed 190 million years ago

(Carroll, 1987). The flowering plants emerged 135 million years ago. The eutherian mammals became a distinct group 58 million years ago. There is a clear-cut interval whose length remains undetermined. This is a problem to be elucidated by future research in molecular biology. Moreover, the fact that these groups have lived on Earth for this length of time does not mean that their placenta was necessarily formed at the time they appeared. However, in the case of the placental mammals it is evident that the placenta evolved concomitantly with their emergence (Macdonald, 1984). Thus, we are facing the same situation as we did when periodicity was first recognized among the chemical elements. Order and regularity became evident but no well defined interval could be established. Even today there are many irregularities in the periodicity present at the chemical level.

Formation of the same organ by different tissues. The placenta is formed in animals by many different tissues. These are: (1) the uterus, (2) the skin of the pouch, (3) the gills, and (4) the tails. This enables us to understand how it is formed without difficulty in plants, which also possess vascular vessels that carry nutrients to the different tissues of the plant, just as blood vessels do in animals.

The emergence of the organ is not directly related to the general environment. In amphibians, *Pipa americana* has a well-developed placenta, whereas *Xenopus laevis* lays eggs which develop freely; yet, both species live in water. The same is the case with the placental fishes. These occur in the same environment inhabited by the fishes which release their eggs directly into the water. The conditions under which the marsupial and placental mammals evolved were essentially the same, yet they differ sharply in the possession of this organ.

The molecular mechanism responsible for the emergence of the placenta. Like other organs, the placenta is not the product of a single tissue or of a single function, but results from a coherent differentiation package consisting of several tissues coordinated with associated functions. The question, then, to be asked is: What internal mechanisms are involved in creating suddenly and repeatedly this ordered package? In the case of the placenta an answer is available. This is because every phenomenon that occurs in humans is considered important, and as a consequence is more extensively investigated than those solely present in animals or plants.

Studies of human and animal physiology reveal that the formation of the placenta is due to the action of two distinct cascades of hormonal signals (Figure 6). Molecular cascades have been discovered recently to be the cause of important differentiation processes. What characterizes a cascade is the production of a specific chemical which obliges the emergence of an ordered sequence of other chemicals that cause the formation and affect the activity of different organs. What characterizes the molecular cascade is that its pathway is strictly *canalized* and has a multitude of structural and functional side effects. In the case of the placenta, the follicle-stimulating hormone (FSH) initiates the first cascade of molecular processes by increasing the cellular concentration of cyclic AMP. As a result phosphorylated proteins are produced that affect the transcription of DNA. The second cascade is started by the hormone oestrogen. This is secreted by the egg follicles and their maturation is also activated by FSH. The oestrogen acts on the walls of the uterus where the placenta develops and leads to the surge of the luteinizing hormone. In a third step, this last hormone provokes the secretion of progesterone that induces the pregnancy of the fertilized egg. Other hormones, such as prostaglandins also participate in the molecular cascade (Challis, 1979; Dorrington, 1979; Friday and Ingram, 1985).

Such an internal and highly ordered sequence of chemical signals can now be seen as the cause of the coherent package of structures and functions that results in the formation of the placenta. Once the production of a given chemical in the cell reaches a threshold, a morphogenetic process ensues that cannot be stopped. It results in a novel and harmoniously constructed organ.

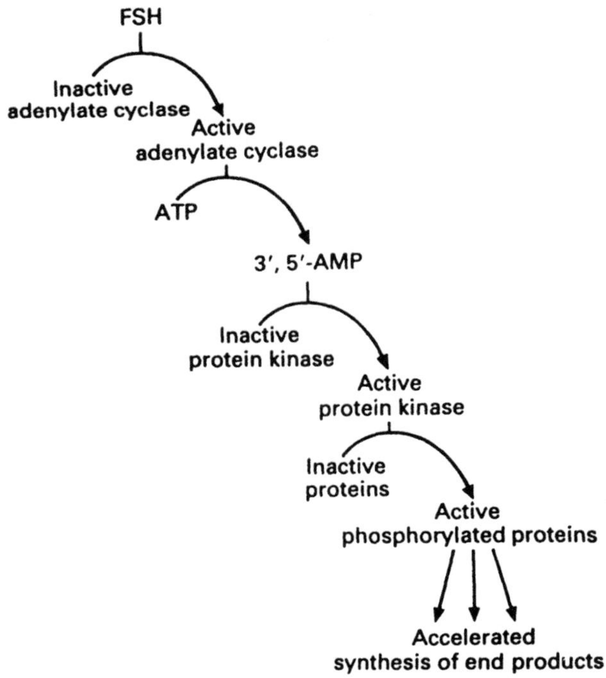

Figure 6. Cascade of chemical events involved in the mechanism of action of the follicle stimulating hormone (FSH). (From Dorrington, 1979).

The cascade of hormones furnishes the mechanism by which the organ is formed so suddenly with anatomical and physiological coherence, but it does not provide information about another aspect that is equally relevant for evolution.

What is the genetic mechanism that allows the DNA sequences of the chromosomes to produce a molecular signal that had not been available for millions of years? In other words, how does the genetic constellation of the organism suddenly produce the same protein or the same hormone that had not been available for a long period. The answer is now becoming apparent. Most genes in eukaryotic organisms consist of a large number of DNA sequences called introns and exons that alternate with one another. Both types of sequences are transcribed into RNA. However, the final RNA that is translated into the protein is only formed by the sequences coming from the exons. That is to say, the RNA from the introns is not incorporated into the final messenger. More recently it has been found that exon-intron recombination occurs, and in this way novel proteins, called mosaic proteins, are produced. The exons can change in number and position along the gene. By the same mechanism old proteins that had not been produced for a long time can suddenly be made available for differentiation, creating the periodic reappearance of patterns and functions (Gilbert, 1986; Sudhof et al., 1985a,b).

These mechanisms will be discussed in detail at the end of this chapter.

The Formation of the Penis

The penis is usually considered to be a feature of the more complex vertebrates—the mammals. The reason is that the lower vertebrates such as fishes do not possess this organ; their genital canals terminate in a cloaca. The same is true of the amphibians, reptiles, and birds. Only a few species have a rudimentary penis.

Fishes and amphibians release their eggs and sperm, usually into the water. As such, efficient copulatory organs that allow the introduction of the sperm into the female genitals may be considered superfluous. But in reptiles and birds that like the mammals live on land, the penis would seem to be a most needed organ. However, the organ is not of general occurrence in these classes. What throws new light on the evolution of the penis is the fact that it existed before the vertebrates emerged. Indeed, the lower invertebrates had developed this organ, which is nearly as complex as that of humans.

Hermaphrodite animals seem to be those that would have the least need of a penis. Yet in the Gastrotrichs (simple invertebrates) one animal behaves as male and the other as female when they copulate. The animal that functions as a male has a penis that is used to transfer sperm to the one that behaves as female. Worms, such as the Acanthocephalans, are bisexual and possess a penis that is introduced into the vagina of the female, whereby sperm is released. Other worms have a penis that is highly complex. In *Fasciola hepatica* (flatworms), this organ consists of: (1) a seminal vesicle, (2) a prostate gland, and (3) an ejaculatory duct (Figure 7). In humans, the penis has an ejaculatory canal, accompanied by a seminal vesicle and

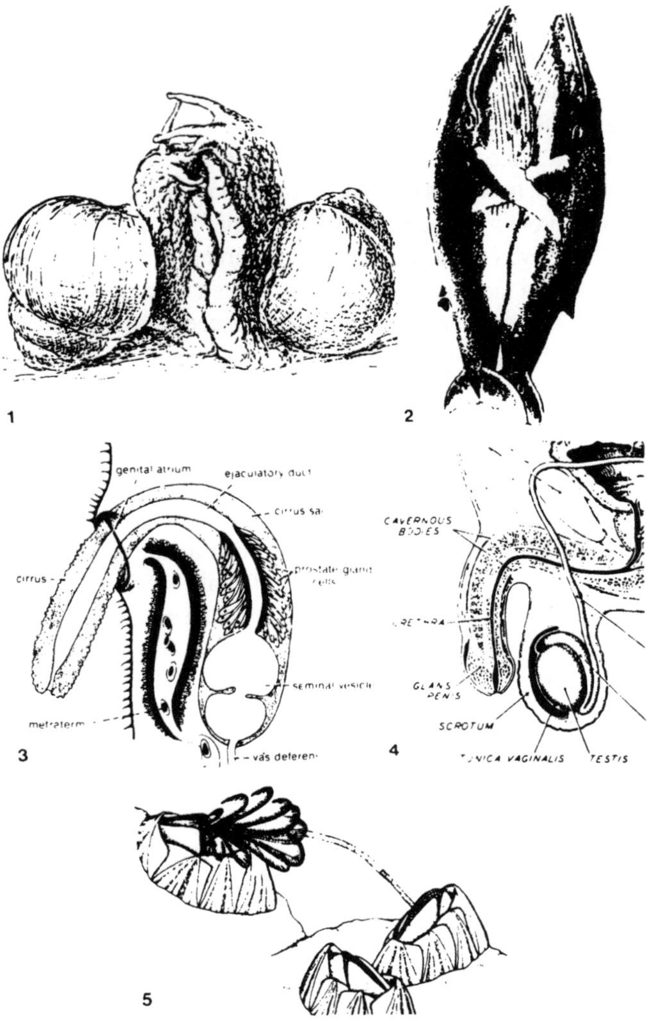

Figure 7. The periodicity of the penis. (**1**) Copulation in snails: these molluscs are hermaphrodites but one may function as male (extruding its penis) and the other as female. (**2**) A pair of mating humpback whales which, like other mammals, have a long penis. (**3**) The protruding penis of the flatworm *Fasciola hepatica*. This organ has a seminal vesicle, a prostate gland, and an ejaculatory duct. (**4**) Penis in the human male with its ejaculatory canal and seminal vesicle which is also connected with a prostate gland. (**5**) The penis of the barnacle, which is an invertebrate, is so long that it extends from one individual (on the left) to another (on the right) during copulation. *Source*: (**1**) Frisch, 1938; (**2**) Orr and Pope, 1986; (**3**) and (**5**) Barnes, 1980; (**4**) Romer and Parsons, 1978.

a prostate gland. *Busycon* species, in molluscs, display a large penis in the frontal part of the body instead of at the rear, as is the case in most animals. *Buccinum undatum*, another mollusc that is a marine species, has an equally long penis (Margulis and Schwartz, 1982).

Insects have small genital organs, but the dragonflies copulate by means of a large penis (Romoser, 1973).

In proportion to their body size, the crustaceans exhibit the largest type of penis that is known. This is present in barnacles of the genus *Balanus*. They are marine animals that are sedentary, being attached to rocks and having the body enclosed in a limestone carapace. Their testes are located in the cephalic region, and the sperm ducts unite within a penis that protrudes from the body. The penis is so long that it reaches another animal located distantly, where it enters the mantle cavity of the female and deposits the sperm into her body (Barnes, 1980).

Marsupials possess a double vagina and a double penis, whereas humans have only one of each (Macdonald, 1984).

The Human Penis Uses the Hydraulic Skeleton Present in Invertebrates

The velvet worm, *Peripatus capensis,* has many legs and moves them with equal efficiency as, for instance, the myriapod *Scolopendra morsitans*. What is of particular interest is that the two groups of animals use different anatomical and physiological solutions. In the myriapod, the legs are supported by a rigid exoskeleton, whereas the velvet worm has hollow legs whose rigidity is maintained by water pressure. In both cases the legs move with equal efficiency as those of vertebrates that are supported by an internal bony skeleton.

Among invertebrates, the penis is sometimes stiffened by a stylet, and in seals and walruses the penis has a long bone that grows with increased sexual maturity (King, 1964).

This bone is absent in most mammalian species; instead under the influence of hormones the boneless penis stiffens and becomes as rigid as any human arm or leg. The human hydroskeleton provides equally effective rigidity and articulation as the one achieved in the body of invertebrates. But the similarity extends further than this. The sensory and locomotor organs of the head and foot of a pulmonate snail are flaccid and crumpled when it copulates. The blood-flow changes within the body lead to the dilatation of the penis of the snail. The same phenomenon exists during human copulation. The penis that was previously flaccid becomes erect due to a redistribution of blood in the body.

The Periodicity of the Penis

Primitive penises occur in some reptile and bird species. This suggests that the genetic constitution necessary to produce this organ is available, but it is not fully realized. In summary, then, internal molecular processes seem to have determined the emergence of the penis since it developed under the following conditions:

1. It emerged in unrelated groups.
2. It appeared in environments in which it did not seem to be needed.
3. Its emergence was precluded in situations in which it could have been favorable to the propagation of the organism. Although birds evolved some 150 million years ago, they did not manage to form this organ.
4. Its complexity, as well as its number, are not directly related to the location of the organism in the evolutionary scale.
5. Different parts of the body may form it. It may protrude from the head or from the rear of the animal.
6. It tends to occur at relatively regular intervals (Figure 8).

The Return to Aquatic Life in Terrestrial Animals

Vertebrates started by being aquatic, while fishes remained in this medium for the rest of their evolution. The conquest of the land took place when the amphibians emerged. Their name derives from their mode of life that is partly in water and partly on land. It is only with the advent of the reptiles that full terrestrial life had become established.

Several phyla of invertebrates are aquatic, but in some groups, such as the molluscs, land forms of the species developed.

What is significant from the point of view of evolution is that:

1. The terrestrial habitat reverted to the aquatic form both in the vertebrates and invertebrates.
2. The return to water occurred in a series of unrelated classes and orders.
3. It led to the formation of the same type of structures and functions, irrespective of whether it was an invertebrate, a reptile, or a mammal that acquired the aquatic habitat.

The repetition of this event, accompanied by the same type of biological solution, provides evidence of the periodicity of the phenomenon.

Independent Orders of Mammals Evolved into Aquatic Forms

DNA–DNA hybridization, enzyme analysis, and other molecular methods indicate that whales (order cetacea) are derived from the ancestors of deer, which belong to the order artiodactyla, the even-toed ungulates (Figure 9) (Janvier, 1984; Arnason, 1985; Goodman et al., 1985; Arnason and Widegren, 1986; Springer and Kirsch, 1993).

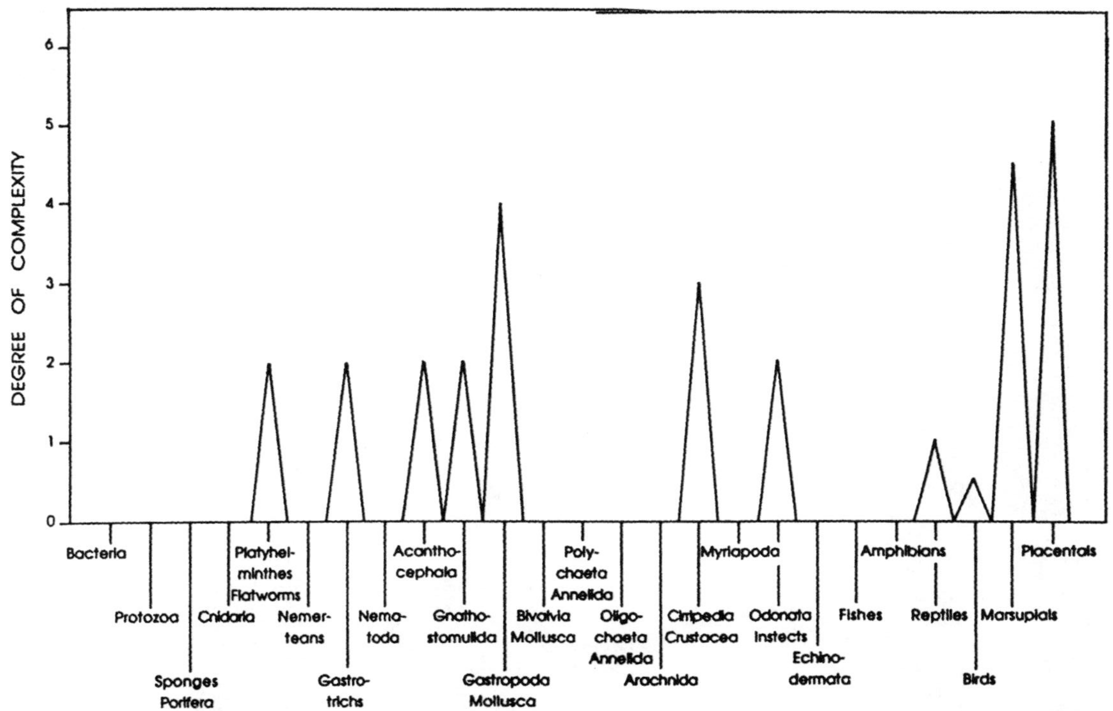

Figure 8. Period penis. The periodicity of the penis is remarkable. It occurs in a highly developed form in Gastropods (Molluscs) and it is well developed in other invertebrate phyla, such as the Crustacea (Cirripedia), flatworms (Platyhelminthes), and insects (Odonata). It is absent in many other invertebrates such as the Echinoderms. In vertebrates, the penis is highly developed in both the Marsupials and the Placentals, but it is nearly absent, or rudimentary in birds and reptiles and totally absent in fishes and amphibians. The emergence of the penis is not directly related to the environment: it occurs in marine forms, such as the Cirripedia and the Gastropods, in which the sperm could be released into the water, and it is hardly present in birds that live on land like mammals.

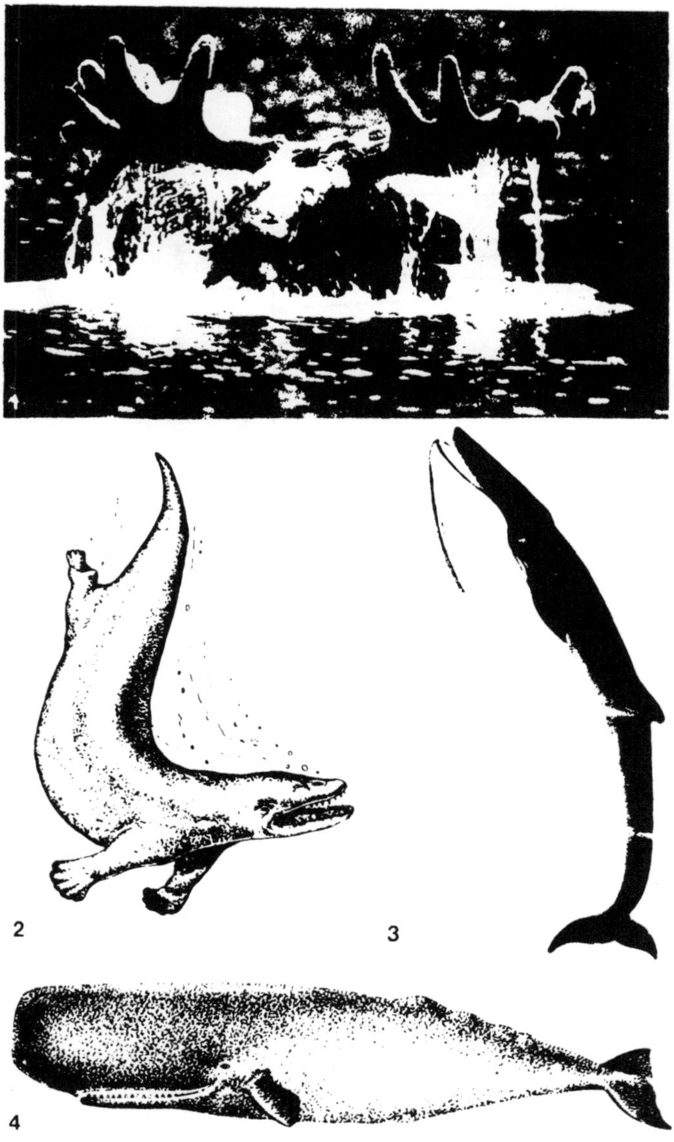

Figure 9. Air-water transformations in whale ancestors and living whales. (**1**) A male elk, *Alces alces*, swimming across a Swedish lake; females also swim and have no horns. DNA studies show that the terrestrial even-toed ungulates, such as the elk, are among the ancestors of the whales. (**2**) Restoration of *Pakicetus*, the most primitive known whale (from early Eocene). (**3**) Sei whale, *Balaenoptera borealis*, a baleen whale with its characteristic mouth plates that allow it to ingest plankton. (**4**) Sperm whale, *Physeter catodon*, a toothed whale which has strong teeth and is a carnivore. *Source*: (**1**) Commercial photograph; (**2**) Savage and Long, 1986; (**3**) Curry-Lindahl and Tinggaard, 1965; (**4**) Freeman, 1972.

Similar methods have established that seals (order Pinnipedia) have had as ancestors the carnivores (order Carnivora). The dog, the bear, and the otter are their close relatives. The eared seals originated about 25 million years ago from a dog-like branch; the walrus came from a bear-like stem; and the true seals appeared

Figure 10. Air-water transformations in seal ancestors, seals and walruses (order Pinnipedia). DNA studies disclose that the terrestrial carnivores had the same ancestors as the Pinnipeds with the otter and the mink being the closest relatives. (**1**) Tiger bathing, (*Panthera tigris,* carnivore); tigers often lie in water to keep cool. (**2**) and (**3**) Polar bear (*Ursus maritimus,* carnivore); like the sea lions, the polar bear swims only with its forelimbs (the hind limbs being used only as a rudder) and closes its nostrils when it dives. (**4**) Swimming southern sea otter (*Enhydra lutris*); otters are truly amphibious, foraging in water but living also on land. (**5**) The typical eared seal *Arctocephalus pusillus.* The eared seals comprise the fur seals and the sea lions. (**6**) A typical true seal *Phoca vitulina.* (**7**) The walrus *Odobenus rosmarus. Source*: (**1**) and (**4**) Beazley, 1980; (**2**) and (**3**) Matthews and Carrington, 1972; (**5**) and (**6**) Macdonald, 1984; (**7**) Freeman, 1972.

15 million years ago, derived from an otter-like group (Macdonald, 1984) (Figure 10).

Sea cows (order sirenia) arose from the ancestors of the elephants (order proboscidea) (Figure 11).

Pigs and peccaries are the close relatives of the hippopotamuses (order artiodactyla), which derive from animals belonging to the same order: the suiforms (Colbert, 1980).

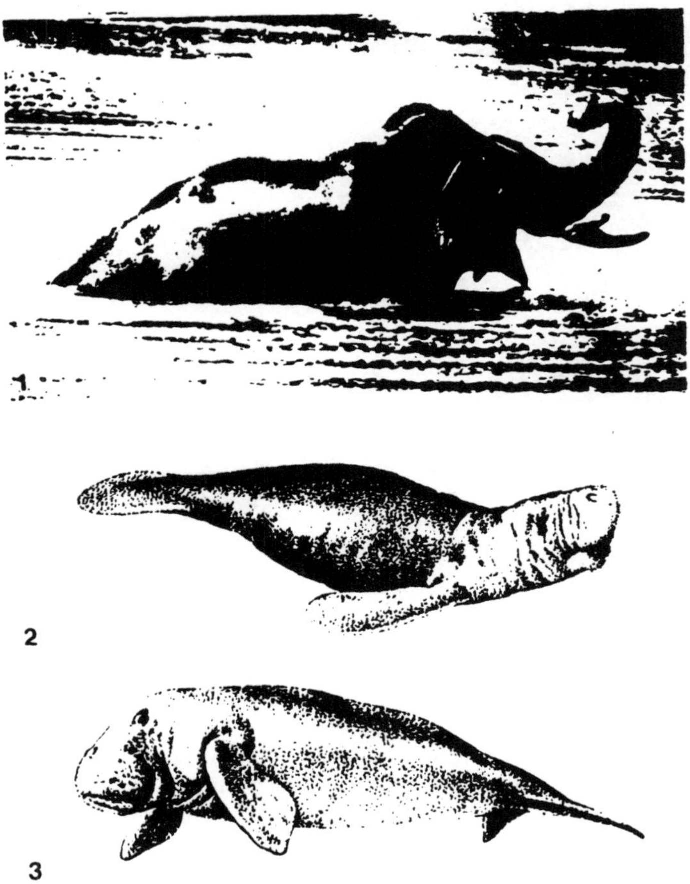

Figure 11. Period return to aquatic life—sea cows. (**1**) An elephant bathing. These animals are good swimmers and eat usually the plants that grow under water. The sea cows are derived from the ancestors of the elephants. (**2**) The North American manatee *Trichechus manatus.* (**3**) The Dugong *Dugong Dugon.* The manatee and the dugong are sea cows. *Source*: (**1**) Matthews and Carrington, 1972; (**2**) and (**3**) Freeman, 1972.

Seals and hippopotamuses live both on land and in water, but whales and sea cows are strictly aquatic. The four groups are derived from three different stocks that have their origin in the artiodactyla, the carnivora, and the proboscidea.

The Reptiles and Amphibians Also Became Fully Aquatic

Two types of reptiles returned to the marine environment: the ichthyosaurs and the plesiosaurs. The first emerged 200 million years ago, while the latter lived between 181 and 135 million years before the present.

Several groups of amphibians returned to a fully water habitat. An example is provided by the stereospondyls which became specialized aquatic forms. They reversed the terrestrial trend of evolution 230 million years ago (Carroll, 1987).

Among the Invertebrates Gastropods and Insects Returned to the Water

The gastropods are molluscs which started as marine animals, but later conquered the land, thereby becoming fully terrestrial. However, they became once again aquatic by invading freshwater habitats.

Land insects are also believed to have populated fresh waters. They developed new organs such as tracheal gills in which the respiratory vessels are ramified and lie under a thin cuticule. Some submerged insects carry an air bubble, as a physical gill, from which they absorb oxygen, and others come to the surface at regular intervals to breathe air, as aquatic mammals do (Withers, 1992).

Same Solution Occurred in the Different Groups

The return to aquatic life could have led to disparate and unrelated solutions if it would have been an accidental process. Instead, a study of the different organisms reveals that the same type of forms and functions emerged, despite the fact that they were separated by millions of years of evolution, as is the case in invertebrates and higher mammals.

The fresh water forms of Gastropods (molluscs) such as *Limnaea* (snails), come to the surface to obtain air, as seals and whales do. The snail has a lung which has an opening through which air is allowed. The opening is closed when the animal is submerged. Whales have lungs and close their nostrils by employing valves when they swim under water. Both the mollusc and the whale can remain submerged in water for more than an hour (Orr and Pope, 1986).

It is interesting that the fossil stereospondyls, which became specialized aquatic types, increased in size as they moved into the water, thus becoming the largest amphibians ever to live. The proportions within the body changed. Their heads became immense in size but the limbs and feet became smaller. Land mammals that returned to water much later underwent the same type of transformation. Whales are the largest animals that ever lived on this planet; their heads are quite large in relation to the body size, while the limbs are greatly reduced (Colbert, 1980).

Early land reptiles did not have streamlined forms, but when the ichthyosaurs became aquatic their body assumed the form of that of a tuna fish. The legs and feet were transformed into paddles giving them the same function as those of seals. The bifurcated tail of fishes had disappeared in amphibians and reptiles long ago, but an "astonishingly" similar one reappeared suddenly in the ichthyosaurs.

Fishes have a dorsal fin that is supported, like their other fins, by an inner bony skeleton. What is astounding is that reptiles also acquired a dorsal fin, which is practically identical in form to that of fishes, but had no skeleton. The whales repeated this feat many millions of years later. They have an identical dorsal fin that was formed without bones (Orr and Pope, 1986; Carroll, 1987).

Some bird species returned partly to water, as is the case with penguins, and their body features and physiology show similarities with those of the other groups. For instance: their wings became paddles; when they dive they draw oxygen from oxyhemoglobin; and they acquired layers of fat that insulate them from the cold water. The whales developed a similar fat layer that gives them insulation.

When the mammals returned to the aquatic life, whales, sea cows, and seals acquired a fully streamlined body. A remarkable feature is the flat tail of whales, which is a new formation without bones akin to their dorsal fin. In fishes, the tail is part of the bony skeleton. Without a skeleton the tail of a whale can be more than two meters long and still has the same function as in other marine animals.

This means that the dorsal fin and tail of fishes have a different tissue origin than the dorsal fin of the ichthyosaurs and the dorsal fin and the tail of the whales, which are all fleshy structures. As in the case of the placenta, different tissues can give rise to the same organ.

Other body transformations were of the same type in the ichthyosaurs and whales. The four legs were transformed into paddles in the marine reptiles and in primitive whales. Their arm bones and fingers also changed in a parallel way.

In seals, the four feet are transformed into paddles, and they have webbed toes similar to those of aquatic birds. Hippopotamuses have also partial webbing between their toes. Seals lack a dorsal fin and a propulsive tail, but close their nostrils on immersion. They are only partly aquatic.

Like whales, sea cows (the manatees and dugongs) are totally aquatic, but like seals have not developed a dorsal fin. They have, however, a streamlined body, have lost their hindlimbs, have a tail, and possess valved nostrils.

No Change in Genetic Constitution is Necessary to Produce a Streamlined Form

Algae are marine and fresh water organisms. Their later descendants conquered the land in the form of the flowering plants, which are mainly terrestrial. Yet among them, some returned to water, thereby becoming amphibious.

What is interesting from the point of view of evolution is that no change in genetic constitution is necessary to acquire the streamlined shape, which is characteristic of aquatic plants and animals. In species such as *Potamogeton gramineus*, the submerged leaves have a streamlined form being highly dissected and thin, whereas the aerial ones are thick and entire. The functions of the leaves lying under water are also modified. Their photosynthesis and respiration processes are adjusted to different wavelengths and oxygen concentrations. Since the structural and functional changes occur within the same individual, it is evident that we are dealing with the same genetic constitution in both situations. Accordingly, one can conjecture that the ancestors of the aquatic reptiles and mammals did not need to change their genetic constitution to obtain the streamlined form. Supporting this is the fact that the whales appeared suddenly in tertiary times. Colbert (1980) points out that they enjoyed at the outset a series of extraordinarily rapid evolutionary changes. The first fossils already look like whales. This is not due to the lack of fossils. The same is the case in reptiles that have become aquatic (the ichthyosaurs) and in mammals that suddenly started to fly (e.g., bats). Few or no intermediate forms exist.

Moreover, seals and whales did not need to change their genetic constitution to acquire vital aquatic functions. Both mammals dive for periods as long as 60 minutes, a long period of time to withstand asphyxia. In this connection, a study has been carried out on the hormones that regulate the asphyxia to which the human embryo is subjected during prolonged birth. And, simultaneously, a comparative analysis was made of diving seals. The results thus obtained show that the human fetus and the seal use the same hormone, namely catecholamine. When its level increases, there is a redistribution of blood flow to vital organs, and an increase in heartbeat and blood pressure. All three processes follow the same pattern in both organisms when oxygen is unavailable (Lagerkrantz and Slotkin, 1986).

One can safely state that most mammals go through the same type of birth, as humans and seals do. This is taken to mean that marine mammals had no need to develop a new mechanism to tolerate asphyxia during diving. They only used the one that they already had to counteract a long and difficult birth.

Experimental Demonstration that Water Governs the Streamlined Pattern

The submerged leaves of the water buttercup (*Ranunculus aquatilis*) are narrow and dissected, whereas the floating ones are broad and entire. Experiments show that the leaf-type produced is determined by the contact of the plant crown with

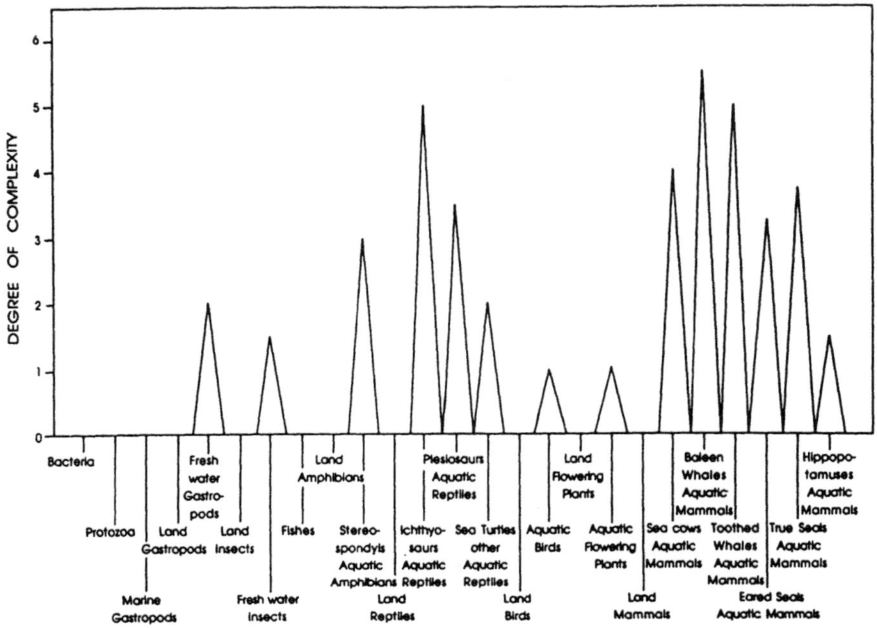

Figure 12. Period return to aquatic life. The majority of invertebrates live in the sea or freshwater, but some phyla became terrestrial such as the insects. The early vertebrates were also marine, such as the fishes, but subsequently became terrestrial starting with the amphibians. In both groups the terrestrial species returned to the water environment and did this periodically. The land gastropods returned to aquatic life; the terrestrial insects returned to fresh-water; the land amphibians became aquatic as they transformed into the stereospondyls; the land reptiles got transformed into the marine reptiles called ichthyosaurs and plesiosaurs. Some birds and some flowering plants also became aquatic. The periodicity is particularly evident in the return of the land mammals to water which occurred in several orders that are not closely related: the sea cows, the whales (baleen and toothed), the seals (eared and true), and the hippopotamuses.

water. When this organ is under water, dissected leaves are formed, and when in air, entire leaves appear. If, however, the experiment is reversed and the crown is again immersed in water, the growth of broad leaves stops and the plant resumes the formation of linear ones. It is now known that the shape of leaves is determined by hormones that move along the stem (Dale, 1982).

The buttercups are flowering plants that were terrestrial from the beginning and that, like the mammals, became partly or wholly aquatic.

The Periodicity of the Return to Aquatic Life

In summary, biological periodicity results from the combination of the following processes:

1. The phenomenon is recurrent. It appears in: (a) terrestrial plants, (b) land invertebrates, (c) amphibians, (d) reptiles, (e) land birds, and (f) terrestrial mammals. All have developed forms that became partially or fully aquatic (Figure 12).
2. The structural and functional changes that occur in these organisms are essentially alike, and they affect a whole variety of organs from the skeleton to the lungs.
3. Most important is that these similar solutions appeared independently of the degree of complexity and of the phylogenetic position of the organisms affected. Millions of years of evolutionary separation failed to diminish this capacity.
4. The transformation that led to aquatic life occurred only in specific orders, families, or genera, and emerged at relatively regular intervals since it appeared not only in the major groups of vertebrates but also in invertebrates and plants.
5. The new solutions arose as a coherent package of properties including (a) streamlined body shape, (b) closing of openings connected with lungs during submersion, (c) the ability to remain for long periods of time under water, (d) the modification of limbs and other body parts, (e) the creation of new dorsal fins and new tails, (f) the webbing of toes, and (g) insulation of the body by thick layers of fat.
6. No change in genetic constitution needs to occur to create the streamlined form in plants or the ability to remain under water for long periods of time in mammals. This means that the transformations could have been relatively rapid as is indicated by the fossil record.
7. Coherent packages of new structures and functions that affect the whole body are produced in frogs within days. Tadpoles change into adults due to a tenfold increase in thyroid hormone levels. The result is (a) the reabsortion of the tail, (b) the change from gill to lung respiration, (c) the formation of a skeleton with adjoining muscles, and (d) the reshaping of the intestines and blood circulation (Fox, 1981; White and Nicoll, 1981). This cascade of organ transformations caused by a single small chemical has now been studied at the molecular level. The hormone binds to receptor proteins. The genes for these proteins have been characterized and cloned in *Xenopus laevis*. Thyroid hormone changes gene expression in every tissue of the tadpole (Shi and Brown, 1990; Yaoita et al., 1990).

Atomic, Molecular, and Biological Mimicry

Mimicry is a concept familiar to zoologists and botanists but in the last few years it has become of general application among biochemists. This is especially true in immunology where one is confronted with a multitude of molecules that closely resemble each other.

Animals belonging to different species or even to unrelated orders display similar properties. Animals may also resemble plants and vice versa. An example of this is the flower of the orchid *Ophrys* that is strikingly similar to a wasp which copulates with the flower (Mossberg and Nilsson, 1977). Two molecular phenomena are expected to be mainly responsible for the biological mimicry. First, it is known that many genes have been conserved for millions and even billions of years. This is the case with the DNA sequences for cytochrome c, chlorophyll, hemoglobin, ribosomal RNA, histone IV, and others. These genes have not changed appreciably since the bacteria were formed (3 billion years ago), as is the case of the ribosomal genes, or were only slightly modified since eukaryotes emerged (1.2 billion years ago), as it happened in cytochrome c and histone sequences (Rimington and Kennedy, 1962; Granick, 1965; Margoliash and Smith, 1965; Dickerson, 1980). And, second, genes that are not necessarily identical code for proteins that contain key atoms, with similar electronic configurations to those of other macromolecules. These would be expected to lead also to the repetition of a pattern or function. For example, proteins that have different amino acid sequences can possess similar properties due to the existence of a particular atomic configuration present in all of them (Murzin, 1993). This is better understood when it is recognized that biological mimicry was preceded by atomic and molecular mimicry.

Atoms may contain different numbers of protons and neutrons in their nuclei, as well as different number of electrons in their surrounding shells, yet they may mimic each other in their properties. It is the number of electrons of the outer shell that governs this similarity of properties. Beryllium, magnesium, calcium, strontium, barium, and radium all have similar properties due to the presence of two electrons in the most peripheral shell of their atoms (Greenwood and Earnshaw, 1989). It is this phenomenon that led to the establishment of periodicity at the atom level.

As one examines minerals many examples of mimicry are found. Zincite (ZnO), bromellite (BeO), greenockite (CdS), wurtzite (ZnS) and swedenborgite ($NaBe_4SbO_7$) produce crystals with the same symmetry and which belong to the same crystal class, yet they have very different chemical compositions (Bloss, 1971).

Not less than 15 different compounds are isostructural with the mineral fluorite. The mechanism responsible for the mineral mimicry has to do with the arrangement

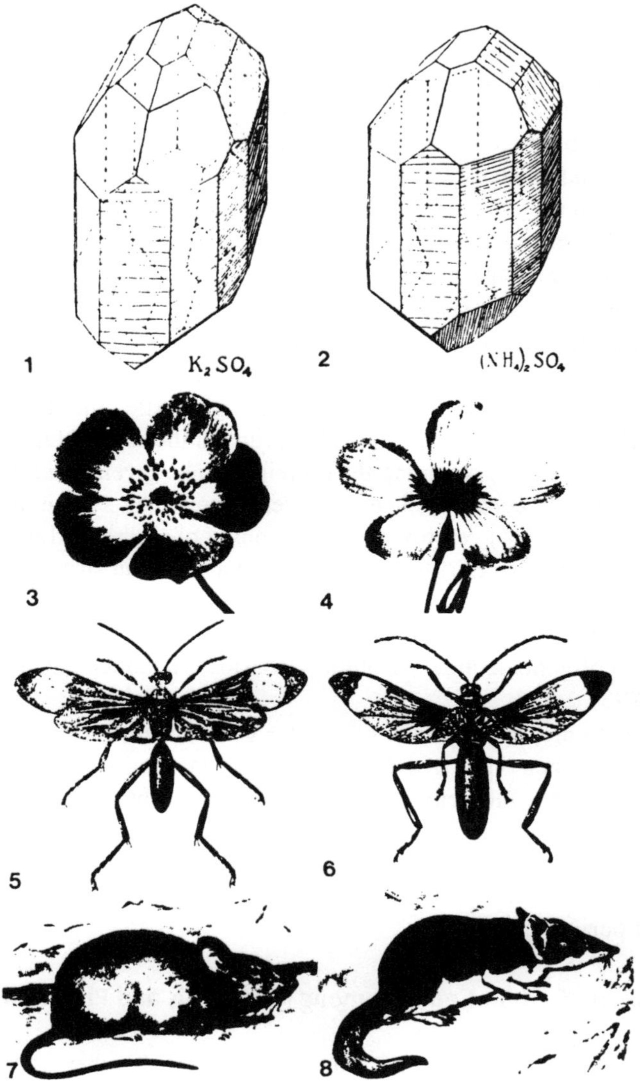

Figure 13. Mimicry in crystals, plants, invertebrates and vertebrates. (**1**) Crystal of potassium sulfate, K_2SO_4. (**2**) Crystal of ammonium sulfate $(NH_4)_2SO_4$. Both crystals have essentially the same form and crystallize in the orthorhombic system. (**3**) The flower of *Rosa pendulina* with 5 petals distributed radially (Family Rosaceae). (**4**) Flower of *Oxalis adenophylla*, with 5 petals also distributed radially, but belonging to a different family (Oxalidaceae). (**5**) The wasp *Mygnimia aviculus* of the order Hymenoptera. (**6**) A most similar species *Coloborhombus fasciatipennis* belonging to the order Coleoptera. (**7**) The house mouse *Mus musculus* is a placental mammal. (**8**) The marsupial mouse *Dasycercus* belongs to the non-placental mammals but is strikingly similar to the house mouse. *Source*: (**1**) and (**2**) Pauling, 1949; (**3**) and (**4**) Heywood, 1978; (**5**) and (**6**) Galiano, 1929; (**7**) Curry-Lindahl and Tinggaard, 1965; (**8**) Halstead, 1978.

of the atoms in the crystal lattice, and with their electronic configurations (Figure 13) (Pauling, 1960; Dickerson and Geis, 1979; Sharp, 1988).

Molecules exhibit the same phenomenon. When a saturated solution of iodoform (CHI_3, triiodomethane) is allowed to crystallize it produces the same type of crystals, as when water turns into snow flakes. Although they have only one atom in common, hydrogen, the final result is in both cases the production of a flat six-branched crystal with secondary ramifications forming the size gradient characteristic of water crystals (Nakaya, 1954).

Proteins are built by only 5 atoms: carbon (50%), oxygen (25%), nitrogen (15%), hydrogen (7%) and sulfur (in small quantities). Their assembly results in large macromolecules, but these atoms maintain the properties that they had in the minerals. Their electron and proton effects remain the same. Numerous examples of molecular mimicry among the proteins can be given. Four proteins will suffice: (1) staphylococcal nuclease, (2) yeast asp-tRNA synthetase, (3) cholera-like cytoxin, and (4) verotoxin-1. They consist of amino acid sequences that show no significant similarities, yet they have a common structural and functional solution that consists in their capacity to bind oligomers. They share a folding structural motif, called the five-stranded beta-barrel, which obliges them to bind oligonucleotides or oligosaccharides (Branden and Tooze, 1991; Murzin, 1993).

Hence, not only the presence of similar gene constellations, but also the occurrence of other types of DNA sequences that lead to the production of proteins, and other molecules, containing atom groups with similar electronic configurations, ought to be considered as the primary cause of the emergence of structural and functional mimicry among living organisms (see Figure 13). This is considered to be the basis of the emergence of biological periodicity.

Production of "Carbon Copies" among Marsupials and Placental Mammals

Just as atoms, molecules, and minerals mimic each other, thus producing "carbon copies," it is not surprising that the same situation arises at the biological level. When a comparison is made between the marsupial and the placental mammals, a long array of species, belonging to different orders turn out to be similar and, in some cases, nearly identical in the two groups of mammals. This copying process has been described long ago by zoologists, but it has not been understood in terms of periodicity simply because it has not been considered as the product of internally coordinated cellular processes.

"Carbon copies" are found for members of the following orders: Carnivora, Primates, Rodentia, Dermoptera, Insectivora, Edentata, Perissodactyla, and Artiodactyla (Table 1 and Figure 14) (Halstead, 1978; Savage and Long, 1986).

The large sabretooth cats or tigers are known from the fossil record. They had very long upper canines that protruded from their mouths. This type of animal

Figure 14. Some of the species that are "carbon copies" of one another in placental and marsupial mammals. (**1**) The sabretooth large cat *Smilodon* (placental), and (**2**) the sabretooth cat *Thylacosmylus* (jaguar sized marsupial); both are fossils. (**3**) Scaly-tailed squirrel *Anomalurus peli* (placental), and (**4**) the greater glider *Petauroides volans* (marsupial). (**5**) The common mole *Talpa europaea* (placental), and (**6**) the pouched mole *Notoryctes typhlops* (marsupial). *Source*: (**1**) Halstead, 1978; (**2**), (**4**), and (**6**) Macdonald, 1984; (**3**) Freeman, 1972; (**5**) Sandars, 1937.

emerged several times among the placentals. Two well known genera are *Smilodon* and *Eusmilus*. A practically identical sabretooth appears among the fossil marsupials called *Thylacosmylus*. The teeth had the same shape, it lived at the same time, and had the same carnivore habit as the placental species.

The living lion of the placentals (*Panthera leo*) has an equivalent in a marsupial fossil, the lion *Thylacoleo carnifex*. Other carnivore copies are found among the hyena, the wolf, the wild cat, and the otter. This last species and the water opossum (marsupial) have the same body shape, the same aquatic habitat, swim and catch food in a similar way, and are able to close certain organs when they become immersed.

Within the primates there are monkeys that resemble opossum species. And in the rodents the similarity is extreme. Placental species of mice are difficult to distinguish in body features and habits from the marsupial mice.

Among the most striking "carbon copies" are the gliding squirrels. The placental scaly tailed squirrel and the marsupial greater glider have the same type of

Table 1. Equivalence between Placentals and Marsupials

Placentals			*Marsupials*		
Order	*Family*	*Species*	*Order*	*Family*	*Species*
Carnivora	—	Sabretooth *Smilodon* (fossil) and *Eusmilus* (fossil)	Marsupialia	—	Marsupial Sabretooth *Thylacosmylus* (fossil). Upper canines had same form and shape as those of placental sabretooth, lived at the same time and had same carnivore habit as placental equivalents.
Carnivora	Felidae	Lion *Panthera leo*	Marsupialia	Thylacoleonidae	Marsupial lion *Thylacoleo carnifex* (fossil). Strong stabbing and shearing teeth, lion- like in body and head shape, well equipped for hunting and carnivore nutrition.
Carnivora	Hyaenidae	Spotted hyena *Crocuta crocuta*	Marsupialia	Borhyaenidae	Pouched hyena *Borhyaena* (fossil). Hyena-like.
Carnivora	Mustelidae	European otter *Lutra lutra*. Aquatic habit, paws are webbed in most species, catches food with forefeet or mouth, feeds mainly on crustaceans and fish, closes ears and nostrils under water.	Marsupialia	Didelphidae	Water opossum *Chironectes minimus*. Aquatic habit, hindfeed are webbed, swims with hindfeet, forefeet used to catch food, feeds mainly on crustaceans and fish, closes pouch under water.
Carnivora	Canidae	Common wolf *Canis lupus lupus*	Marsupialia	Thylacinidae	Tasmanian wolf *Thylacinus cynocephalus*. Wolf like in overall form and carnivore nutrition.
Carnivora	Felidae	Wild cat *Felis silvestris*	Marsupialia	Dasyuridae	Tasmanian cat *Sarcophilus harrisii*. Similar to a cat, has powerful jaws and sharp teeth, carnivore.

Table 1. *(continued)*

Placentals			Marsupials		
Order	*Family*	*Species*	*Order*	*Family*	*Species*
Primates	Tarsiidae	Spectral tarsier *Tarsius spectrum.* Arboreal habit, large eyes directed forward, flattened face and long tail.	Marsupialia	Didelphidae	Black-shouldered opossum *Caluromysiops irrupta.* Arboreal habit, protruding eyes directed forward, face like that of primates and long tail.
Primates	Cebidae	Spider monkey *Ateles paniscus.* Tail prehensile as an extra hand.	Marsupialia	Didelphidae	White-eared opossum *Didelphis albiventris*. Tail prehensile as extra hand.
Rodentia	Muridae	House mouse *Mus musculus*	Marsupialia	Dasyuridae	Marsupial mouse (mulgara) *Darcycercus cristicaudata.* Similar in body shape and habits. Many species in this family resemble placental mice.
Rodentia	Anomaluridae	Scaly-tailed squirrel *Anomalurus peli*. Has a membrane that extends between forefeet and hindfeet, glides well.	Marsupialia	Pseudocheiridae	Greater glider *Petauroides volans.* Has a membrane that extends between forearms and hindfeet, same body solutions and gliding efficiency as placental species.
Dermoptera	Cynoce-phalidae	Philippine colugo *Cynocephalus volans*. It is also called "flying lemur" because it glides well, has a membrane that extends between forefeet, hindfeet and end of tail.	Marsupialia	Petauridae	Sugar glider *Petaurus brevicepts.* Has a membrane between forefeet and hindfeet, same body solutions and gliding capacity as placental species.
Insectivora	Talpidae	Common mole *Talpa europaea.* Subterranean way of life, cylindrical body, feeds on insects and worms, muzzle is long and naked, eyes covered by skin, no external ears, short tail.	Marsupialia	Notoryctidae	Pouched mole *Notoryctes typhlops.* Subterranean way of life, cylindrical body, feeds on insects, hands used to excavate, horny shield on tip of snout, no eyes and no external ears, tail is a stub.

Table 1. *(continued)*

Placentals			*Marsupials*		
Order	*Family*	*Species*	*Order*	*Family*	*Species*
Insectivora	Soricidae	Common shrew *Sorex araneus*	Marsupialia	Dasyuridae	Red-tailed Phascogale *Phascogale calura.* Has long conical snout, insect eater, looks like the shrew in appearance and habits.
Edentata	Myrmeco-phagidae	Lesser anteater *Tamandua tetradactyla.* Digs with clawed fingers, eats ants and termites with long narrow tongue	Marsupialia	Myrmecobiidae	Numbat, banded anteater *Myrmecobius fasciatus.* Digs with clawed fingers, eats termites and ants with long narrow tongue.
Perissodactyla	Rhinocero-tidae	Rhinoceros *Diceros bicornis*	Marsupialia	Diprotodontidae	Marsupial Rhinoceros *Diprotodon* (fossil). A three meter long herbivore with the size, shape and habits of a placental rhinoceros.
Artiodactyla	Suidae	Wild boar *Sus scrofa*	Marsupialia	Peramelidae	Pig-footed bandicoot *Chaeropus ecaudatus.* Forefeet have only two digits that are functional, forming a paired foot pad, like that of artiodactyla.

membrane and gliding efficiency. These species belong to the order Rodentia but gliding animals also appeared in the order Dermoptera. These are the placental Philippine colugo and the marsupial sugar glider. The four species have the following features in common: (1) a large membrane of skin that connects the four limbs; (2) the membrane extends to the toes, fingers (or forearms), and whole or part of the tail; (3) the membrane is furred and attached to the neck and sides of the body; (4) the animals have the proper muscles that control gliding; and (5) there is coordination of movements directed by the nervous system.

Another paradigm is found in the mole (order Insectivora). The pouched mole (marsupial) has the same cylindrical body, subterranean way of life, insect feeding, and absence of external ears as the placental counterpart. The eyes are reduced or absent in both cases.

Other copies are found among the shrew, anteater, rhinoceros, and wild boar.

In all the species described the transformation is not the product of one or a few modifications, but consists of a package of a series of structures and functions that are coherently organized. Otherwise the result could not have been a nearly identical animal (Macdonald, 1984).

This is especially evident in the anteaters, which occur in the placentals, the marsupials, and the monotremes (a group of primitive egg-laying mammals). They are characterized by elongated snouts, small eyes, long tongue, reduced or absent dentition, large claws on the fore feet, and digestive enzymes that interact with the formic acid from the ants that are their regular source of food. Some of the species are: *Tachyglossus aculeatus* (monotreme), the placentals, *Myrmecophaga tridactyla, Manis gigantea, Tamandua tetradactyla*, and the marsupial *Myrmecobius fasciatus*. Some of the species eat both ants and termites.

Nearly all body structures and functions are found to have participated in the establishment of the equivalence between the placentals and the marsupials. The most evident structures are: body configuration, body size, skeletal traits, attachment of muscles, shape of cranium, shape of mouth, type of tongue, and type of teeth. The following functions participated: vision, feeding habits, type of diet, digestive activity, muscle strength, running capacity, and gliding power.

An unexpected parallel turns out when one compares the marsupial–placental periodicity with that of the chemical elements. In the main body of the periodic table there are two series of atoms that stand out because they represent two groups with special properties. These are the lanthanide and actinide series. The first starts with lanthanum (La) and comprises 14 elements. The second begins with actinium (Ac) and consists equally of 14 elements (see Figure 1). All lanthanides have properties in common, the reason why the series is given the same designation. And all actinides are characterized by several properties of their own (for instance, they are all radioactive). Of particular significance is that the two series are aligned by placing the 14 elements of one of them on the top of those of the other. This is possible due to the correspondence in properties, element by element, along the series. This results from the fact that actinium forms very similar compounds to those of lanthanum (Sharpe, 1988). This situation has a parallel at the biological level in the placental–marsupial equivalence. The species of placental mammals build a series mainly characterized by the presence of a placenta. The second series is built by the marsupials that have a pouch as their typical feature. The two series can be aligned and put on top of each other building two rows like those of the elements. This is possible because, not all, but many species display the same essential properties. Animals of the upper row have corresponding ones on the lower sequence (see Table 4, p. 313). The parallel may seem far-fetched, but this is because we are not accustomed to think in terms of periodicity, and are becoming confronted with a novel integration of the properties of living organisms with those of their constituent atoms.

The Emergence of High Mental Ability

One of the characteristics common to atomic and biological periodicity is that atoms and organisms may display a given property, but those which are closely related to them, may not show it.

Anyone who contemplates the behavior of a mosquito or of a beetle can hardly imagine that they should have a high form of mental ability. Even if one considers a bee or an ant, as isolated individuals, nothing in the animals tells us that they can produce one of the most complex societies known outside those created by humans. This is not only true of the insects but also of molluscs. Some of the species of this phylum are among the most inconspicuous invertebrates from a mental point of view. No one is expected to find this ability well developed in a shellfish, such as an oyster, which has minute eyes, a reduced head, and is a prisoner of its shell. However, other molluscs such as the cephalopods, which include the giant squids and the octopods, can recognize specific objects and be trained like dogs.

Octopuses can learn to distinguish between different geometrical figures. Just as in experiments with higher mammals, they were shown a figure at the same time that they got a reward (a crab to eat) or were exposed to an electric shock if they missed the target. They could distinguish between a cross and a square and between 27 forms intermediate between these two extremes. They could, as well, recognize the shapes that were closer to a given form (Figure 15). It could be ascertained that octopuses have short-term and long-term memory similar to that of higher vertebrates (Russell-Hunter, 1979; Wells, 1983).

Another behavior well established in mammals was discovered in *Octopus vulgaris*. The animals were found to be able to perform a task when they were allowed to observe other trained octopuses. When they had seen the behavior of the trained ones, they were able to choose the same object that was picked up by their previous companions. This learning occurred irrespective of the object chosen (Fiorito and Scotto, 1992). Giant nerve cells are present in the brain of cephalopods and these are known to form a complex system of brain connections (Bullock and Horridge, 1965; Katz and Miledi, 1966; Withers, 1992).

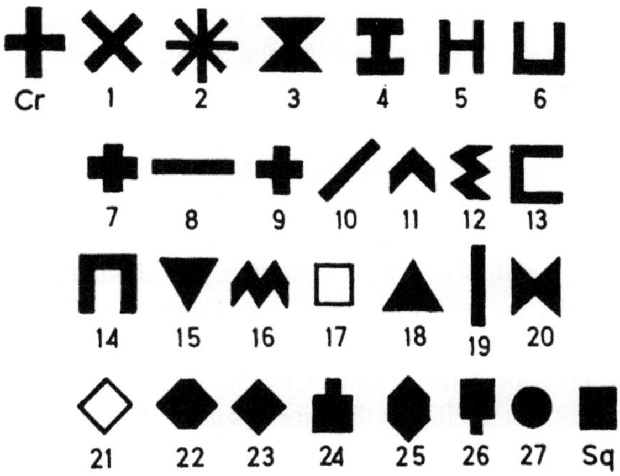

Figure 15. Octopuses can distinguish between the 29 shapes shown above by sight. They can be trained to distinguish between a cross (*above left*) and a square (*below right*) and 27 intermediate forms. They are given rewards (e.g., food) when they attack one of the symbols. The numbers indicate the order in which the octopuses ranked the shapes. This is directly related to the transition from cross to square (From Wells, 1983).

Bees, ants, and termites are social insects that belong to two different orders: Hymenoptera (wasps, bees, and ants) and Isoptera (termites). It is important to realize that there are within the same order other insect species that do not build societies, and are as nonsocial as most other insect orders. An example is the solitary species of wasps (Romoser, 1973). Moreover, social organization in insects has evolved independently eleven times in the Hymenoptera, but only once in all the other orders (the Isoptera). Insects number more than 800,000 species and are divided into 26 orders (Hinde, 1981).

Social behavior in insects is characterized by:

1. Highly organized societies.
2. Well-defined languages, which possess dialects.
3. The ability to decide the fate of their progeny by differential fertilization.
4. Determination of the social rank of a given individual, such as the "queen," by feeding specific chemicals to some of the members of the colony.
5. The development of agriculture consisting of the cultivation of fungi.
6. The raising of other species (such as aphids) and the use of their sugary products, just as humans raise cattle and collect their milk.
7. The capacity to move the animals under their control to plants where they can feed better and, as such, produce higher yields (Hermann, 1979).

Honeybees are said to be the only creatures other than man that communicate by a symbolic language (Dreller and Kirchner, 1993). Moreover, bees and ants, like birds, are known to use celestial navigation. They follow with different degrees of accuracy the movements of the sun and its relationship to the horizon. This type of navigation is a highly complex process since the rate of movement of the sun depends on time of day, time of year, and geographical latitude (Wehner and Müller, 1993).

Ants, wasps, bees, and termites are not directly related to mammals or humans. Moreover, they have a brain that is about one million times smaller than ours. The brain of an insect weighs between 1 milligram and 1 microgram, whereas the human brain weighs on the average 1400 g (Changeux, 1985). The brains of octopuses are also minimal in comparison to our brain. In mammals, the high mental ability is not necessarily displayed by closely related species. Dolphins are considered to be not less intelligent than gorillas or chimpanzees, yet these two last species are our closest relatives.

The periodicity in mental ability also does not show any direct relationship to the general environment. Insects are terrestrial and aerial animals; octopuses live in the ocean, sometimes at great depths; and mammals and humans live mainly on land.

Distribution of Vision in Invertebrates

Most vertebrates have well-developed and complex eyes. However, in invertebrates the situation is different. Eyeless species, rudimentary eyes, and highly complex eyes appear in different phyla.

Two features disclose biological periodicity. First, complex vision emerged during the long evolution of the invertebrates and disappeared not accidentally but according to a recurrent pattern. Second, within a given group there are species without eyes, and others with highly complex vision. The occurrence of the two extreme situations among closely related species also shows periodicity (Figures 16 and 17).

Light sensitivity is an integral part of the cell. Unicellular organisms such as *Amoeba* and *Paramecium* already react to light, although they have no primordial eyes. These appear in other species such as *Erythropsidinium* and turn out to be already relatively complex. These ocelli consist of a lens, a fluid-filled chamber, and a cup covered by light-sensitive pigment. Although this is a single-cell organism, the lens can change its shape, the pigment can move freely in relation to the lens, and the whole ocellus can be protruded from the protozoan body, allowing it to orient in different directions. The dinoflagellates use their primordial eyes to detect the presence of prey just as humans do. The protozoa and the mammals are situated at the two extremes of the evolutionary scale (Figure 16) (Barnes, 1980; Friday and Ingram, 1985).

True complex eyes are formed in the multicellular invertebrates. These consist, among other parts, of a cornea, a lens, an iris, a ciliary muscle, a retina, and optic nerve. This is, for instance, the case in the Cephalopods, such as *Octopus* (molluscs)

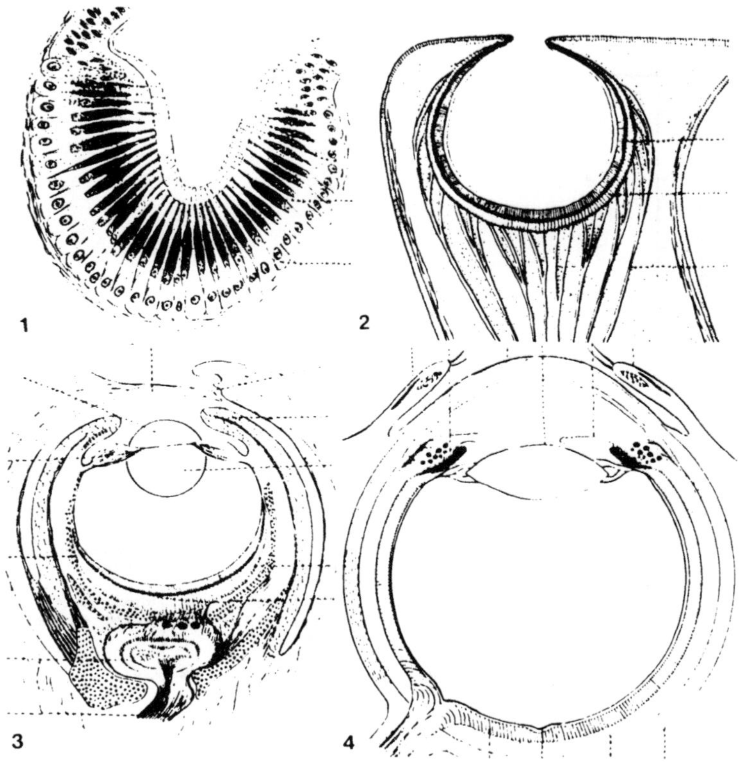

Figure 16. Periodicity of vision. (**1**) Eye of the mollusc *Patella* consisting of a simple depression in contact with the exterior and retina cells. (**2**) Eye of the mollusc *Nautilus* with a deeper depression, pigmented zone, retina, and nerves. (**3**) Eye of the mollusc *Sepia* (cephalopods) consisting of 11 differentiated parts: retina, pigment cells, optic ganglion, optic nerve, cartilaginous capsule, epithelial body, two juxtaposed crystallines, iris, cornea, and a small opening to the surface. (**4**) Human eye, formed by cornea, iris, crystalline lens, retina and optic nerve. *Source*: Rabaud, 1932.

Figure 17a. (*overleaf*) Period vision. (**A**) Protozoa to *Gnathostomulida*. Light sensitivity is inherent in the cell. This is why in the graphs the base-line, representing the degree of complexity, is not set at zero but at the value of 0.2. Simple eyes already appear in the protozoa. Before one reaches the vertebrates, which have complex eyes, there is a periodic eruption of eye complexity among the invertebrates. Moreover, the complex eyes appear within classes or phyla that also consist of animals without eyes. This is the case in the Cnidaria, in which the cubomedusans have complex vision, but the sea anemones are eyeless. This is also the case in the Molluscs, in which the octopods have eyes as well developed as those of vertebrates, yet there are many species within the phylum which have rudimentary eyes or no eyes at all. This situation repeats itself in the Annelida, the Crustacea, and the Insects. Most significant is that the Hemichordata and the early Chordata, such as the Ascidiaceae, which are evolutionarily more advanced than the invertebrates, have no eyes or very rudimentary vision. This is taken to mean that the periodicity is not directly related to the degree of evolutionary complexity. (**B**) Patella to Araneae.

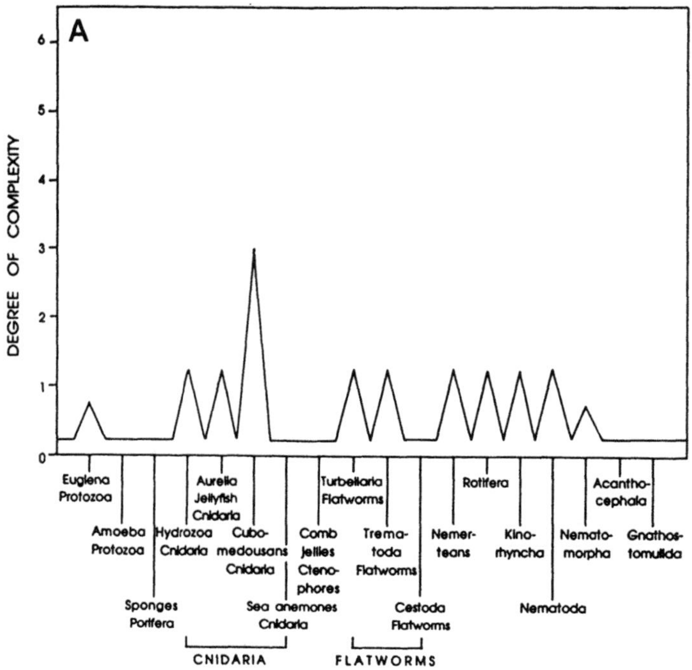
A
DEGREE OF COMPLEXITY
0
1
2
3
4
5
6
Euglena Protozoa
Amoeba Protozoa
Sponges Porifera
Hydrozoa Cnidaria
Aurelia Jellyfish Cnidaria
Cubo-medousans Cnidaria
Sea anemones Cnidaria
Comb jellies Cteno-phores
Turbellaria Flatworms
Trema-toda Flatworms
Cestoda Flatworms
Nemer-teans
Rotifera
Kino-rhyncha
Nematoda
Nemato-morpha
Acantho-cephala
Gnathos-tomulida
CNIDARIA
FLATWORMS

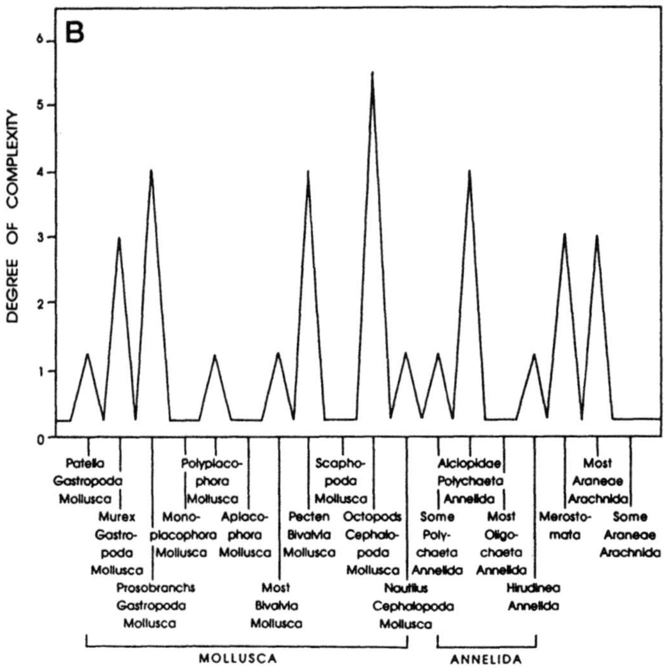
B
DEGREE OF COMPLEXITY
0
1
2
3
4
5
6
Patella Gastropoda Mollusca
Murex Gastro-poda Mollusca
Prosobranchs Gastropoda Mollusca
Mono-placophora Mollusca
Polyplaco-phora Mollusca
Aplaco-phora Mollusca
Most Bivalvia Mollusca
Pecten Bivalvia Mollusca
Scapho-poda Mollusca
Octopods Cephalo-poda Mollusca
Nautilus Cephalopoda Mollusca
Some Poly-chaeta Annelida
Alciopidae Polychaeta Annelida
Most Oligo-chaeta Annelida
Hirudinea Annelida
Merosto-mata
Most Araneae Arachnida
Some Araneae Arachnida
MOLLUSCA
ANNELIDA

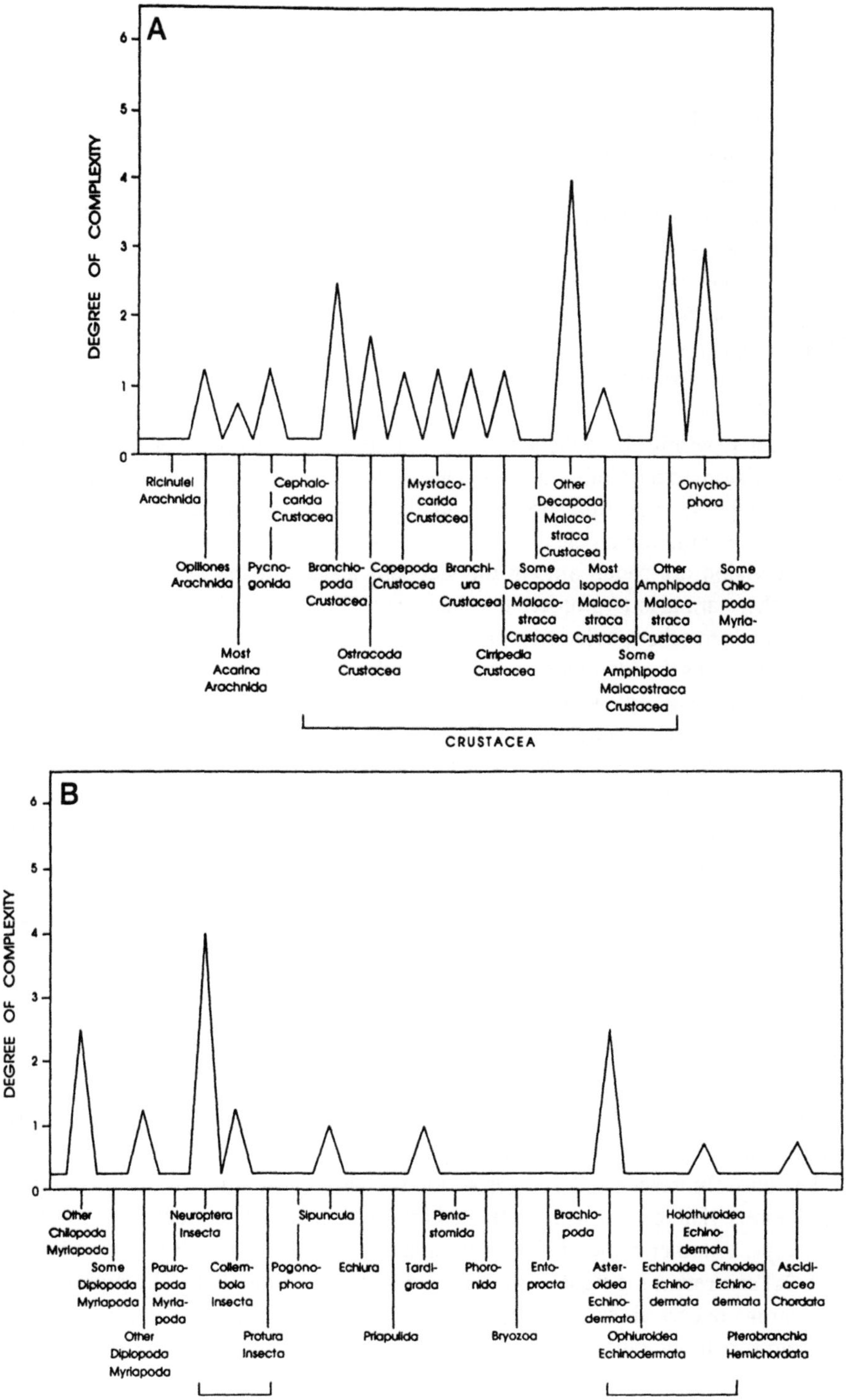

Figure 17b. Period vision. (**A**) Ricinulei to Chilopoda. (**B**) Other Chilopoda to Ascidiaceae.

which are capable of image formation and shape discrimination. The eye of the mollusc *Sepia* is formed by 11 differentiated parts: retina, pigment cells, optic ganglion, optic nerve, cartilaginous capsule, epithelial body, two juxtaposed crystallines, iris, cornea, and a small opening to the surface. This eye is considered to be as complex as the human eye.

The periodic eruption of eye complexity is in line with evidence of the sudden emergence of complex vision in: (1) cubomedusans (Cnidaria); (2) Octopods (Molluscs); (3) Alciopidae (Annelida); (4) lobsters (Crustacea); and (5) the compound eyes of, for instance, the dragonflies (Insects) (Russell-Hunter, 1979; Barnes, 1980).

Similarly, within these distant groups, there are species without eyes or rudimentary vision. In the Cnidaria, the sea anemones are eyeless. Molluscs such as the Monoplacophora, the Aplacophora and the Scaphopoda have no eyes. In the Annelida the situation repeats itself, since most Oligochaeta lack eyes. Crustacea without vision are found among the Cephalocarida and the Decapoda. And in insects there are also two orders that lack eyes (Romoser, 1973). The absence of eyes extends to other phyla of invertebrates.

As one leaves the invertebrates and enters the advanced chordates (to which the vertebrates belong) one would expect to find highly advanced eyes. However, this is not the case. The Hemichordata (Pterobranchia) and the Ascidiacea, which are true Chordates, consist of blind species. The organisms of the first class have no eyes and those of the second have only dispersed photoreceptor cells.

Anatomical studies indicate that the eye has had an independent origin in several animal phyla. An example is the third eye of some lizards that is situated on the forehead (Rabaud, 1932; Romer and Parsons, 1978). Another aspect of the periodicity of vision is that it cannot be easily correlated with the type of environment to which the group is exposed. This is the case of animals that live in shallow waters, yet have no eyes (Echiura, Bryozoa).

Bioluminescence

Bioluminescence is a clear-cut example of a biological function that was inherited directly from the minerals (Figure 18).

The term fluorescence derives from the property of the mineral fluorite (CaF_2) which emits visible light when activated by invisible radiation, such as ultraviolet or X-rays. Another mineral that is luminescent is calcite ($CaCO_3$). Also barite ($BaSO_4$) can retain the light of the sun and reemit it in the dark. The presence of extra atoms in crystals favors the phenomenon. These create structural boundaries that trap subatomic particles, the result being the emission of photons (Campbell, 1988).

At the biological level, several chemicals are required to establish the energy cascade that results in photon emission. They include:

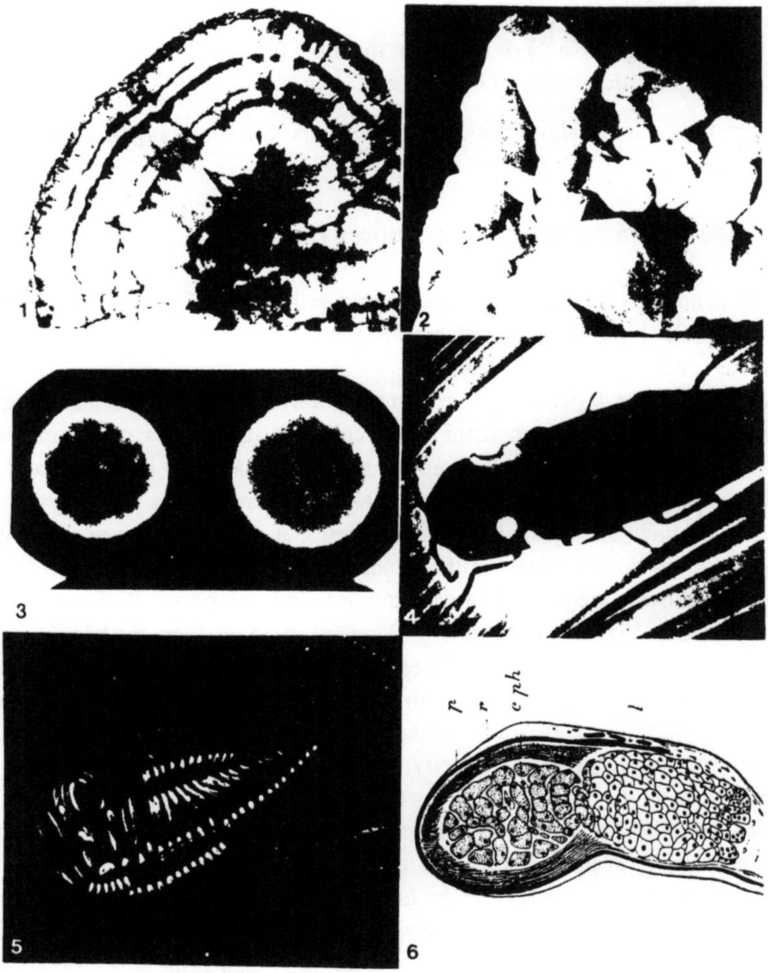

Figure 18. Periodicity of luminescence. The word fluorescence derives from the property of fluorite to emit visible light when activated by invisible radiation such as X-rays. (**1**) The mineral fluorite (CaF_2). (**2**) Group of calcite crystals, another mineral that also fluoresces. (**3**) Two colonies of *Photobacterium* photographed in obscurity. (**4**) The tropical insect *Pyrophorus noctilucus* which produces such a strong light that it can be used to read. (**5**) The abyssal fish *Argyropelecus* with the body covered by luminous organs. (**6**) The luminous organ of the fish *Argyropelecus affinis* showing the cells that produce light, the lens, and the reflecting structure. *Source*: (**1**) and (**2**) Cabrera, 1937; (**3**) Combes, 1938; (**4**) Francé, 1943; (**5**) Frisch, 1938; (**6**) Bohn, 1934b.

- oxygen (or one of its metabolites);
- luciferin (a substrate);
- luciferase (an enzyme);
- cofactors (such as aldehydes);
- cations (Mg^{2+}, Ca^{2+}) or transition metals (Fe^{2+}, Cu^{2+}); and,
- fluor (F) which is bound to a protein.

The bound fluor atoms show how bioluminescence derives its organization from a mineral construction.

A Periodic Function that can be Measured Physically

Light in animals is produced intracellularly in the form of a glandular secretion, or as the result of a symbiosis with luminous bacteria. In these microorganisms it is also the result of a chemical cascade.

The light organs have become accompanied by additional structures that enhance the effect. These are: melanin filters, collagen light guides, chitin reflectors, and lenses. This is a situation that repeats the solutions found in vision, which is the ability to sense light. In both cases, additional structures that improve the phenomenon have been added at different times. A common feature of complex organs involved in vision and luminescence is the presence of a lens.

Few phenomena are so widespread among living organisms. Bioluminescence is represented in 16 phyla and almost 700 genera. It occurs from the bacteria to the fishes being represented in many invertebrates. However, it could not establish itself in algae, flowering plants, amphibians, reptiles, birds, and mammals, as well as in large groups of invertebrates. This is what makes bioluminescence one of the best examples of biological periodicity.

The following information provides more compelling evidence for periodicity. Many bacteria, both marine and fresh water species, are strongly luminous, but the blue-green algae (cyanobacteria), that are also prokaryotes, did not develop this function. In the fungi many species are luminous. However, there are no eukaryotic algae or flowering plants that are luminescent. This situation repeats itself in the invertebrates. Insects, such as fireflies, and molluscs, such as the cephalopods, emit intense light, whereas the nematodes and the brachiopods lack this capacity. As one reaches the vertebrates one is faced with the same clear-cut separation. Not a single mammal, including humans, has been found to emit light, but fishes in deep oceans are usually covered by a series of luminous organs. It should be noted that whales that prey on deep sea fish and giant squids (which are also luminous and live at great depths), are not themselves luminescent.

Another typical feature associated with periodicity is the absence of this function in closely related species. There are many insects and fishes that are not luminous.

Bioluminescence can be measured physically. No other function can be measured with such accuracy. This represents a valuable asset in judging periodicity.

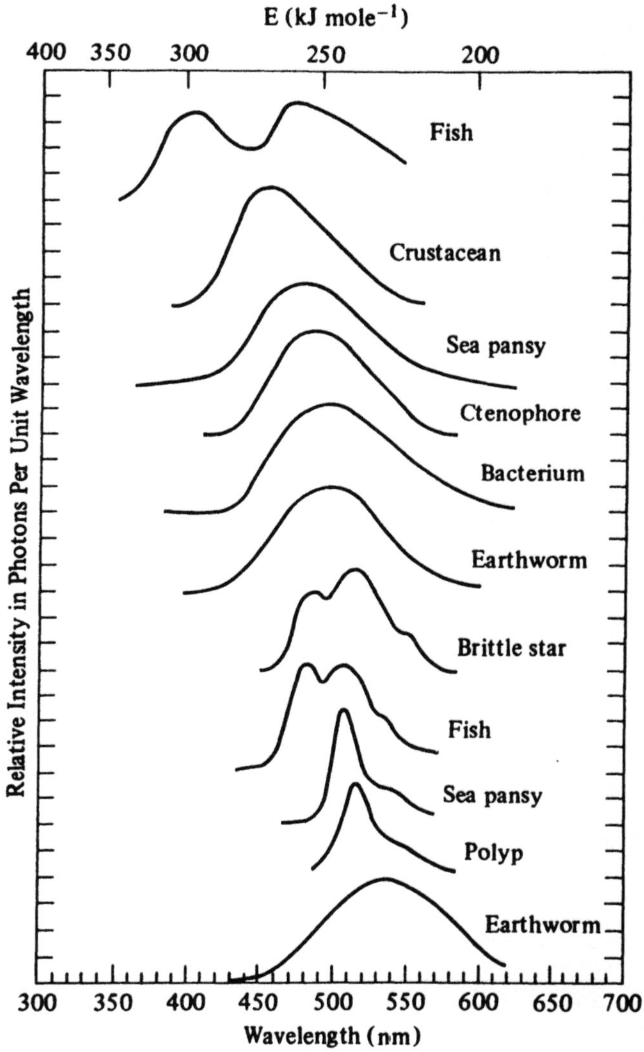

Figure 19. Wavelength (lambda; nm) and energy (kJ mole photons^{-1}) emitted by bioluminescent organisms. Searsid fish exudate; crustacean *Cypridina* reaction *in vitro*; sea pansy *Renilla* reaction *in vitro*; ctenophore *Mnemiopsin* emission *in vivo*; photobacterium emission *in vivo*; earthworm *Diplocardia* emission *in vivo*; brittle star emission *in vivo*; fish *Porichthys* emission *in vivo*; sea pansy *Renilla* emission *in vivo*; hydropolyp *Clytia* emission *in vivo*; earthworm *Diplotrema* emission *in vivo*. (From Wampler, 1978).

The relative intensity in photons per unit wavelength—the light energy emitted by the organism—as well as the wavelength can be recorded. It turns out that the intensity of emitted light and the wavelength vary with the species (Figure 19). A bacterium, a crustacean, and a fish have, however, not very different illumination powers (Wampler, 1978; Withers, 1992). This means that, as in other functions, the reemergence of a given property, and its degree of organization, are not necessarily related to the complexity of the organism.

The sudden recurrence of bioluminescence does not seem to be related to the general environment. Sharks, which are fishes, prey in the darkness of the abyss, yet they do not have luminous organs. Amphibians and reptiles that inhabit caves, where darkness is almost total, were not able to develop light organs.

The bioluminescence occurs at relatively regular intervals (Figure 20) and it may have arisen independently more than 30 times (Campbell, 1988) (Table 2).

Table 2. Period Bioluminescence: Presence and Absence of Bioluminescence

Presence	*Absence*
Photobacterium and other Species	Cyanobacteria (Blue-Green Algae)
Dinoflagellates (Protozoa)	Protozoa (many species)
Eumycota (Fungi)	Myoxomycota (Fungi)
Porifera (Sponges)	All Algae
Cnidaria (Jellyfishes)	Flowering Plants
Ctenophora	Platyhelmintha (Flatworms)
Nemerteans	Rotifera
	Gastrotricha (Aschelminths)
Mollusca —Bivalvia (Pholas)	Kinorhyncha (Pseudocoelomates)
—Pulmonata (Land Snails)	
—Gastropoda (Limpets)	Nematoda (Roundworms)
—Cephalopoda (Squids)	
Annelida —Oligochaeta	Acanthocephala
—Polychaeta	Entoprocta
Insects —Fire Flies	Priapulida
—Beetles	
Diplopoda (Millipedes)	Sipuncula
Crustacea —Ostracoda (Vargula)	Echiura
—Decapoda (Shrimp)	Tardigrada
Echinodermata—Asteroidea (Star Fish)	Pentastomida
—Holothuroidea (Sea Cucumbers)	Onychophora
Hemichordata	Phoronida
Chordata —Fishes	Gnathostomulida
	Loricifera
	Brachiopoda
	Chaetognatha
	Chordata —Amphibians
	—Reptiles
	—Birds
	—Mammals

Notes 1. Only main groups are represented.
2. "Presence" means that the group contains some or many luminous species.
3. "Absence" denotes that luminescence has not been recorded.
4. Based on data from Campbell, 1988.

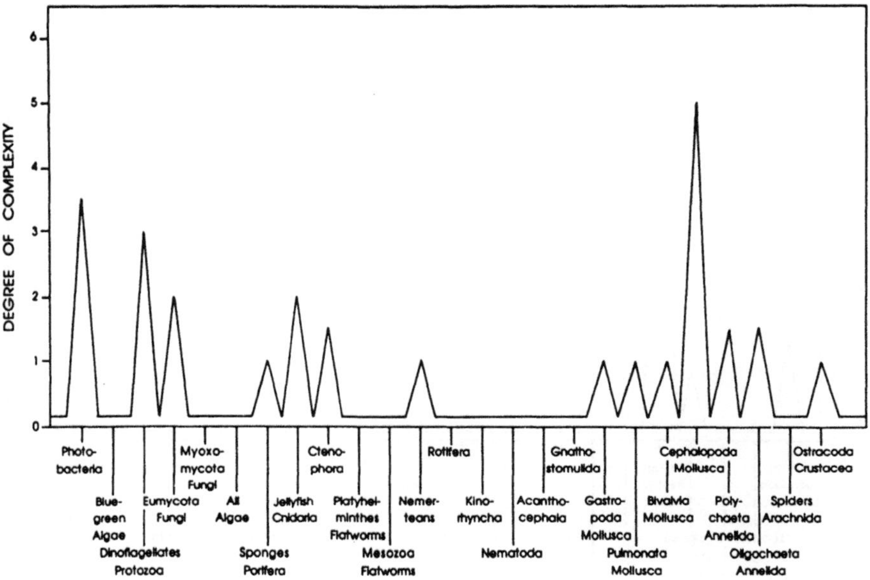

Figure 20a. Period luminescence. Photobacteria to Ostracoda. Very weak luminescence can be detected in most cells; for this reason the degree of complexity is represented by a baseline that is set at 0.1 in the graph. The ability to emit light emerges at the dawn of life with nearly the same intensity as is acquired much later in the vertebrates. The photobacteria are strongly luminescent. The periodicity of the phenomenon is evident in several ways: (1) The close relatives of the photobacteria, the cyanobacteria (blue-green algae) could as well be luminescent but are not; (2) as the eukaryotes evolve there are whole phyla that do not exhibit this phenomenon but it erupts at regular intervals occurring in the Dinoflagellates (Protozoa), the Cephalopods (Molluscs), the fireflies (Insects), and the fishes; and (3) after the fishes, luminescence is extinguished in the vertebrates. Even marine mammals such as the whales that inhabit the same dark depths of the ocean as the fishes and the cephalopods have not developed luminescence. Lack of luminescence is also found in all eukaryotic algae which live in deep sea waters. Hence, the emergence of this property is not necessarily correlated with a given environment.

Bioluminescence in Humans and Its Transfer to Other Organisms

As mentioned above, mammals including humans, have not developed light organs. However, human cells are able to produce an ultraweak chemiluminescence that is invisible to the naked eye, but that is detectable by photomultipliers. The human cells produce less than one photon per cell per second, whereas bacteria emit 1000 photons during the same period. The production of photons appears as a phenomenon inherent to the cell, which under certain genetic conditions attains canalized forms that result in high intensity.

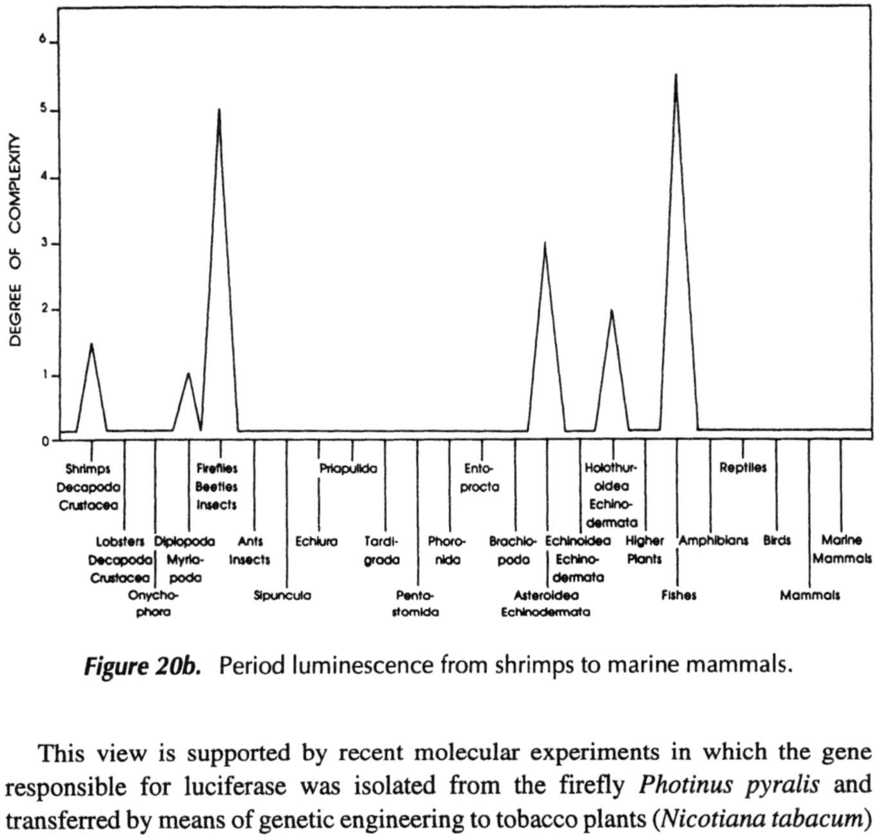

Figure 20b. Period luminescence from shrimps to marine mammals.

This view is supported by recent molecular experiments in which the gene responsible for luciferase was isolated from the firefly *Photinus pyralis* and transferred by means of genetic engineering to tobacco plants (*Nicotiana tabacum*) that, as all other flowering plants, lack luminescence. The leaves, stems, and roots of the transgenic plants became luminous (Ow et al., 1986; Millar et al., 1992).

Hence, it is possible to create luminescence in organisms that did not possess it beforehand. Such experiments open the way to the understanding of the molecular mechanisms that led to the establishment of periodicity.

The Capacity to Fly

Humans cannot fly, but other mammals are able to do so with high efficiency. They are the bats (order Chiroptera). It is significant for the understanding of biological periodicity that no other mammalian order comprises organisms that fly. This represents a rather unique situation since bats suddenly acquired a function that was not present in their terrestrial ancestors. Some mammalian species glide but that represents a different situation.

The rapid development of biotechnology, combined with an understanding of the molecular mechanisms responsible for periodicity will open the way for the understanding of new biological patterns. When these mechanisms become better understood, one may be in a position to find out what gene sequences, and what gene coordination are necessary to produce a wing, with its adjoining structures and functions in a terrestrial mammal. A key question which then arises is: Will it be possible to transfer the gene constellation responsible for flight in bats to the human genome? The result could be a human with "natural" wings. The transfer of genes among mammals, and the introduction of new genes into plants, animals, and humans is now a procedure that is being used (Ow et al., 1986; Hiatt et al., 1989; Millar et al., 1992).

The Punctuated Occurrence of Flight

There are 30 phyla of invertebrates (Margulis and Schwartz, 1982) but in only one of them, the Arthropoda, did flight emerge. Within this phylum only the class insecta developed this ability. No other invertebrate group had shown this phenomenon before the insects, and no other invertebrate group displayed it in the future, although the predecessors of the vertebrates have had at least 600 million years at their disposal to transform their patterns. This demonstrates the punctuated appearance of this function, a feature characteristic of biological periodicity.

Moreover, flight in insects (which originated about 310 million years ago) led from the outset to wings of large dimensions. The fossils of the giant dragonfly *Meganeura*, which inhabited the forests of the Carboniferous period (350 to 270 million years ago), had wings that measured 73 cm from tip to tip. This is the length of the wings of many birds, and no insect that lives today comes close to such a size. Hence, from the very beginning flight attained full capacity.

The second time flight appeared occurred much later. As the vertebrates emerged, fishes, amphibians, and most reptiles could not fly. Without any direct relationship to the flying insects, arose the flying reptiles: the pterosaurs (230 to 195 million years ago). The pterosaurs became extinct during the Cretaceous period leaving no survivors (Walker, 1974).

Birds emerged (180 to 135 million years ago) from flightless reptiles, such as the thecodonts (related to the dinosaurs) which were dominant during the Triassic period (Whitten and Brooks, 1988).

All of a sudden in the Eocene Age (only 40 million years ago) bats derived from flightless mammals appeared. No other mammals can fly: fossils having intermediate stages have thus far not been found. Those that exist are well preserved and indicate the presence of the stomach contents of the first bats (Colbert, 1980).

Another sudden event was the emergence of flight in fishes. Most fishes do not fly, but among the teleosts, which appeared 190 million years ago, a few species fly (Figures 21 and 22) (Hanstrom and Johnels, 1962; Beazley, 1980).

Five times in unrelated groups and during different periods, animals that did not fly gave rise to species that became aerial.

Figure 21. Periodicity of flight: structural similarity. (**1**) Insect *Chrysops discalis*. (**2**) Extinct flying reptile *Rhamphorhyngue*. (**3**) The bird, giant kingfisher, *Megaceryle maxima*. (**4**) Flying fish *Dactylopterus orientalis* (gurnard). (**5**) A bat (flying mammal). *Source*: (**1**) Brumpt, 1936; (**2**) Boule and Piveteau, 1935; (**3**) Beazley, 1974; (**4**) Freeman, 1972; (**5**) Perrins, 1976.

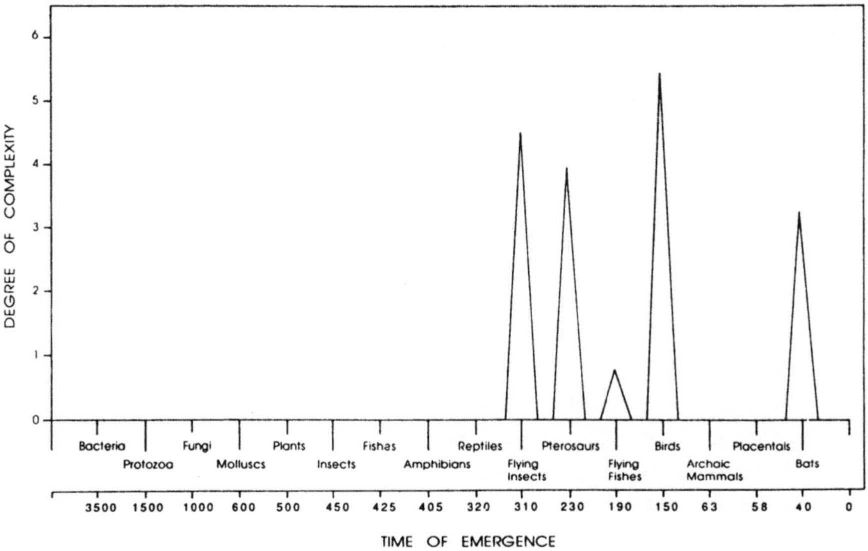

Figure 22. Period flight. Flight is a phenomenon that arises from nowhere. There are 30 phyla of invertebrates but in only one of them did the ability to fly develop. Moreover, it attained in insects such a capacity that it was seldom exceeded later in evolution. It appears again suddenly in the reptiles called pterosaurs (pteron = wing), in flying fishes, birds, and bats (which are mammals). None of these groups is directly related to the preceding group that was able to fly but have arisen from nonflying relatives. Flying fishes are teleosts; they developed after their emergence which was 190 million years ago.

Structures and Functions Necessary to Produce Flight

The wing is only one of the many structural and functional components that must be present for an animal to become aerial. Several bird and insect species have wings but do not fly. The most important components are:

1. The animal's skeleton which may consist of bones or other types of tissues.
2. Articulations necessary for the change of direction.
3. Powerful muscles that control the various movements.
4. Additional energy that is used during flight.
5. Great quantities of blood and air that are necessary during wing movements.
6. Control by the nervous system of direction, stability, and speed.

The Same Coherent Solutions Appeared in Different Groups

The separation in time is so great (millions of years) that it could be expected that different biological solutions would have led to the flying capacity. This is, however, not the case. If one compares insects, pterosaurs, birds, fishes, and bats, it turns out that the same basic solutions were used by them. This is regarded as another typical feature of biological periodicity. No accidental solutions were allowed; rather, the following rigid and coherent pattern is repeated:

1. A fusiform body is present in most groups.
2. The wings are located above the center of gravity on the dorsal surface both in invertebrates and vertebrates.
3. A tail is not absolutely necessary but its presence in most groups improves stability.
4. Air filled bones and other organs are present in several groups.
5. Antagonistic muscles and articulations are obligatory components.
6. Glycogen is the source of high energy in insects and birds.
7. The nervous system participates all the time in this activity, in which balance and vision are the main functions involved (McFarland, 1981).

The flight of fishes is not a passive gliding process, as is the case, for instance, in gliding squirrels, but involves strong muscular contractions that oblige the body and the wings to vibrate like those of insects. For the fish to be able to lift the body out of the water a strong force is necessary that is provided by the powerful muscles of the caudal fin, which is capable of 70 contractions per second. Fishes can change direction during flight such as circling around a boat and can stop in air, if necessary. They are flying in a medium which is adverse to their respiration. This is why they move in air for short distances and must return quickly to water in order to breathe regularly (Hanström and Johnels, 1962).

Flight is a Paradigm of Biological Periodicity

Few biological phenomena exhibit periodicity so well as the ability to become aerial. Some typical examples are:

1. Flight arises without immediate predecessors in five major groups.
2. It emerges with few intermediate forms or suddenly.
3. It tends to occur in a restricted class or order.
4. A coherent package of structures and functions arises in every case, which consists of essentially similar components.
5. The degree of complexity of flight is not directly related to the evolutionary position of the organism; for example, the flight of bats is not more complex than that of insects.
6. The number of wings per animal may range from two in birds to four in insects and fishes.

7. The general environment does not seem to have been mainly responsible for its emergence. For example, fishes become aerial in an adverse medium, and bats had no obvious need for flying.
8. Its occurrence seems to have taken place at relatively regular intervals, irrespective of the evolutionary situation in which the organism happened to find itself (Figures 21 and 22).

MOLECULAR MECHANISMS INVOLVED IN BIOLOGICAL PERIODICITY

The question that is commonly asked when a new phenomenon is described is: What is the mechanism responsible for its occurrence? A ready answer is often unavailable because it is only when the phenomenon is discovered that a search can start. This is particularly true of the periodicity of the chemical elements. Its mechanism took half a century to become established.

The processes responsible for biological periodicity have yet to be elucidated. However, since molecular biology and molecular cytogenetics have made great progress in the last two decades, some data are available that shed light on the molecular pathways that may be involved in the production of biological periodicity.

The most characteristic features of biological periodicity are threefold. First, the punctuated emergence of the same form and function at given intervals. Second, these structures and functions occur as packages that are coherently organized. Third is the correlation of their appearance with internal and external molecular signals.

We will now describe the molecular evidence that elucidates these three situations.

Chemical Properties of DNA Lead to the Formation of New Mosaic Proteins and the Reappearance of Old Ones

Many structural genes in eukaryotes are interrupted structures consisting of intervening DNA sequences termed introns, and of other sequences coding for proteins, termed exons. In most structural genes the exons alternate regularly with the introns along the whole length of the gene. Both types of sequences are transcribed into RNA, but only the exons become part of the messenger RNA that is translated into the polypeptide sequence that builds the protein. Since RNA of the introns is cleaved and deleted, it does not participate in the messenger RNA.

The number of introns may be as high as 25 and their length may extend from 50 to 50,000 base pairs. They are usually larger than the exons. An example is the human gene that encodes for the blood-clotting factor VIII; it spans 186,000 base pairs while the coding sequence for its protein is only 9000 base pairs (Gitschier et al., 1984).

Gilbert (1978) and Blake (1978) proposed that the change in number and position of the introns and exons along the gene could play an important role in evolution. That is, it would lead to the creation of new genes and the recreation of old ones. Examples of these situations in plants, invertebrates, and vertebrates are now available.

Intron-exon recombination has led to the formation of new mosaic proteins (Doolittle et al., 1986). They have been given this designation because they are considered to be made up of several exons that were originally located in other DNA sequences. Two examples are low-density lipoprotein (LDL)-receptor, a cholesterol transport protein which performs multiple functions (Südhof et al., 1985a,b), and the human tissue-type plasminogen activator (Ny et al., 1984).

There are three other mechanisms responsible for protein diversity which are now well established: (1) DNA duplication, (2) gene rearrangements, and (3) alternative RNA splicing (Smith et al., 1989; Benne and Van der Spek, 1992). Milk lactalbumin is derived from lysozyme by the first process (Doolittle et al., 1986). Protein diversity arises from separated genes that become united in the somatic cells, as is the case in human and mouse immunoglobulin genes (Hozumi and Tonegawa, 1976; Rabbitts, 1978). The most interesting mechanism is the third one, because alternative RNA splicing does not change the DNA sequences of the cell. It only leads to modification of the messenger RNA that is finally produced, the result being the formation of different proteins. The myosin heavy-chain gene of *Drosophila*, the GTP-binding protein Gs alpha gene, and sarcomeric contractile protein genes, are cases in which exons are known to be alternatively spliced. This phenomenon, moreover, is reversible (Breitbart et al., 1987; Smith et al., 1989; Van der Ploeg, 1990) (Figure 23).

The chemical properties of DNA and RNA offer at least four different mechanisms that lead in an obligatory manner to the formation of compound proteins, with new or old features, as a result of the reversibility of these processes.

Internally Directed Cascades of Molecules: The Source of Coherent Packages of Structures and Functions

Most macromolecules present in the cell have an *address*. This is achieved by specific parts of the molecule that are only recognized by other molecular edifices. Many proteins have short segments of amino acids which function as targeting signals. Interestingly, the Golgi complex also participates in the implementation of order in the interaction of molecules within the cell (Pugsley, 1989).

This order extends to other molecular pathways that display even more rigid canalizations. These are the so-called molecular cascades that have been found to be the key components of internally organized functional and structural processes (see Figure 6).

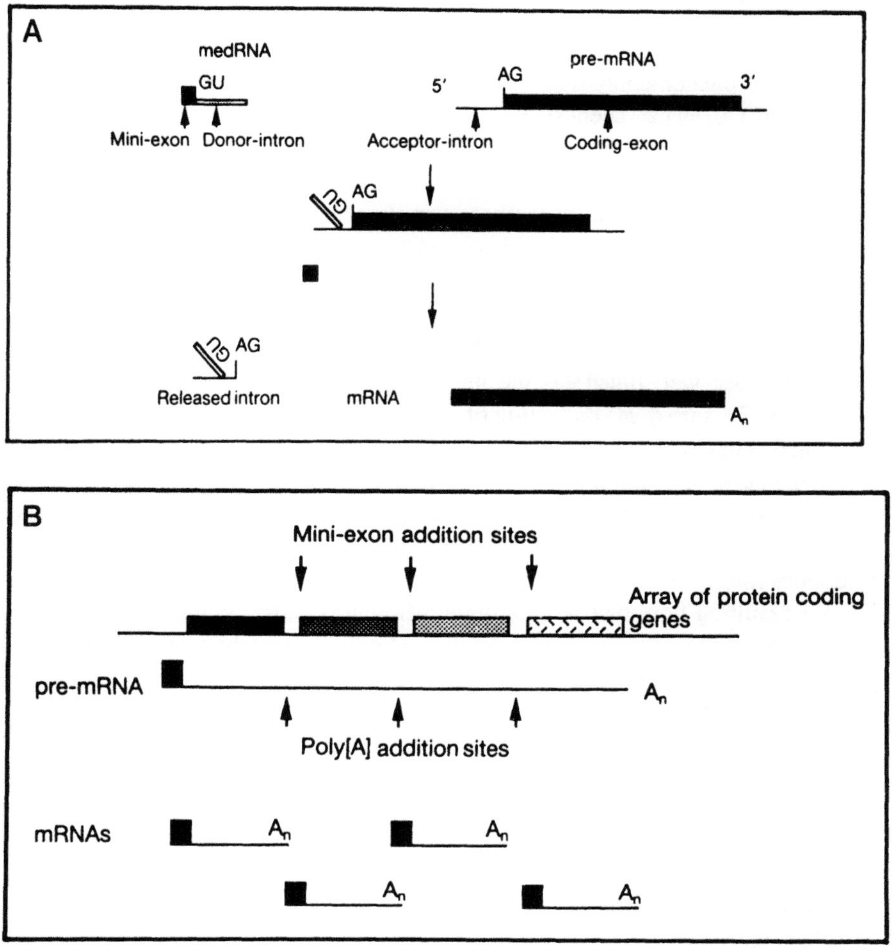

Figure 23. (**A**) Representation of the process of trans-splicing in the genome of the trypanosome *Trypanosoma brucei* (and related species) occurring in the genes that code for the variant cell surface glycoproteins (VSG genes). Trans-splicing leads to the joining together of different RNAs into a single messenger RNA. Two separately transcribed RNA precursors unite to form the final messenger (mRNA). The first precursor is transcribed from a mini-exon which is coupled to an intron that functions as a donor (med RNA). The second precursor derives from a coding-exon coupled to an acceptor–intron and is a pre-messenger RNA (pre-mRNA). The splicing takes place at the terminal nucleotides (GU and AG) and is accompanied by the release of the introns. The final product is a compound messenger RNA (mRNA) that is used in the translation of a new protein. (**B**) Trans-splicing occurring with the help of a donor–intron which adds mini-exon sites to an array of protein coding genes. This generates multiple mature messenger RNAs (mRNAs) from a single pre-messenger RNA (pre-mRNA). Each final messenger is a compound structure which contains, in addition to one of the segments of the pre-messenger RNA, the sequence from the mini-exon (From Van der Ploeg, 1990).

The stimulation of liver cells by the hormones glucagon and epinephrine results in a cascade phenomenon. The first step is the transformation of adenosine 5′-triphosphate (ATP) into adenosine 5′-monophosphate (cyclic AMP) by an enzyme. This reaction stimulates the activation of two other enzymes that follow in succession, the final result being the conversion of glycogen into glucose.

Blood clotting in humans is another example which involves a cascade of more than 10 proteins. There is an intrinsic and an extrinsic pathway. Both are canalized by proenzymes and activator proteins (Factors III, VII, IX, X, XI, and XII). The sequence is determined by intrinsic molecular events that were created by the evolution of the proteins themselves (Ryden and Hunt, 1993).

A third case well illustrates the internal coherence that becomes evident in periodicity. This involves the action of a single molecule which results in radical functional changes in a series of different body organs. The anterior pituitary is an endocrine gland situated in the brain. It produces a polypeptide called melanocorticotropin. Following three successive cleavages, the polypeptide is carved into six hormones. These are amino acid sequences of various lengths (the smallest being 5 peptides long). The cleavage occurs at specific sites along the mother molecule. The six resulting hormones act on quite different organs: (1) they stimulate the adrenal cortex adjacent to the kidneys; (2) act on lipid metabolism; (3) affect the pigmented cells of the skin; and (4) influence brain function in a way similar to morphine (Figure 24).

This is a situation similar to that caused by thyroxine in the frog embryo. This hormone leads to the production of a fully integrated adult that is totally different from the larva. The larva has no lungs or bones. After an increase in thyroxine, the heart, digestive apparatus, and muscles are radically changed and new organs are formed, such as the lungs and a vertebral column with accompanying bony members (Alberts et al., 1983).

What characterizes the molecular cascade are the following processes:

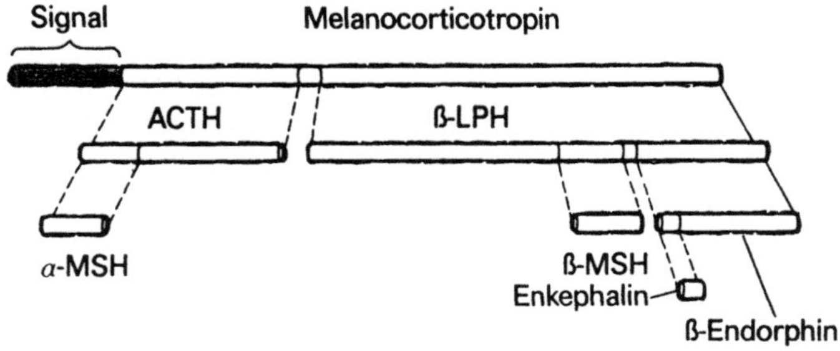

Figure 24. One polypeptide gives rise to six hormones that activate different organs. The melanocorticotropic hormone is carved into adrenocorticotrophin (ACTH) and beta-LPH (lipotropin). A second cleavage occurs releasing alpha- and beta-MSH (melanocyte stimulating hormones) and beta endorphin. The third cleavage produces enkephalin. (From de Duve, 1984b).

1. Many steps involving enzyme catalysis.
2. Every molecule is part of a sequence that cannot be altered.
3. Once the cascade starts, different cells or organs are affected by the various products.
4. There often is an amplification of the original signal. For every glucagon molecule bound to a receptor in a cell, 100 million molecules of glucose are produced.

Molecular cascades provide insight into the mechanisms that produce integrated transformations affecting simultaneously a series of organs. These are considered as one of the main features of biological periodicity.

The Role Played by External Signals

The changes caused by the chemical properties of DNA and by the cascades of molecules represent internal factors. But there are also external factors that may cause the phenomenon of periodicity. One such chemical agent is water. The role of this chemical in the air–water transformation, described earlier, reveals the following:

1. No change in genetic constitution is necessary to produce a hydrodynamic form in plants, since within the same individual the part that is immersed in water shows streamlined leaves, whereas that which is in contact with air retains the aerial leaf shape.
2. The agent of the hydrodynamic form is water, as demonstrated by experiments in which the crown of the plant was immersed or was placed outside water. Only when the crown was under water did streamlined leaves appear.
3. In mammals no change in genetic constitution seemed necessary when immersed for long periods of time, since the hormonal processes that allow prolonged asphyxia already existed in mammals as a normal device that allows a prolonged birth without serious brain damage.
4. Invertebrates, such as snails, and vertebrates, such as whales, developed similar physiological solutions when they returned to water. These are taken as an indication that similar genes and similar external factors (water) have been the main decisive factors in the development of periodicity.

However, the relationship with the environment is not a simple one and only future experiments will elucidate this situation. Take, for example, vision. Some animals that inhabit caves are usually blind or have small eyes. Remarkably there are whole phyla among the invertebrates that are exposed to strong sunlight—example is coastal water species—yet these invertebrates are blind or have only rudimentary eyes.

Flight in fishes occurs in an external environment that is fully adverse to their respiration capacity and, as a consequence, obliges them to return to water within minutes. Yet the flight shows the characteristics found in other groups. Clearly, one needs to await more molecular information before this relationship is clarified.

ATOMIC PERIODICITY COULD NOT BE ERASED: BIOLOGICAL PERIODICITY MANIFESTS MANY FEATURES OF CHEMICAL PERIODICITY

There is an enormous gap in evolutionary time and level of organization between simple chemical elements and complex living organisms. Most chemical elements were formed at the dawn of the Universe, that is, about 15 billion years ago (Von Baeyer, 1986). As such, one should not expect to find a relationship between the periodicity present at the chemical level and that found at the biological one. Such a long evolution could have erased any trace of similarity. This is, surprisingly, not the case. The reason may be simple. We are dealing with exactly the same atoms at both levels. These have not acquired different properties as they became assembled into macromolecules, cells, and organisms. An atom of nitrogen or of cobalt maintains the same electronic configuration and has the same properties, whether it is an isolated structure or part of an enzyme, or a cell organelle. Nothing has been modified in the atoms in the course of evolution of living organisms, and this is how periodicity perpetuated.

The relationship between the two levels is characterized by the preservation of 10 similarities (see Figure 1 and Table 3).

1. There is a recurrence of properties at both levels.

2. This recurrence occurs at relatively regular intervals in atoms and organisms.

3. Mimicry is displayed by atoms with different electronic configurations, as well as by macromolecules with different chemical compositions, and by organisms with different gene constellations. The peripherally located electrons, specific atoms, and particular DNA sequences appear to be the key components of the similarity of forms and functions at the three stages of organization.

4. Elements located nearby in the periodic table have different properties. Closely related species, within the same class or order may also have quite different properties. This is the case, for instance, in vision, in which the most complex eyes and the simplest ones appear in closely related species.

5. Properties which are common to chemical elements are not isolated events, but occur in packages which may include as many as 10 separate ones. At the biological level, periodicity also involves a package composed of a series of properties.

Table 3. Chart of Functional Periodicity

	Functions						
	Bioluminescence	*Advanced Vision in Invertebrates*	*Placenta*	*Penis*	*Return to Aquatic Life*	*Flight*	*High Mental Ability*
	Photobacteria	Cubomedusae (Cnidaria)	Onychophorans	Platyhelminthes (Flatworms)	Fresh Water Gastropods	Insects	Insects
	Dinoflagellates (Protozoa)	Prosobranchia (Mollusca)	Sharks[3] (Fishes)	Gastrotrichs	Fresh Water Insects	Pterosaurs (Reptiles)	Cephalopoda (Mollusca)
	Cephalopoda (Mollusca)	Pecten Bivalvia (Mollusca)	Amphibians	Acanthocephala	Stereospondyls (Aquatic Amphibians)	Teleosts (Fishes)	Mammals
General Increase in Organism Complexity	Fireflies Beetles (Insects)	Cephalopoda (Mollusca)	Reptiles	Gnathostomulida	Ichthyosaurs (Aquatic Reptiles)	Birds	
	Asteroidea (Echinodermata)	Alciopidae (Annelida)	Teleosts (Fishes)	Gastropoda (Mollusca)	Plesiousaurs (Aquatic Reptiles)	Bats (Mammals)	
	Fishes[1]	Decapoda[2] (Crustacea)	Flowering Plants	Cirripedia (Crustacea)	Sea Cows (Aquatic Mammals)		
		Amphipoda[2] (Crustacea)	Placental Mammals	Odonata (Insects)	Whales (Aquatic Mammals)		
		Neuroptera (Insects[2])		Marsupials	Seals (Mainly Aquatic Mammals)		
				Placentals	Hippotamuses (Mainly Aquatic Mammals)		

Notes: [1]The distribution of bioluminescence is found in Table 2. Here are only included the groups in which it is most prominent.
[2]Only some Decapoda and some Amphipoda have advanced eyes. Neuroptera are only mentioned as an example, other insect orders have advanced eyes.
[3]Some functions, such as the placenta, are only present in some species of a group. Detailed information is found in the text.

6. It is significant that chemists were forced to locate two series of chemicals outside the periodic table. These are the lanthanide and actinide series, composed of 14 elements each. This special location resulted from the following:

- Each element within the lanthanides has the same basic properties (this is why they retain the same name).
- Each element within the actinides has also the same basic properties, as the others within the group.
- Moreover, there is a full correspondence, element by element, between the 14 elements that constitute the two series. This is why they are placed in two rows, one over the other.

A comparable situation of two separate series occurs with respect to the marsupial and placental mammals (Table 4). Each species within the marsupials is usually characterized by having a pouch, as well as other features, and each placental species possesses a placenta and has other properties in common. Moreover, the species of the marsupials and of the mammals can be aligned in two rows, one over the other, like the lanthanides and the actinides. This is possible because there is a correspondence between many of the species of both series. Flying squirrels, mice, moles, sabretooth cats, and others have full correspondence in forms and functions among the placentals and marsupials.

7. The construction of a periodic table was initially a difficult task for chemists. There were too many exceptions and too many empty spaces in the table that could not be easily explained. This led to doubt about the existence of periodicity and order among the chemical elements. The intervals connected with the periodicity seemed too irregular and difficult to define. Fifty years had to elapse before additional knowledge led to the filling of the empty spaces and the sorting out of other irregularities. At present, we are facing the same situation with respect to biological periodicity. There are empty spaces that are not easy to explain similar to what was experienced at the chemical level. It took a long time for the chemists to find most of the solutions, but it may well take a shorter time for the biologists in view of the information explosion in molecular biology.

8. The periodicity among the chemical elements is evident for simple, as well as complex properties. The same is true of living organisms. Such a simple phenomenon as luminescence (which already occurs in minerals), and complex mental activity display periodicity.

9. Hierarchy is present at both levels. Not all electrons that circle in orbits around the nucleus of an atom are equally important in determining its properties. The same is true of living organisms. The atoms present in nucleic acids and proteins have a preponderant role in determining their properties as compared, for instance, with those that constitute polysaccharides.

10. Complexity is not a prerequisite for periodicity to emerge at both levels. Very simple and very complex atoms may show the same type of properties. Simple

Table 4. Equivalence between Placentals and Marsupials[1]

Placental Series													
Sabre-tooth	Lion	Hyena	Otter	Wolf	Cat	Mouse	Gliding Squirrel	Gliding Colugo	Mole	Shrew	Ant-eater	Rhino-ceros	Wild-boar
Smil-odon	*Pant-hera*	*Cro-cuta*	*Lutra*	*Canis*	*Felis*	*Mus*	*Anoma-lurus*	*Cynoce-phalus*	*Talpa*	*Sorex*	*Taman-dua*	*Dice-ros*	*Sus*
Marsupial Series													
Sabre-tooth	Lion	Hyena	Water Opossum	Wolf	Cat	Mouse	Greater Glider	Sugar Glider	Mole	Phasco-gale	Ant-eater	Rhino-ceros	Pig-footed Bandi-coot
Thyla-cosmy-lus	*Thyla-coleo*	*Borhy-aena*	*Chiro-nectes*	*Thyla-cinus*	*Sarco-philus*	*Darcy-cercus*	*Petau-roides*	*Petau-rus*	*Notor-yctes*	*Phasco-gale*	*Myrme-cobius*	*Dipro-todon*	*Chaer-opus*

Note: [1]The equivalence between the placentals and the marsupials is displayed as two parallel series, because their parallelism is compared to that of the chemical elements that build the lanthanum series and the actinium series (Figure 1). Both at the chemical and the biological levels the components within each series have common characteristics, yet at the same time, they show properties common to both series that have an exact correspondence.

invertebrates with little DNA, and higher vertebrates with excessive amounts of DNA show the same package of functions and structures that manifest periodicity.

The conclusion to be drawn is that no fundamental characteristics could be erased. At the two extremes of the evolutionary process, the simple atoms and the most complex mammals were obliged to follow similar pathways for the simple reason that the constitutive atoms have not changed with time. The billions of years that have elapsed since the elementary particles assembled into atoms, were not enough to change their original construction. We all retain this feature of the original assembly. Canalization was framed by the pattern of the atoms, but at the same time this has allowed the formation of a plethora of diversified living organisms. Rigidity and plasticity are the two antithetical components of inert and living matter.

RULE OF PERIODICITY

From the data that have been presented it becomes possible to formulate a rule of biological periodicity: *A coherent package of forms and functions arises in an organism independently of the main properties of its immediate predecessors, recurs at relatively regular intervals irrespective of its phylogenetic position, and appears without necessarily being related to the general environment.*

This statement demands clarification.

The properties of the immediate predecessors do not need to be directly correlated with the emergence of periodicity, yet they must possess particular genetic constellations that favor its emergence.

The phylogenetic position is not a limiting factor in the emergence of a given structure or function. A protozoan does not have the same possibilities as a multicellular organism. However, as one reaches the invertebrates, this limitation seems to cease.

The internal and external environments may not by themselves decide the emergence of the periodical properties. However, they may contribute to trigger them, once they act on a genetic constitution that may respond to their intervention.

A detailed treatment of biological periodicity is to be found in Lima-de-Faria (1995).

SUMMARY

When chemical elements were aligned after their increasing atomic weight they turned out to exhibit properties that recurred at relatively regular intervals. It was this unexpected periodicity that led to the establishment of the periodic chart of the elements. Later, as more information became available atomic weights were substituted by atomic numbers. This is especially evident in: helium (He), neon

(Ne), argon (Ar), krypton (Kr), xenon (Xe) and radon (Rn) that have the atomic numbers: 2, 10, 18, 36, 54, and 86, respectively. Yet, they all have the same properties; that is, they are inert gases. The length of the period that marks the repetition of a certain property may vary from 2 to 32 chemical elements. Chemical periodicity has turned out to be a most fruitful finding that allows us to explain many of the phenomena encountered today in chemistry.

Biological periodicity is characterized by an equally unexpected recurrence of groups of properties that emerge at relatively regular intervals, as the biological complexity increases throughout the evolutionary scale. An example is flight. This function originates in insects; does not show up in any other invertebrate group; reappears in the fossil reptiles; later becomes well established in birds; it occurs in certain fishes; and finally, reappears suddenly in the bats, but not in any other mammalian order.

In this survey attention is drawn to the periodicity of functions that are relevant to the higher mammals and humans. These are: (1) the activity of the placenta; (2) the function of the penis; (3) the features associated with the return to aquatic life in terrestrial mammals; (4) the occurrence among the placental mammals of structural and physiological "carbon copies" of the marsupial species; (5) the emergence of high mental ability among not closely related organisms; (6) the formation of complex vision organs in invertebrates; (7) the presence of bioluminescence from bacteria to high eukaryotes; and (8) the capacity for flight.

In all these cases there are well-defined groups of species that display a function that is usually not shown by their closest relatives. However, the same function had appeared long ago in unrelated organisms, and had emerged several times at given intervals.

The molecular mechanism behind biological periodicity is to be searched for in the information available from molecular cytogenetics. DNA segments can rearrange themselves inside split genes, and reverse the order of introns and exons. In this way they allow the reemergence of ancient proteins. Compound RNA messengers can be formed that produce new proteins. Moreover, proteins with different amino acid sequences are able to exhibit the same function, showing that different DNA sequences may lead to the production of similar physiological solutions.

Biological periodicity permits us to understand the evolution of higher mammals, and especially of the human species, from a rather novel point of view.

REFERENCES

Alberts, B., Bray, D., Lewis, J., Raff, M., Roberts, K., & Watson, J. D. (1983). Molecular Biology of the Cell. Garland Publishing, New York.

Arnason, U. (1985). Valen–mera hjort än lejon? Forskning och Framsteg 6, 16–21.

Arnason, U. & Widegren, B. (1986). Pinniped phylogeny enlightened by molecular hybridizations using highly repetitive DNA. Mol. Biol. Evol. 3(4), 356–365.

Barnes, R.D. (1980). Invertebrate Zoology (4th ed.), pp. 1–1089. Saunders College, Philadelphia.

Beazley, M. (Ed.) (1974). The World Atlas of Birds. Mitchell Beazley Publishers, London.

Beazley, M. (1980). The Atlas of World Wildlife. Rand McNally, printed in the Netherlands.

Benne, R. & Van der Spek, H. (1992). L'Editing des messages génétiques. La Recherche - 245 Juillet-Août 23, 846–854.

Blake, C. C. F. (1978). Do genes-in-pieces imply proteins-in-pieces? Nature 273, 267.

Bloss, F. D. (1971). Crystallography and Crystal Chemistry: An Introduction, pp. 1–545. Holt, Rinehart and Winston, New York.

Bohn, G. (1934b). Leçons de Zoologie et Biologie Générale. VI. Vertébrés Inférieurs (Poissons, Batraciens, Reptiles). Actualités Scientifiques et Industrielles 183. Hermann et Cie, Éditeurs, Paris.

Boule, M. & Piveteau, J. (1935). Les Fossiles. Éléments de Paléontologie. Masson & Cie, Éditeurs, Paris.

Branden, C. & Tooze, J. (1991). Introduction to Protein Structure, pp. 1–302. Garland Publishing, New York and London.

Breitbart, R. E., Andreadis, A., & Nadal-Ginard, B. (1987). Alternative splicing: A ubiquitous mechanism for the generation of multiple protein isoforms from single genes. Annu. Rev. Biochem. 56, 467–495.

Brumpt, E. (1936). Précis de Parasitologie. II. Collection de Précis Médicaux, Masson & Cie.

Bullock, T. H. & Horridge, G. A. (1965). Structure and Function in the Nervous Systems of Invertebrates, Vols. 1 and 2. Freeman and Company, San Francisco.

Burton, R. (1987). Egg. Nature's Miracle of Packaging, pp. 1–158. William Collins, London.

Cabrera, A. (1937). Historia Natural, Geologia, Vol. 4, pp. 620. Instituto Gallach, Barcelona.

Calvin, M. (1983). The path of carbon: From stratosphere to cell. In: Advances in Gene Technology: Molecular Genetics of Plants and Animals (Downey, K., Voellmy, R. W., Ahmad, F., & Schultz, J., Eds.), Vol. 20, pp. 1–35. Winter Symposia. Academic Press, New York.

Campbell, A. K. (1988). Chemiluminescence. Principles and Applications in Biology and Medicine. Ellis Horwood, VCH.

Carroll, R. L. (1987). Vertebrate Paleontology and Evolution, pp. 1–698. W. H. Freeman, New York.

Challis, J. R. G. (1979). Prostaglandins. In: Mechanisms of Hormone Action, Reproduction in Mammals (Austin, C. R. & Short, R. V., Eds.), Book 7, pp. 81–116. Cambridge University Press, Cambridge.

Changeux, J.-P. (1985). Neuronal Man. The Biology of Mind, pp. 1–348. Oxford University Press, Oxford.

Colbert, E. H. (1980). Evolution of the Vertebrates. A History of the Backboned Animals through Time. (3rd ed.), pp. 1–510. A Wiley-Interscience Publication, John Wiley & Sons, New York.

Combes, R. (1938). La biologie végétale. In: La Biologie, pp. 9–22. Palais de la Découverte, Masson et Cie, Editeurs, Paris.

Curry-Lindahl, K. & Tinggaard, K. A. (1965). Djuren i Färg. Däggdjur, Kräldjur, Groddjur. Almqvist & Wiksell, Stockholm.

Dale, J. E. (1982). The Growth of Leaves. The Institute of Biology's Studies in Biology No. 137, pp. 1–60. Edward Arnold, London.

De Duve, C. (1984a). A Guided Tour of the Living Cell, Vol. 1. Scientific American Library, Scientific American Books, New York.

De Duve, C. (1984b). A Guided Tour of the Living Cell, Vol. 2. Scientific American Library, Scientific American Books, New York.

Denffer, D. von, Schumacher, W., Mägdefrau, K., & Ehrendorfer, F. (1971). Strasburger's Textbook of Botany (New English Edition), pp. 1–877. Longman, London and New York.

Dickerson, R. E. (1980). Cytochrome c and the evolution of energy metabolism. Sci. Am. 242(3), 99–110.

Dickerson, R. E. & Geis, I. (1979). Chemistry, Matter, and the Universe. An Integrated Approach to General Chemistry, pp. 1–669. Benjamin/Cummings, Menlo Park, CA.

Doolittle, R. F., Feng, D. F., Johnson, M. S., & McClure, M. A. (1986). Relationships of human protein sequences to those of other organisms. Cold Spring Harbor Symposia on Quantitative Biology, Vol. LI. Molecular Biology of *Homo sapiens*, pp. 447–455. Cold Spring Harbor Laboratory.

Dorrington, J. (1979). Pituitary and placental hormones. In: Mechanisms of Hormone Action, Reproduction in Mammals (Austin, C. R. & Short, R. V., Eds.), Book 7, pp. 53–80. Cambridge University Press, Cambridge.

Dreller, C. & Kirchner, W. H. (1993). How honeybees perceive the information of the dance language. Naturwissenschaften 80, 319–321.

Fiorito, G. & Scotto, P. (1992). Observational learning in *Octopus vulgaris*. Science 256, 545–547.

Fox, H. (1981). Cytological and morphological changes during amphibian metamorphosis. In: Metamorphosis: A problem in Developmental Biology (Gilbert, L. I. & Frieden, E., Eds.), pp. 327–362. Plenum Press, New York.

Francé, R. H. (1943). La Maravillosa Vida de los Animales. Una Zoologia Para Todos. Editorial Labor, S.A., Barcelona.

Freeman, R. (1972). Classification of the Animal Kingdom. An Illustrated Guide, pp. 1–55. Hodder and Stoughton, The Reader's Digest Association.

Friday, A. & Ingram, D. S. (1985). The Cambridge Encyclopedia of Life Sciences, pp. 1–432. Cambridge University Press, Cambridge.

Frisch, K. V. (1938). Noi e la Vita. Biologia Moderna per Tutti. Editore Ulrico Hoepli Milano.

Galiano, E. F. (1929). Los Fundamentos de la Biología. Editorial Labor, S.A., Barcelona, Buenos Aires.

Gilbert, W. (1978). Why genes in pieces? Nature 271, 501.

Gilbert, W. (1986). The RNA world–origin of life. Nature 319, 618.

Gitschier, J., Wood, W. I., Goralka, T. M., Wion, K. L., Chen, E. Y., Eaton, D. H., Vehar, G. A., Capon, D. J., & Lawn, R. M. (1984). Characterization of the human factor VIII gene. Nature 312, 326–330.

Goodman, M., Czelusniak, J., & Beeber, J. E. (1985). Phylogeny of primates and other eutherian orders: A cladistic analysis using amino acid and nucleotide sequence data. Cladistics 1(2), 171–185.

Granick, S. (1965). Evolution of heme and chlorophyll. In: Evolving Genes and Proteins (Bryson, V. & Vogel, H. J., Eds.), pp. 67–88. Academic Press, New York.

Greenwood, N. N. & Earnshaw, A. (1989). Chemistry of the Elements, pp. 1–1542. Pergamon Press, Oxford, New York, Beijing.

Halstead, L. B. (1978). The Evolution of the Mammals. Peter Lowe, Oxon, U.K.

Hanström, B. & Johnels, A. G. (1962). Benfiskar. In: Djurens Värld, Band 6, Fiskar: 2 (Hanström, B., Ed.). Förlagshuset Norden AB, Malmö.

Hermann, H. R. (1979). Insect sociality—an introduction. In: Social Insects. (Hermann, H. R., Ed.), pp. 1–33. Academic Press, New York, San Francisco, London.

Heywood, V. H. (1978). Flowering Plants of the World. Oxford University Press, Oxford.

Hiatt, A., Cafferkey, R., & Bowdish, K. (1989). Production of antibodies in transgenic plants. Nature 342, 76–78.

Hinde, R. (1981). Social organization. In: The Oxford Companion to Animal Behaviour (McFarland, D., Ed.), pp. 518–527. Oxford University Press, Oxford, New York.

Hozumi, N. & Tonegawa, S. (1976). Evidence for somatic rearrangement of immunoglobulin genes coding for variable and constant regions. Proc. Nat. Acad. Sci. 73, 3628–3632.

Jaffe, H. W. (1988). Introduction to Crystal Chemistry. Student Edition, pp. 1–161. Cambridge University Press, Cambridge, New York.

Janvier, Ph. (1984). Cladistics: theory, purpose, and evolutionary implications. In: Evolutionary Theory: Paths into the Future (Pollard, J. W., Ed.), pp. 39–75. John Wiley and Sons, New York.

Katz B. & Miledi, R. (1966). Input/output relation of a single synapse. Nature 212, 1242–1245.

King, J. E. (1964). Seals of the World. The British Museum (Natural History), London.

Klein, C. & Hurlbut, C. S. Jr., (1985). Manual of Mineralogy (20th ed.). John Wiley & Sons, New York, Chichester.

Lagercrantz, H. & Slotkin, A. (1986). The "stress" of being born. Sci. Am. 254(4), 92–102.

Lima-de-Faria, A. (1983). Molecular Evolution and Organization of the Chromosome, pp. XLIV & 1186. Elsevier Science Publishers, Amsterdam and New York.

Lima-de-Faria, A. (1988). Evolution without Selection. Form and Function by Autoevolution, pp. 372. Elsevier, Amsterdam, New York, Oxford.

Lima-de-Faria, A. (1995). Biological Periodicity. Its Molecular Mechanism and Evolutionary Implications. JAI Press, Greenwich, CT (in press).

Macdonald, D. (Ed.) (1984). The Encyclopaedia of Mammals, Vols. 1 and 2, pp. 1–895. George Allen & Unwin, London, Sydney.

Margoliash, E. & Smith, E. L. (1965). Structural and functional aspects of cytochrome C in relation to evolution. In: Evolving Genes and Proteins (Bryson, V. & Vogel, H. J., Eds.), pp. 221–242. Academic Press, New York.

Margulis, L. & Schwartz, K. V. (1982). Five Kingdoms. An Illustrated Guide to the Phyla of Life on Earth, pp. 1–338. W. H. Freeman, San Francisco.

Matthews, L. H. & Carrington, R. (1972). Världens Djurliv. Reader's Digest AB, Stockholm, Sweden.

Mazurs, E. G. (1974). Graphic Representation of the Periodic System during One Hundred Years. University of Alabama Press, Tuscaloosa, Alabama.

McFarland, D. (1981). The Oxford Companion to Animal Behaviour. (McFarland, D., Ed.), pp. 1–657. Oxford University Press, Oxford, New York.

Millar, A. J., Short, S. R., Chua, N.-H., & Kay, S. A. (1992). A novel circadian phenotype based on firefly luciferase expression in transgenic plants, The Plant Cell 4, 1075–1087.

Mossberg, B. & Nilsson, S. (1977). Nordens orkidéer, pp. 1–128. J.W. Cappelens Förlag A.S.

Murzin, A. G. (1993). OB(oligonucleotide/oligosaccharide binding)-fold: Common structural and functional solution for non-homologous sequences. The EMBO Journal 12(3), 861–867.

Nagl, W. (1973). The angiosperm suspensor and the mammalian trophoblast: Organs with similar cell structure and function? Soc. Bot. Fr. Mém. Coll. Morphol. 289–302.

Nagl, W. (1976). Zellkern und Zellzyklen, pp. 1–486. Verlag Eugen Ulmer, Stuttgart.

Nakaya, U. (1954). Snow Crystals. Natural and Artificial. Harvard University Press, Cambridge.

Needham, J. (1950). Biochemistry and Morphogenesis. Cambridge University Press, Cambridge.

Ny, T., Elgh, F., & Lund, B. (1984). The structure of the human tissue-type plasminogen activator gene: Correlation of intron and exon structures to functional and structural domains. Proc. Nat. Acad. Sci. USA 81, 5355–5359.

Orr, R. & Pope, J. (1986). Mammals of Britain & Europe. Peerage Books, Godalming, UK.

Ow, D. W., Wood, K. V. DeLuca, M., De Wet, J. R., Helinski, D. R., & Howell, S. H. (1986). Transient and stable expression of the firefly luciferase gene in plant cells and transgenic plants. Science 234, 856–859.

Pauling, L. (1949). General Chemistry, pp. 1–618. W. H. Freeman, San Francisco.

Pauling, L. (1960). The Nature of the Chemical Bond (3rd ed.). Cornell University Press, Ithaca, NY.

Perrins, C. (1976). Bird Life. An Introduction to the World of Birds. Elsevier-Phaidon.

Pugsley, A. P. (1989). Protein Targeting, pp. 1–279. Academic Press, San Diego, New York.

Rabaud, E. (1932). Zoologie Biologique. Fascicule I: Morphologie Generale et Systeme Nerveux. Gauthier-Villars et Cie, Paris.

Rabaud, E. (1934). Zoologie Biologique. Fascicule III: Les Phénomènes de Reproduction. Gauthier-Villars, Paris.

Rabbitts, T. H. (1978). Evidence for splicing of interrupted immunoglobulin variable and constant region sequences in nuclear RNA. Nature 275, 291–296.

Rimington, C. & Kennedy, G. Y. (1962). Porphyrins: structure, distribution and metabolism. In: Comparative Biochemistry (Florkin, M. & Mason, H. S., Eds.), Vol. 4, pp. 557–607. Academic Press, New York.

Romer, A. S. & Parsons T. S. (1978). The Vertebrate Body. Shorter Version. Fifth Edition, pp. 1–476. W.B. Saunders, Philadelphia.

Romoser, W. S. (1973). The Science of Entomology, pp. 1–449. Macmillan Publishing, New York; Collier Macmillan Publishers, London.

Russell-Hunter, W. D. (1979). A Life of Invertebrates, pp. 1–650. Macmillan Publishing, New York; Collier Macmillan Publishers, London.

Rydén, L. G. & Hunt, L. T. (1993). Evolution of protein complexity: The blue copper-containing oxidases and related proteins. J. Mol. Evol. 36, 41–66.

Sandars, E. (1937). A Beast Book for the Pocket. The Vertebrates of Britain Wild and Domestic other than Birds and Fishes. Oxford University Press, Humphrey Milford, London.

Sanderson, R.T (1967). Inorganic Chemistry. Reinhold Publishing, New York.

Savage, R. J. G. & Long, M. R. (1986). Mammal Evolution. An Illustrated Guide. British Museum (Natural History), London.

Sharp, D. W. A. (1988). The Penguin Dictionary of Chemistry (Sharp, D. W. A., Ed.), pp. 1–434. Penguin Books, London.

Shi, Y.-B. & Brown, D. D. (1990). Developmental and thyroid hormone-dependent regulation of pancreatic genes in *Xenopus laevis*. Genes and Development 4, 1107–1113; Cold Spring Harbor Laboratory Press.

Smith, C. W. J., Patton, J. G., & Nadal-Ginard, B. (1989). Alternative splicing in the control of gene expression. Annu. Rev. Genet. 23, 527–577.

Springer, M. S. & Kirsch, J. A. W. (1993). A molecular perspective on the phylogeny of placental mammals based on mitochondrial 12S rDNA sequences, with special reference to the problem of the Paenungulata. Journal of Mammalian Evolution 1(2), 149–166.

Stubbings, H. G. (1975). *Balanus balanoides* (175 pp.) Liverpool University Press.

Südhof, T. C., Goldstein, J. L., Brown, M. S., & Russel, D. W. (1985a). The LDL receptor gene: A mosaic of exons shared with different proteins. Science 228, 815–822.

Südhof, T. C., Russel, D. W., Goldstein, J. L., & Brown, M. S. (1985b). Casette of eight exons shared by genes for LDL receptor and EGF precursor. Science 228, 893–895.

Van der Ploeg, L. H. T. (1990). Antigenic variation in African trypanosomes: genetic recombination and transcriptional control of VSG genes. In: Gene Rearrangement (Hames, B. D. & Glover, D. M., Eds.), pp. 51–97. IRL Press, Oxford University Press, Oxford, New York.

Von Baeyer, H. C. (1986). Physics finds a childhood. The Sciences Jul./Aug., 8–10.

Walker, C. (1974). Introduction: the world of birds. In: The World Atlas of Birds, pp. 10–33. Mitchell Beazley Publishers, London.

Wampler, J. E. (1978). Measurements and physical characteristics of luminescence. In: Bioluminescence in Action (Herring, P. J., Ed.), pp. 1–48. Academic Press, New York.

Wehner, R. & Müller, M. (1993). How do ants acquire their celestial ephemeris function? Naturwissenschaften 80, 331–333.

Weinberg, E. D. (1989). Cellular regulation of iron assimilation. The Quarterly Review of Biology 64(3), 261–290.

Wells, M. (1983). Cephalopods do it differently. New Scientist 100, 332–337.

White, B. A. & Nicoll, C. S. (1981). Hormonal control of amphibian metamorphosis. In: Metamorphosis: A Problem in Developmental Biology (Gilbert, L. I. & Frieden, E., Eds.), pp. 363–396. Plenum Press, New York.

Whitten, D. G. A. & Brooks, J. R. V. (1988). The Penguin Dictionary of Geology, pp. 1–493. Penguin Books, London.

Withers, P. C. (1992). Comparative Animal Physiology, pp. 1–949. Saunders College Publishing. Harcourt Brace Jovanovich College Publishers, Fort Worth, Philadelphia, San Diego.

Yaoita, Y., Shi, Y.-B., & Brown, D. D. (1990). *Xenopus laevis* alpha and beta thyroid hormone receptors. Proc. Natl. Acad. Sci. USA 87, 7090–7094.

Chapter 11

The Origin and Evolution of Man

MICHAEL H. DAY

Principles of Medical Biology, Volume 1B
Evolutionary Biology, pages 321–351

ISBN: 1-55938-802-1

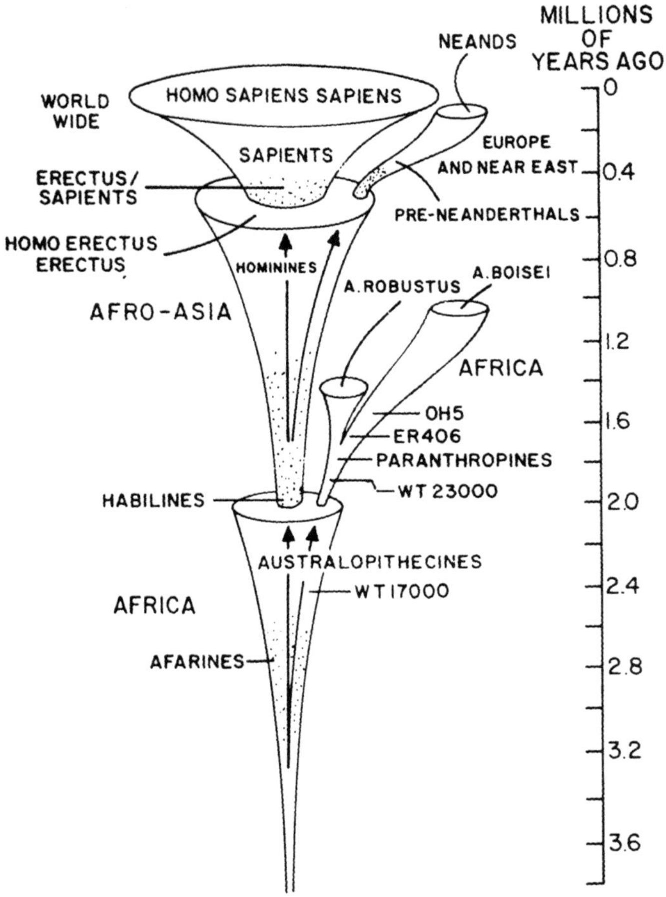

INTRODUCTION

The study of human evolution has a long history. Indeed, early civilization produced the first questionings of human origins based on the observations and thinking of Greek scholars. Almost every ethnic group, tribe or religion has its own creation myth as a response to an apparent human need for answers to eternal questions. Who are we, where did we come from, how long have we been here?

These are universal questions posed by sentient man to himself, to the old by the young, and to the learned by those who seek enlightenment. The answers given are as varied and unproven as the systems of faith and belief propounded by their followers. The Judeo–Christian belief is recorded in the Bible as the Genesis story of Adam and Eve, although many contemporary believers regard this as an allegory rather than a literal account of what took place.

Scientific appraisal of the question of human origins is based upon

1. The observations of comparative anatomy that show the clear relationships between man and other groups of animals such as vertebrates, mammals, and primates.
2. The fact of evolution as demonstrated originally by the work of Darwin and later reproduced experimentally.
3. The recovery of fossil evidence for this process such as fossil primates, fossil hominids (man-like creatures), and fossil man himself in a time sequence that is clearly defined and increasingly well dated by modern techniques.

We now know how, when, and where man originated and evolved on Earth. We also know the mechanism that brought it about (normal genetic variation acted upon by natural selection), and perhaps in a few years we shall even know the fully sequenced human genome.

The purpose of this chapter is to outline the primate background to human evolution, the fossil record of the evolving hominids, and the emergence of modern man as a single worldwide species that we call humanity and which forms the basis for the humanitarian calling of doctors, paramedics, and nurses wherever they may work.

PRIMATES AND THEIR EVOLUTION

The Order Primates

Man is a primate. That means that he belongs with the prosimians, monkeys, and apes in an Order of the Class Mammalia and shares with them the mammalian advantages of warm blood and temperature control, internal reproduction and placentation, the ability to suckle and nurture relatively helpless young, the possession of hair and two sets of teeth (milk and permanent), as well as a repertoire of behavior more complex than that of other vertebrates such as fishes, reptiles, and birds.

The overriding feature of the primates, however, is that all living primates are capable of climbing trees at some point in their lives, or have an ancestor that was truly arboreal. This fact alone has had the most significant bearing on the evolution of the Order and upon the anatomy of individual primate species. Each and every

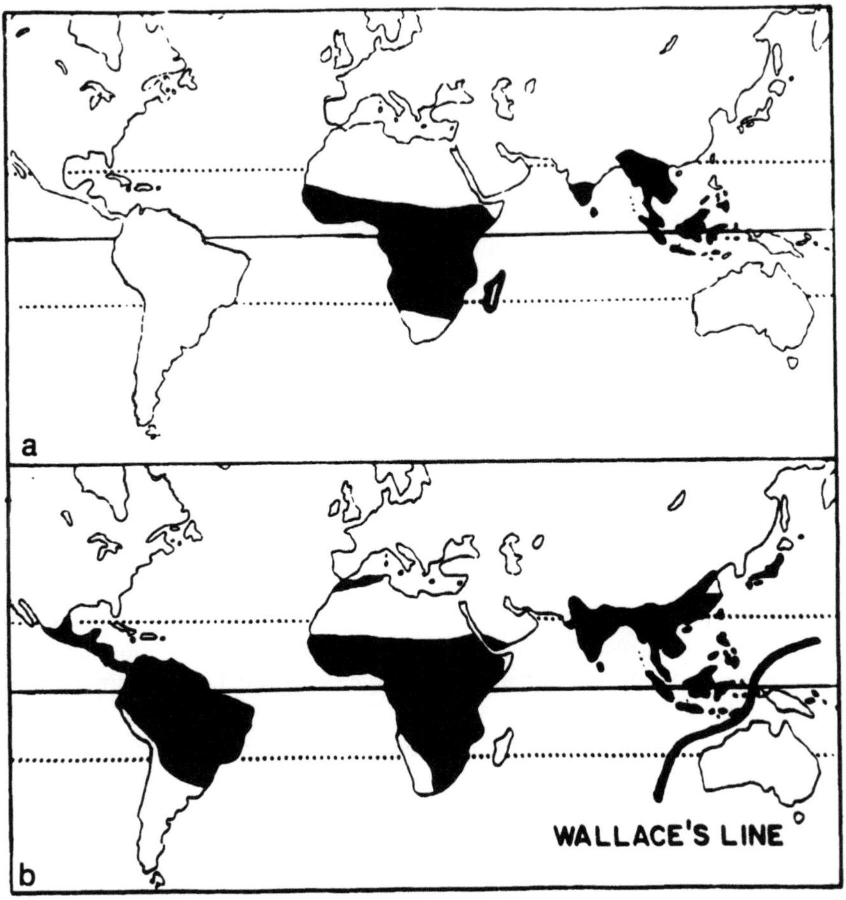

Figure 1. The distribution of modern primates. (**a**) Lemurs, lorises, and tursiers. (**b**) Monkeys and apes. There are none in the wild to the east of Wallace's Line. (Redrawn after R. D. Martin).

species of primate has anatomical specializations that relate to a lifestyle in a "three-dimensional" environment—features that can be shown to be preadaptive to many human attributes.

Native primates are tropical animals and are found in both the New and Old Worlds within an irregular band that circles the Earth roughly between the Tropics of Cancer and Capricorn, but only as far to the east as Wallace's Line. There are no primates in the wild in New Guinea, Australia, New Zealand, or the intervening islands. There are no native primates in Europe; the so-called "Barbary Apes" of Gibralter were introduced from North Africa in the nineteenth century.

The major groups of primates are the small prosimians such as the galagos and lorises of Africa and Asia; the lemurs of Madagascar; the monkeys of Africa, Asia, and South America; the gibbons and siamangs of the Far East; the chimpanzees and gorillas of Africa; and mankind whose distribution is worldwide. A number of primate species are endangered by the destruction of habitats such as the rain forests of South America and the Far East, and the encroachment of herding and agriculture on natural areas in Africa where primates live.

Primate Characteristics

Arborealism and the Skull

Most mammals are terrestrial and are adapted to life on the ground. This sets a high premium on an acute sense of smell that is used for seeking prey by following scent trails, or for seeking a mate by marking behavior, or for scenting danger in the wind. Olfactory dominance of the senses is reflected in the long noses of dogs, antelopes, horses, and many other mammals. Housing a long nose in the skull places the eyes to the sides and allows maximum range of vision, another terrestrial advantage. Living in trees reduces the value of olfactory dominance since no "trail" can be reasonably followed, thus the advantage of binocular, stereoscopic vision is paramount. Stereoscopic vision gives appreciation of distance, which is vitally important when leaping or running from branch to branch 100–200 feet above the ground.

Nasal reduction and orbital shift to the front begin to mold the evolving primate skull. At the same time, snout reduction lessens room for the dentition, while the brain expansion needed for the evolving cerebellum (an important part of the brain concerned with posture and balance) adds to the transformation of the primate skull. The prognathism of the skull typical of the mammal is progressively reduced until the orthognathism of man is reached.

Teeth

Dental reduction leads to the primate dental formula of two incisors, one canine, two premolars, and three molars (2123). The preceding deciduous dental formula is two incisors, one canine, and two milk molars (212). The dentition is therefore heterodont (varied tooth forms) and diphyodont (coming in two sets). An arrangement of the cusps and fissures on the crowns of the molar teeth (the so-called Y5 arrangement) is typical of the dryopithecines—a group of hominoids that date back to the Miocene—and is to be found in both modern apes and man. It is anatomical similarities in homologous structures of this type between man and the primates that places us unequivocally within the order and shows that our closest relatives to be the great apes. In recent years this long-known comparative anatomical evidence has been immensely strengthened by immunological studies and the discovery of DNA.

Brain

The brain expansion that has occurred during the evolution of the primates has resulted from an enormous increase in the relative size of the cerebral hemispheres so that in higher primates they overlap the cerebellar hemispheres by a considerable amount. The massive relative increase of the cerebral hemispheres has necessitated the development of a new system of blood supply, the carotid system, bringing blood through the cranial base to the Circle of Willis, a distribution ring circuit at the base of the brain. Terrestrial mammals rely on the vertebral arteries alone for the nourishment of the brain. While the skull is principally concerned with housing the brain and the senses that collect information about the environment (eyes, ears, smell), communication (vocalization), and food preparation (mastication), the limbs and trunk are concerned with locomotion.

Primate Locomotion

Examination of the primate axial and appendicular skeleton, including that of man, discloses that it is relatively unspecialized and that there have been no dramatic losses of digits as in the horse and no extraordinary exaggerations such as the neck of the giraffe. Indeed, the basic vertebrate pattern is almost intact. In the upper limb, the proximal segment has a single bone, the distal segment two bones, the hand is pentadactyl, and the small bones of the wrist relate to the five digital rays and the lower ends of the forearm bones. The two forearm bones rotate freely in the movements of pronation and supination, and the elbow flexes and extends almost without limitation. The shoulder joint is a ball and socket with a wide range of freedom. In fact, the whole jointed structure has evolved to allow the placement of the grasping organ almost anywhere in the sphere of reach of the individual. All of the above is true of man and has been preadaptive to the incredible range of human manipulative activity from typing this script, to playing a piano sonata, to the skills of the carpenter, the sculptor, and the watchmaker.

Similarly with the lower limb the proximal segment has a single bone, the distal segment two bones, a number at the ankle followed by a pentadactyl foot. Most primates have a grasping foot again free to be placed in space to maximum effect; here, however, man shows a clear specialization for upright stance and bipedal gait.

There have been several classifications of primate locomotion that include categories such as terrestrial quadrupeds (baboons), arboreal quadrupeds (monkeys), brachiators (gibbons), semi-brachiators (langurs), and knucklewalkers (chimpanzees and gorillas). Only man is a true upright biped and the hallmarks of this extraordinary method of progression can be seen in the anatomy of the vertebral column (the lumbar lordosis), the pelvis (the position of the sacrum that restricts the pelvic cavity and requires a birth mechanism unique to man), the hip (the iliofemoral ligament strap mechanism), the knee (the valgus angulation between femur and tibia), and the specialized form of the foot that has lost the ability to grasp but gained a short lever system that is propulsive during the "controlled fall" that we term walking.

Primate Life Strategies

Specialism versus Generalism

Darwinian theory suggests that the way to survive as a species is to produce more offspring than another group that is competing for the same territory and/or resources. The survival strategy commonly adopted is that of specialism to fit the environment in terms of diet, locomotion, anatomy, or behavior. Thus, fishes are specialized for life in water, herbivores for grassland, and birds and bats for flight; many other examples come easily to mind. This evolutionary strategy is commonly very successful, but does have one serious flaw. What happens if the environment changes radically? The species concerned may be trapped by the constraints of its specialized diet, anatomy, or ecological needs, and extinction may follow. The primates have avoided this trap since they have almost specialized in generalism. Throughout the order the keynote has been adaptability, and it is this strategy that has enabled the primates to survive and evolve rapidly through the 70 million years of their history, a period of frequent climatic and ecological change, particularly in Africa.

Family Planning?

Some mammals, such as rodents, have large and frequent litters, short gestation periods, short lactation periods, and little parental investment in the offspring—the altricial reproductive complex has a high reproductive potential. Primates have adopted the alternative reproductive strategy: the precocial reproductive complex which has a lower reproductive potential but one with maternalism. Primates have long gestation periods, long oestrous cycles with spontaneous ovulation, long lactation, long dependent infancy, and late sexual maturity coupled with long lifespans. The precocial complex is associated with all living primates and, of course, with man. Parental investment and maternal care are key elements in the precocial system and these will link with the sexual and group behavior of the species as well as with the habitat. What is appropriate for a tropical rain forest may not be so for an open savannah environment.

Habitat, Diet, and Activity Patterns

Primates are found in a wide variety of habitats from tropical rain forests to virtual deserts within their geographic range. The majority of primates live in forests ranging from primary rain forest, to secondary rain forest, to gallery forest along rivers, to woodland savannah. Within each there are smaller habitats depending upon level from top canopy to forest floor.

The dietary preferences of primates vary widely but the fruits, seeds, and young shoots of forest trees and plants are widely taken. There are some primates, such as the langurs, who are specialized leaf eaters, while baboons will forage on the ground for edible roots and grubs.

Most primates are active during the day but indulge in a rest period around midday. Many of the smaller prosimians are nocturnal, and this may be the primitive condition for the order. Only one true monkey is nocturnal (the owl monkey of South America). Primates divide their days between feeding, travelling, and resting; about half their time is spent in feeding, and the rest divided between their other activities that include sexual behavior, mutual grooming, and interactions with infants.

Social Life

Primates are gregarious social animals who live in groups. The simplest of these is a mother and her offspring, common among the prosimians. Monogamous family groups occur, as well as polyandrous groups and one-male groups. The latter usually consists of one dominant male and a number of females and offspring. Groups of this type are subject to "take-over" attempts by outside males and, if successful, may result in the death of the infants followed by the newcomer mating with the females. Multimale groups are also known in which males and females and their offspring live and forage together. It may be that the secondary males in these groups are not yet sufficiently mature to pose a serious threat to the leader of the troop.

Clearly, living in groups is advantageous in terms of the defense of a territory (howler monkeys, gibbons), feeding and foraging (baboons), care of offspring and the educational advantages of the interactions of the young with older troop members (gorillas and chimpanzees). In addition, there is always safety in numbers and the defense of the troop is likely to be better served by banding together. Mate selection and successful recognition of sexual signaling is facilitated by group interactions.

TIME AND FOSSILS

The Ancient History of Human Evolution

The writings of the ancient Greeks and Romans tells us that they were well aware of the existence of monkeys and apes and of their similarity in so many ways to man, but it was Aristotle (c303 B.C.) who noted that groups of animals share features that enable them to be classified together. It was he who, in his *Historia Animalium*, assigned animals into two groups: those with red blood (the vertebrates), and those whom he thought lacked blood (the invertebrates). Aristotle also wrote of a "tree of life" with man at the top, although he does not appear to have

Table 1. Linnaean Classification of Modern Man

KINGDOM	ANIMALIA
PHYLUM	CHORDATA
CLASS	MAMMALIA
ORDER	PRIMATES
SUBORDER	ANTHROPOIDEA
SUPERFAMILY	HOMINOIDEA
FAMILY	HOMINIDAE
SUBFAMILY	HOMININAE
GENUS	HOMO
SPECIES	SAPIENS

grasped the notion of evolutionary change with time. Heraclitus (c500 B.C.) thought that animals could change and Anaximander of Miletus (c570 B.C.) first used the term "to evolve." These first glimmerings of progressive biological thought were stifled by the authority and power of Plato, the philosopher, whose view of an ordered and unchanging universe prevailed for centuries. It was not until the birth of scientific anatomy at the time of the Renaissance that the concepts of Aristotle, Plato, and Galen were challenged.

Leonardo da Vinci dissected human bodies, drew what he saw, and noted the close similarities between the anatomy of man and other animals. These observations were extended by Vesalius in 1543 in the book, *De humani corporis fabrica*, that founded human anatomy as a science. Vesalius established the principle of observation in anatomy and recorded what is there to be seen and not what should be there according to some preconceived notion or theory. The first dissection of a chimpanzee was recorded in 1699 by Edward Tyson who termed it *Homo sylvestris*: the man of the woods. The links between man and the apes were becoming more firmly established, and by the mid-eighteenth century, Linnaeus, the great Swedish biologist and taxonomist, classified man and apes together in his highest category of the Class Mammalia *Homo sapiens* (man, the wise) at the pinnacle of the living animal world in his book, *Systema Naturae*.

The classification system proposed by Linnaeus (Table 1) has stood the test of time. His hierarchical binomial system of taxonomy is still used by biologists worldwide, and Latin names are given to all living things by biological scientists according to strict codes of nomenclatural practice that avoid confusion and duplication. Linnaeus devised his system for the classification of all living plants and animals, but it is apparent if the remains of living creatures of the past can be identified in the rocks as fossils, then the same principles of classification can, and do, apply. A fish is a fish if it is alive or if it is a fossil fish.

The Earth and Fossils

Until the late eighteenth century, the general belief was that the Earth was basically volcanic in origin and that sedimentary rocks were the result of catastrophic events, such as volcanic eruptions and floods like the one recorded in the Bible. In 1788, this view was challenged by William Hutton, an Edinburgh physician and scientist, who observed geological unconformities (adjacent layers of differing composition) in his native Scotland, and one in particular on the island of Arran in the Firth of Clyde. This and other unconformities were not explained by the current view that the Earth was an unchanging structure. Hutton also knew of the cycle of cloud formation; rainfall on land, rivers, and oceans; and evaporation to form clouds once again. Hutton proposed that this cycle broke down older rocks by weathering to form new sediments over the millenia in a uniform fashion, and his theory of "Uniformitarianism" was the explanation of stratification in the rocks and of unconformities as they are found with young rocks in direct contact with older rocks (Playfair, 1802). The inclusion of the remains of animals and plants as fossils in these strata made nonsense of the single catastrophe theory of the biblical flood, which was the current explanation of fossils. The science of palaeontology was about to be born. Explanations of the fossil remains of extinct animals and an appreciation of the time that has elapsed between geologic events became possible.

Finally from the work of "Strata" Smith (a British geologist) and others it was appreciated that older rocks normally underlie younger rocks, and the "Law of Superposition" allows the construction of a sequence of geologic events. This layering of the rocks is paralleled by the recognition of fossil remains at successive levels and changes through time in animals and plants. Biostratigraphy, the recognition of layers by their contained animal and plant fossils, has helped in the construction of the geologic column and in the reconstruction of the ecology of any given time level represented by that flora and fauna. It is also a valuable means of the relative dating of fossils, such as those of fossil man and contemporary animals and plants.

Fossils are formed when a creature dies and the carcass putrifies with the soft parts usually rotting away. The hard parts—bones and teeth—may be preserved undamaged if they are covered by sediments and thus avoid the attentions of scavengers. The bones and teeth will preserve only if the groundwaters are alkaline since acid will dissolve the minerals in bones and teeth. Burial in an alkaline environment with the exclusion of oxygen, such as under a volcanic ashfall, will result in mineralization of the bone at molecular level so that the anatomy is perfectly preserved. In time, the fossil contained in the sediment may erode to the surface and disclose the fossil bearing layer. Redeposition of the fossil in a new layer of sediment is possible and may confuse the supposed dating of a find. It is essential, therefore, to link the fossil with the layer in which it is found before it can be dated with assurance.

Methods of Dating Fossils

Two forms of dating fossils are recognized: relative dating and dating in years. Relative dating simply determines which of two fossils is the older and which the

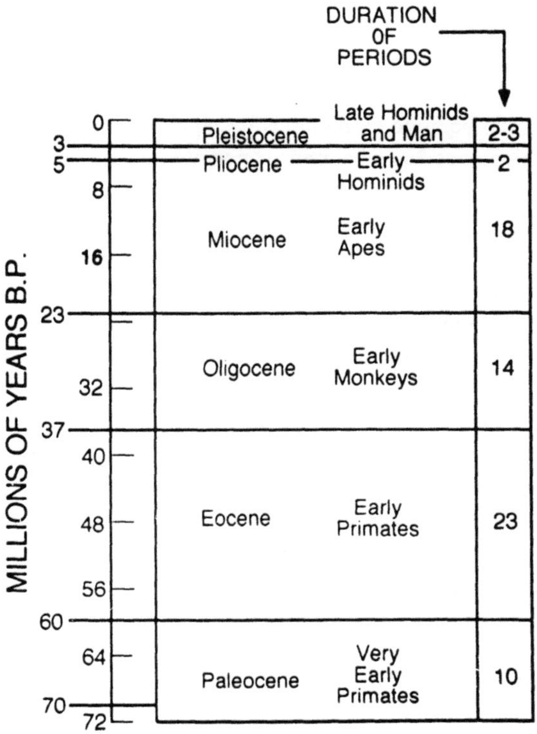

Figure 2. A geological time scale of the last 70 million years showing primate evolution and the appearance of major groups.

younger, or, put another way, determines the contemporaneity of the fossil with the deposit in which it was found by comparison with others from the same layer. Once it has been established that a fossil truly belongs with the layer in which it was found, then the relationship of that layer with others above and below will establish its stratigraphic position. Other fossils both above and below, as well as pollen grains, allow paleontologists to place a layer and its fossils in a local sequence and perhaps a continental sequence. This can be assisted by a determination of the residual paleomagnetic polarity of the rocks in a sequence that may match with the world polarity reversals known from the wanderings of the magnetic poles in the history of the Earth.

Relative dating can be established by reference to the stratigraphy and the biostratigraphy of a fossil, but it is also possible to use chemical analysis to determine its position in a sequence. Fluorine accumulates in buried bone and its level tends to increase with time. It does not accumulate at a universally known and regular rate since it will vary with concentration in the groundwaters. It is, however, a very useful guide to contemporaneity or the relative dating of a sequence of fossils.

Estimation of the levels of nitrogen in fossil bones can also give an idea of the recency of burial, and is again a useful relative dating technique.

Dating fossils in years before the present (y B.P.) or millions of years before the present (my B.P.) can be done by methods based on radioactivity. As some rocks are formed, radioisotopes may be imprisoned within and decay by giving off radiation. The daughter isotope will form at a known rate that can be measured in years. Potassium–argon dating and radiocarbon dating are examples of this method: the former works best at ages of above 250,000 B.P. and the latter at ages less than 60,000 y B.P. Another method that is similar is fission-track dating that relies on the contamination of natural glasses with uranium 238. The uranium decays leaving traces that can be detected and quantified. Newer methods include thermoluminescence dating (TD) and electron-spin-resonance (ESR) dating. By selecting and combining suitable methods it is possible nowadays to get a good estimate of the date at which a fossil find was laid down in the deposits, and this is an essential step in determining the position of a fossil in the sequence of its evolutionary development and its possible role as an ancestor or a descendent. The Primate Order has an evolutionary history of about 70 million years so the importance of accurate dating of fossil primates is paramount in understanding the evolutionary history of the order (Zimmerman and Angel, 1986).

Darwin and Man

While the ancient thinkers of both Greece and Rome had considered it possible that animals might change through time, their ideas were lost in the Dark Ages when religious dogma dominated the intellectual scene and the Biblical story of the individual creation of both man and animals, and their immutability as species, was unchallenged until the beginning of the nineteenth century. In Paris, Buffon's pupil, Jean Baptiste Lamarck, suggested that changes take place gradually in animals through time by the appearance of new characters that are acquired during their lifetime and are passed on to their progeny to the advantage of the species. The inheritance of these acquired characteristics was his explanation of evolutionary change. This idea is wrong because the environment does not directly affect inheritable characteristics (the genes), but it failed to gain acceptance because the leading zoologist of the day, Cuvier, would not entertain the idea of evolution of one species from another or its implication that man had evolved from lower forms.

Charles Darwin was the son of a country doctor from Shrewsbury in England and the grandson of Erasmus Darwin, another doctor and scientist whose views were not unlike those of Lamarck. Charles Darwin joined a naval expedition that was going around the world for five years, and while on the trip he collected thousands of specimens as the official naturalist to the expedition. During that time he became convinced that species had evolved through time particularly by his observations in the Galapagos, a remote group of islands in the Pacific some 600 miles west of Ecuador, whose fauna was clearly associated with that of South

America but was not quite the same. It was clear to him that the ancestors of the Galapagos fauna had arrived there from South America in the past, but in the course of time had varied giving rise to new but closely related forms. His genius was to identify the elegantly simple solution to the problem as to how this had come about. By a mass of evidence Darwin showed in his book, *The Origin of Species*, that animals and plants varied and that some variations were advantageous. In the "struggle for life," those with a favorable variation would survive and produce more offspring. This differential mechanism he termed, "natural selection," or the "survival of the fittest." It was a simple mechanism that accounted for the cumulative effects of thousands of small changes over time, and it was propounded without knowledge of genetics—a science not born in 1859. It was not until 1871, however, that Darwin ventured in print to apply his theory to man in a book entitled, *The Descent of Man*. The history of the period has been covered in detail by a number of Darwin's biographers and makes fascinating reading. The generation of the greatest biological idea that there has ever been, and the battle for its acceptance, is basic to an understanding of scientific medicine which is in itself simply an aspect of applied biology.

PRIMATES IN ROCKS

The First Primates

Normally only the hard parts of a creature (bones, teeth) fossilize, and of them the teeth are the hardest. It is not surprising, therefore, that the earliest fossil primates are known principally from dental remains. The earliest primates evolved from insectivore-like mammals in the Late Cretaceous period just as the dinosaurs were disappearing from the scene. They are not easy to identify in the fossil record for the very reasons that led to their success; they are very generalized and have no special or diagnostic features that set them apart clearly from other small evolving mammals. The earliest recognized primate is *Purgatorius* from several sites in North America; a small creature from the Early Paleocene (60–70 my B.P.) represented solely by a few teeth and jaws. The Middle and Later Paleocene displays a radiation of Plesiadapidae showing a diversity of forms but basically small, tailed, furry, almost squirrel-like forms occupying a variety of econiches including both the ground and the trees.

The succeeding Eocene epoch (57–37 my B.P.) is a period of change in that the fauna took on a more modern aspect and recognizable rodents and ungulates appeared. The Plesiadapidae disappeared and were replaced by primates that began to look like modern prosimians. They already had snout reduction, rounded crania, orbits completed by a bony ring, an auditory bulla, and the beginnings of a carotid system. In the locomotor system they had a divergent and grasping hallux and nails, rather than claws, on many digits. Some of these early primates were like modern lemurs, and others like tarsiers or galagos. The fossil prosimians of the Eocene of North America and Europe were again a widely diverse group who occupied a wide range of econiches. Among them the Adapidae went in for larger size and diurnal leaf-eating, while the Omomyidae were perhaps nocturnal and fruit-eating forms similar to galagos.

Fossil Monkeys

By the Oligocene (37–23 my B.P.) the fossil record of primates that, up to that time, had come from North America and Europe now produced primates from Africa and South America. It also produced the first records of higher primates. The Fayum desert region of western Egypt has an eroded area known as the Jebel Qatrani Formation, dated at between 37–31 my B.P., that contains fossils including those of rodents, insectivores, bats, carnivores, hyraces, and elephants, as well as many fossil primates—a fauna that has a real African feel to it. There are several groups of fossil primates in the Fayum fauna: two prosimians, one lorisoid, one monkey-like, and one ape-like. The monkey-like are the Parapithecidae, a group of at least six species who possess the primitive dental formula of 2133 like the New World monkeys, and for this reason may lie near the divergence of the New World monkeys and the Old World monkeys, apes, and man. How the New World monkeys reached South America is still a vexed question. Did the ancestors of the present monkeys of South America come from North America or directly from Africa? There is no clear answer at present.

The final group of early anthropoids from the Fayum desert are the family Proplipithecidae who have an ape-like dental formula (2123) and who do not have the bilophodont molars typical of the cercopithecoid monkeys. They are more primitive than any living Old World monkeys and apes. One example of this family is well known from the Fayum and this is *Aegyptopithecus*. It is a medium-sized animal, perhaps frugivorous from its teeth, and has skull features that resemble those of other anthropoids but a relatively small brain. The skull in males has a crest, the auditory bulla is like that of the New World monkeys, and it has a tympanic ring rather than a bony auditory meatus. It appears to have been an arboreal quadruped from its limb bones. This mixture of features shows the aegyptopithecines to be neither monkeys nor apes in the true sense of either word, but they are often placed in the "ape" category despite having preceded what is usually considered the monkey–ape divergence.

Fossil Apes

The Miocene is a long epoch that lasted about 18 million years from 23–15 my B.P. In East Africa there was a lot of volcanic activity associated with the development of the East African Rift Valley and the volcanic deposits of this area have disclosed the fossils of some monkeys but also a large Early Miocene radiation of fossil apes, members of the Hominoidea. Europe and Asia show no fossil apes at this time but some do appear in the Middle and Late Miocene.

The best known African Early Miocene hominoids are the family Proconsulidae in which there are a number of genera and species. *Proconsul africanus, Proconsul nyanzae,* and *Proconsul major* are represented by teeth, jaws, and some postcranial bones, but no complete skull is known for any of the Miocene apes of East Africa. They are clearly apes in that they possess the ape dental formula and the Y5 cusp and fissure molar dental pattern; also, none of them have tails as do monkeys. *P. africanus* is the smallest with an almost monkey-like size and proportion while *P. nyanzae* is larger and more terrestrially adapted. *Proconsul major* is much larger but known principally from dental remains alone. Others include a very large ape called *Afropithecus*, a medium-sized ape *Turkanapithecus*, and several of smaller size. This adaptive radiation of apes clearly spread in the East African environment and occupied a variety of vacant niches. They were diurnal; most were fruit eaters but some ate leaves; all were capable of climbing trees, but some were better adapted for arborealism than others. Indeed, some were gibbon-like in their adaptations and were likely to be arboreal specialists; an example might be an ancestor of *Oreopithecus* that is known from the Pliocene soft coal deposits of northern Italy.

There are also an increasing number of Miocene apes known from Eurasia including *Dryopithecus* (the oak ape) and *Sivapithecus*. The latter is a genus that contains a wide range of species some of which are as large as an orangutan. Another closely related genus is *Gigantopithecus* which contains two species, *G. giganteus* and *G. blacki*. These are the largest primates in the whole of the fossil record. The former is known from the Miocene of Pakistan and is modern gorilla-sized, but the latter is known from the Pleistocene of China and, based on the size of its teeth and jaws, may have weighed more than twice as much. Needless to say, these creatures were predominantly ground dwellers.

There has been much speculation about which, if any, of the known Miocene apes gave rise to the modern Hominoidea (in particular gorillas, chimpanzees, orangutans, and man-like creatures that succeeded them). While we do not lack for candidates for this honor, the only one that seems to be well established is *Sivapithecus*. The Asian sivapithecines seem to have given rise to two lines, one that led to the extinct *Gigantopithecus* and the other to the living orangutan (*Pongo pygmaeus*). The resemblance between the Miocene *Sivapithecus* and the modern orangutan are quite striking.

The situation in Africa is less well understood. It was believed at first that two of the *Proconsul* species, *P. africanus*, and *P. major* gave rise respectively to the chimpanzee and the gorilla, but that idea is less well supported in recent years. There is no certainty in the origin of the African apes from the known Miocene radiation of apes at present. The origin of the Hominidae is no less shrouded in mystery.

There is no doubt that the family of man evolved from some form of Miocene ape, but to identify which one from the known range has so far proved impossible.

Ramapithecus was for many years cited as the most likely candidate on account of the supposed rounded form of the dental arcade; once this was shown not to be the case, *Ramapithecus* fell from favor as an ancestor. It is now considered to be the same as *Sivapithecus*. The problem is complicated by the fact that most of the Miocene material is related to the teeth and not to the skeleton, and it is likely that the earliest signs of evolving hominids will show themselves in the evolution of upright stance and bipedalism rather than in dental features. Simply put, at present we have no idea which, if any, of the known Miocene fossil apes gave rise to the hominid line.

THE APE-MEN OF AFRICA

The Walking Ape-Men of Laetoli

In 1976, a chance find at Laetoli near Olduvai Gorge, Tanzania, was a "floor" of consolidated volcanic ash, dated securely at 3.6 my B.P., that had retained the footprints of dozens of mammals and some reptiles and birds that had walked on the wet ash before it set like concrete. There were the prints of giraffes, elephants, baboons, hares, and many others making up a record of a fauna and flora that is remarkably like that of today in the same region—a semi-arid thorn tree scrub (Leakey and Hay, 1979).

The most exciting finds were some footprint trails of bipedal creatures whose gait and foot impressions were very similar to those of modern human unshod people (Day and Wickens, 1980).

The implications of these finds were that hominids (man-like creatures) were living in East Africa 3.6 my B.P. and that their form of locomotion (bipedalism) was well established at that time. The only creatures that they could have been were australopithecines. These ape-men are known from South and East Africa and whose skeletal remains had disclosed bipedalism as their form of locomotion. Later fossil finds of jaws and teeth from Laetoli confirmed that australopithecines were indeed part of the fauna 3.6 my B.P. and they have been attributed to *Australopithecus afarensis*, a species best known from Ethiopia.

The Ape-Men of South Africa

The australopithecines were discovered in 1925 by Raymond Dart, an anatomist, from Taung in South Africa. A single infant skull was declared by him to be neither man nor ape by its dental and cranial anatomy, and also likely to be bipedal from the position of the foramen magnum at the base of the skull. Dart's claim was greeted with skepticism at first, but he was proved to be right when other finds were made at Sterkfontein, Swartkrans, and Kromdraai near Johannesburg and from Makapansgat further away. The finds were difficult to date since they were from limestone cave deposits, but they showed that two principal types of australopithecines were present: one small called *Australopithecus africanus*, and another that is known as *Australopithecus robustus*, a more powerful creature with larger jaws and teeth that is sometimes referred to as the genus *Paranthropus*.

For many years the taxonomic status and significance of these South African finds was a matter of debate. Were they really from two genera? Were they really bipedal and hominid? The answer to the second question seemed to be yes, but the first question is still unanswered.

The Ape-Men of East Africa

In 1959, the late Louis Leakey and his wife Mary recovered a hyper-robust skull from Bed I, Olduvai Gorge, Tanzania (Leakey, 1959). It has massive molar teeth, a heavily buttressed cranium, and a cranial capacity of about 530 cc (Tobias, 1967). It was the first example of an australopithecine to be recovered from East Africa in well-dated deposits associated with the Rift Valley. This hyper-robust skull came from Bed I dated at 1.7 my B.P. by means of the potassium–argon method and by fission-track dating. It was named *Australopithecus boisei* (see Figure 3).

Many more specimens of this species were recovered later by Richard Leakey from the area east of Lake Turkana in Kenya (Wood, 1991), and others from the Omo region of southern Ethiopia (Howell, 1978).

An expedition to the Hadar region led by Don Johanson, Maurice Taieb, and later Yves Coppens, was successful in recovering australopithecines that were all

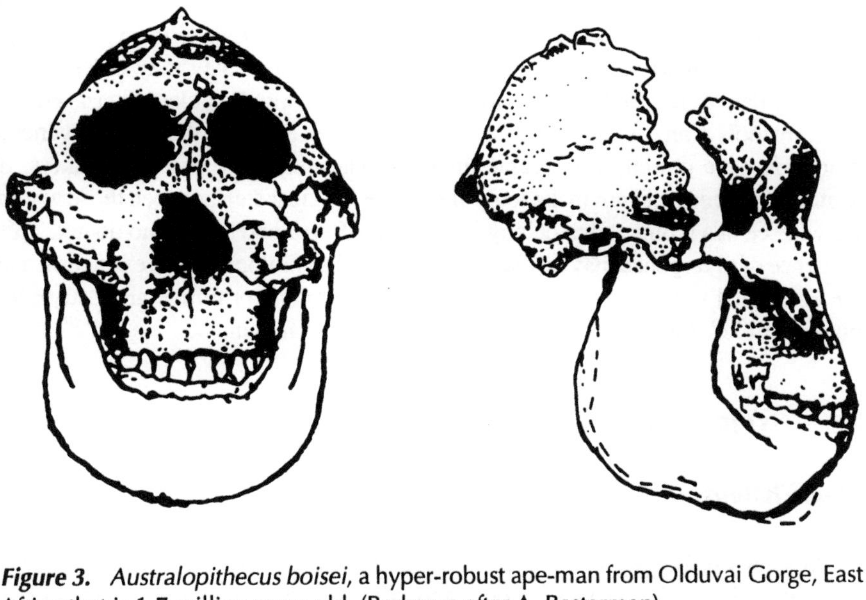

Figure 3. *Australopithecus boisei*, a hyper-robust ape-man from Olduvai Gorge, East Africa that is 1.7 million years old. (Redrawn after A. Besterman).

attributed to a new species, *Australopithecus afarensis*. These finds include the skeleton of a female popularly known as "Lucy," an extremely delicate and gracile creature whose postcranial bones are nearly all represented but whose skull is badly damaged. These skeletal remains dated at about 3 my BP show similarities to the *Australopithecus africanus* material from South Africa, in particular, in the anatomy of the pelvis. This similarity of the pelvic anatomy of two early australopithecines coupled with the evidence of the Laetoli footprints confirmed that bipedalism and upright posture had preceded brain expansion in hominid evolution.

The remains from the Hadar site proved to be widely sexually dimorphic, indeed to such an extent that there is still debate as to the unity of the Hadar finds in one species (*Australopithecus afarensis*) and of the relationship of these finds to the *Australopithecus africanus* finds from South Africa and perhaps the Omo region.

In general terms, the more robust australopithecines have been found in more recent deposits from about 2 to 1 my B.P.; none are younger than that, and none have been found outside Africa. Recently, a very robust skull indeed has been recovered from the west side of Lake Turkana by Richard Leakey and dated at about 2.5 my B.P. This find has caused a radical rethink of australopithecine relationships (Grine, 1988; Walker and Leakey, 1988).

Man and the Ape-Men

Questions that come to mind are, of course, how do these creatures relate not only to each other but how, if at all, do they relate to our ancestry? First we must consider that another hallmark of humanity that can be found in the record in rocks is the evidence of culture in the form of stone tools. It is not absolutely certain, but it seems that the australopithecines did not fashion stone tools with any degree of skill that is regularly recognizable from their living sites; they may indeed have been regular "tool-users" but not "tool-makers" to a set and regular pattern. Where there are tools at an australopithecine site there is often evidence that more advanced hominids were present at the same time and who may have been the makers. If australopithecines did not make tools, then there is a clear step between those whom we will see in a moment that belong to the genus *Homo* and those who do not. This evidence of divergence may be the clue to eventual australopithecine extinction and the continued success of the genus *Homo* from its most primitive and "australopithecine-like" form, through *Homo erectus* to modern man.

THE GENUS *HOMO* AND THE FIRST TRUE MEN

Homo habilis

At Olduvai Gorge, in the years immediately following the discovery of the *Australopithecus boisei* skull, Louis and Mary Leakey recovered a number of hominid remains from the same sites and of the same age that were clearly not of robust australopithecines. They were from smaller creatures, in some ways not unlike the *Australopithecus africanus* remains from South Africa but with some crucial differences. These anatomical differences were noted in dental morphology, cranial capacity, and locomotor and manipulative capability. The molar teeth were narrower and the cranial capacity was larger than that of the smaller australopithecines (Tobias, 1991). The hand and foot bones disclosed anatomical features that indicated manipulative skills, as well as a foot that was incontrovertibly adapted for propulsive bipedalism (Napier, 1962; Day and Napier, 1964). This group of features led to the attribution of these remains to an early form of our own genus as *Homo habilis* (Leakey, Tobias, and Napier, 1964), an astonishing claim that was controversial when it was made. The anatomical claim for manipulative skills in the hand was backed up by the find of numerous stone tools at the site of a primitive, but recognizable culture known as the Oldowan. This was "man the tool-maker" at his earliest appearance at that time, 2.1–1.7 my B.P. The anatomy shown by the

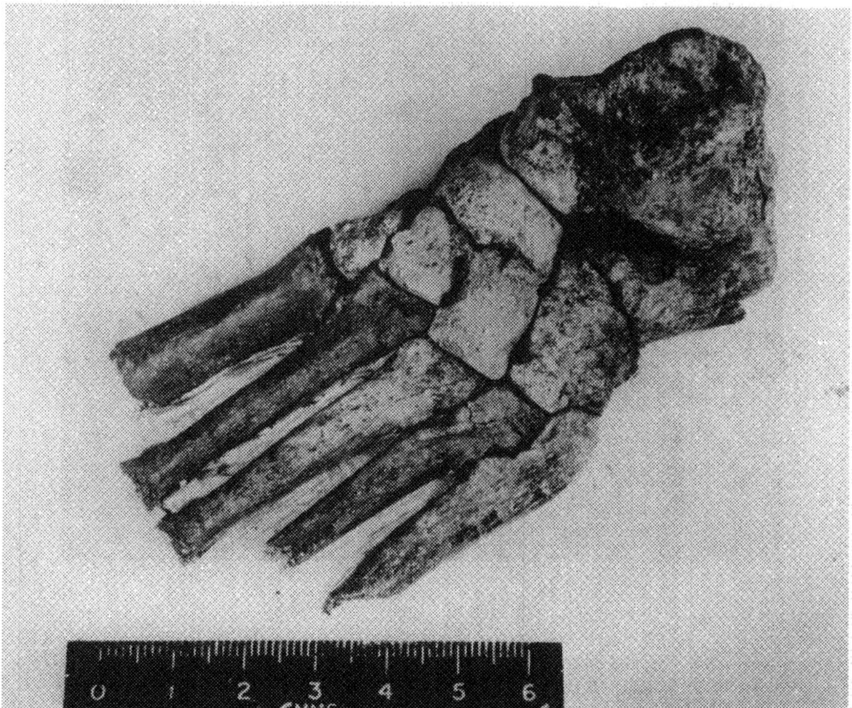

Figure 4. Articulations of the *Homo habilis* foot (*above*) and hand (*overleaf*) from Olduvai Gorge, East Africa dated at 1.7 million years before the present. These are the original articulations made by the author and demonstrate manipulative and bipedal capability in this the earliest member of our own genus *Homo*.(*continued*)

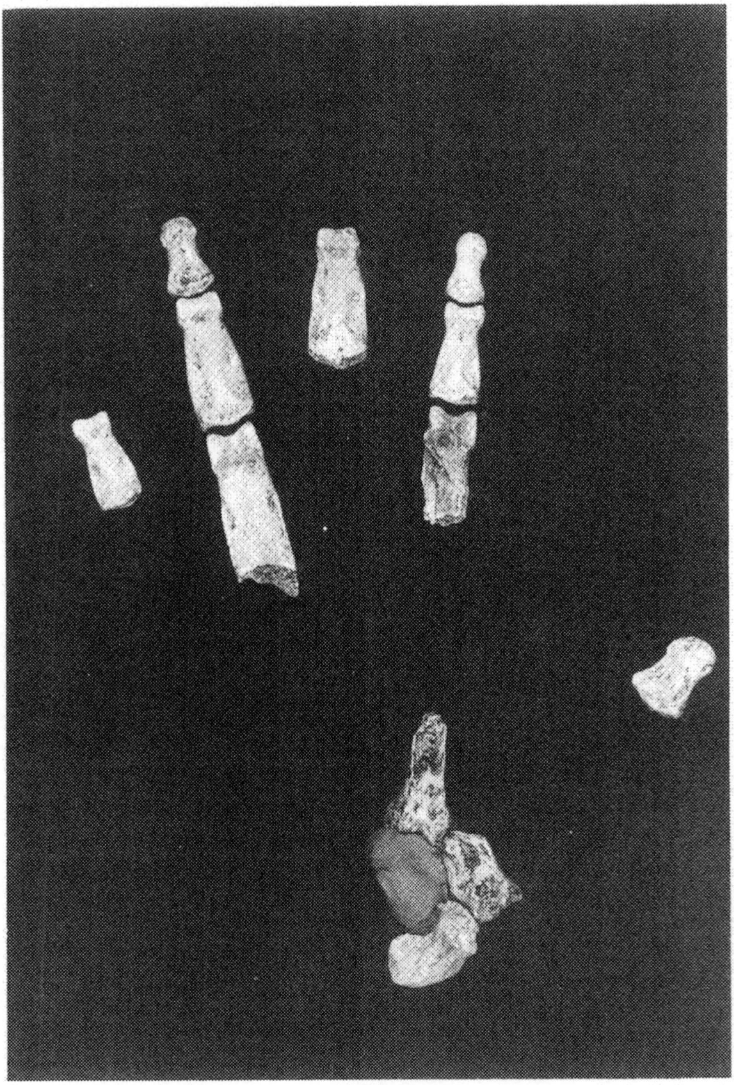

Figure 4. (*continued*)

Olduvai foot was clear in that it was not a grasping foot but one in which the great toe lay alongside the others and acted as part of a short lever system in human-type walking and running. This discovery preceded the Laetoli footprint discovery and was the most persuasive evidence of hominid bipedalism as a precursor to brain expansion that had so far emerged. *Homo habilis* was a small creature between 3 to 4 feet in height whose brain size was about 680 cc. This is larger than the average australopithecine at about 500 cc, but smaller than a later species of man *Homo erectus* at about 800 to 1000 cc and modern man at about 1400 cc.

In 1972, Richard Leakey recovered a skull from Koobi Fora, east of Lake Turkana, Kenya, now known as "1470" (Leakey, 1973; Day et al., 1974). It has been attributed to *Homo habilis* by some authorities, but others have expressed doubts. This skull bears a mixture of features both australopithecine and hominine in nature. After an initial error in the dating was corrected the skull was dated at about 1.8 my B.P., about the same as the Olduvai type specimen. Both Olduvai and Koobi Fora continued to produce fossils that were attributed to *Homo habilis*, but the variability shown by these remains have left the taxon with the question of its validity unresolved. In 1986, Don Johanson excavated at Olduvai Gorge and recovered an almost complete skeleton from deposits dated at 1.8 my B.P. It is a small female between 3 feet, 4 inches and 3 feet, 8 inches in stature, with long arms and some features of the skull that are like *Homo habilis*, but some characters of the limbs that recall *Australopithecus afarensis*. This demonstrable variability in the sample that makes up *Homo habilis* from several African sites is adding to the doubts that have been expressed as to the validity of the taxon.

The crucial question that arises is: What is the relationship of *Homo habilis* to the later hominines represented by *Homo erectus* from Africa, the Far East, and possibly Europe? A straight line relationship from *Australopithecus afarensis—Homo habilis—Homo erectus—Homo sapiens* is an attractive and simple scenario favored by those who believe that normal biological variation is commonly underestimated in fossil samples. Such an underestimation can lead to splitting and the creation and naming of numerous taxa that later have to be lumped together when more material becomes available to scholars. Successive evolution clearly occurred within the hominid evolutionary line, but branching is by no means excluded as shown by the example of the robust and hyper-robust australopithecines. Neither of these forms is found after 1 my B.P. and must be considered extinct and to have contributed little to the hominine gene pool.

Homo erectus

At the turn of the century one of the most important finds in human evolutionary history was made in what is now Indonesia. A Dutch army doctor and former anatomist, Eugene Dubois, found a skull cap and a femur at a place called Trinil. They were heavily fossilized and the skull fragment was very thick and primitive. The frontal region was constricted and the cranial capacity was clearly well below that of modern man. The juxtaposition of the skull cap and the femur, which looked remarkably modern, led Dubois to dub his fossil *Pithecanthropus erectus* (later *Homo erectus*) since, in his judgement, the femur displayed bipedal characters. Not one anatomist disagreed with his view of the femur, but the supposed primitive skull and the modern femur did seem an odd combination. Since the find in 1892 there have been a number of investigations into the dating and contemporaneity of

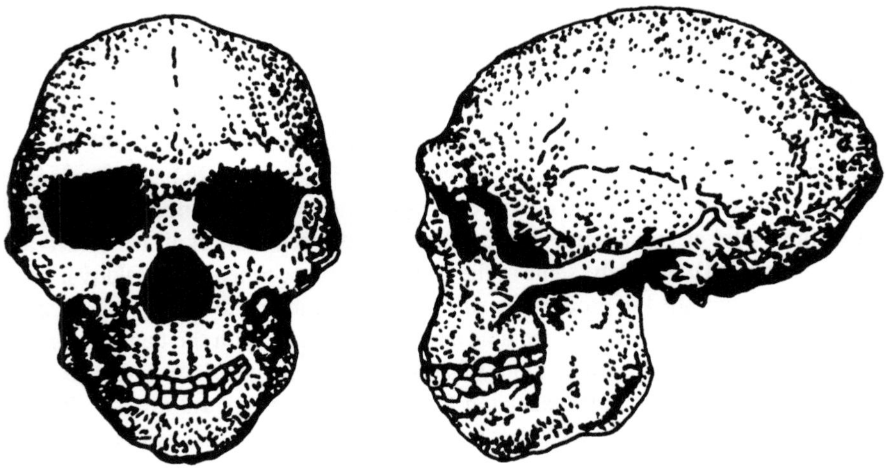

Figure 5. *Homo erectus* from China. Low vault, lack of chin, heavy brow ridge. (Redrawn after Besterman).

the two finds and it seems clear now that the femur came from a later level and belongs to modern man (Bartsiokas and Day, 1993). It is ironic indeed that Dubois was right for the wrong reasons in his view of bipedalism in *Homo erectus*, for since that time further finds of *Homo erectus* from China, Africa, and nearby at Sangiran in Indonesia have confirmed that *Homo erectus* was indeed erect in posture and one of the first true men in the fossil record.

The finds from China came from Choukoutien near Beijing and are preserved in limestone in collapsed caves. They were recovered during the 1920s–1930s and again intermittently ever since. The skulls that came from the Choukoutien site were well preserved and permitted reconstruction and casting; this is very important since almost all of the human fossils from this site were lost during the Second World War. The *Homo erectus* skulls from China are distinctive in that they are extremely thick with a low frontal region, a heavy continuous frontal supraorbital torus, an angulated occiput, and a lack of parietal expansion that leave the maximum breadth of the skull low on the vault at the level of the base. The teeth are large with shovel-shaped incisors and Y5 molar cusp, and fissure patterns that have a tendency to translation into the +4 pattern common in modern man. There were a few postcranial fragments from the site that showed the femur to be of distinctive shaft formation, a detail that was ignored for many years and not accepted as a reason to doubt the Trinil femur.

Africa produced a skull from Olduvai Gorge (Bed II) that had much in common with the Far Eastern remains, and also North Africa contributed several jaws with teeth very similar in morphology to those from China. In addition, the Koobi Fora site in northern Kenya has produced several groups of fossils that almost certainly

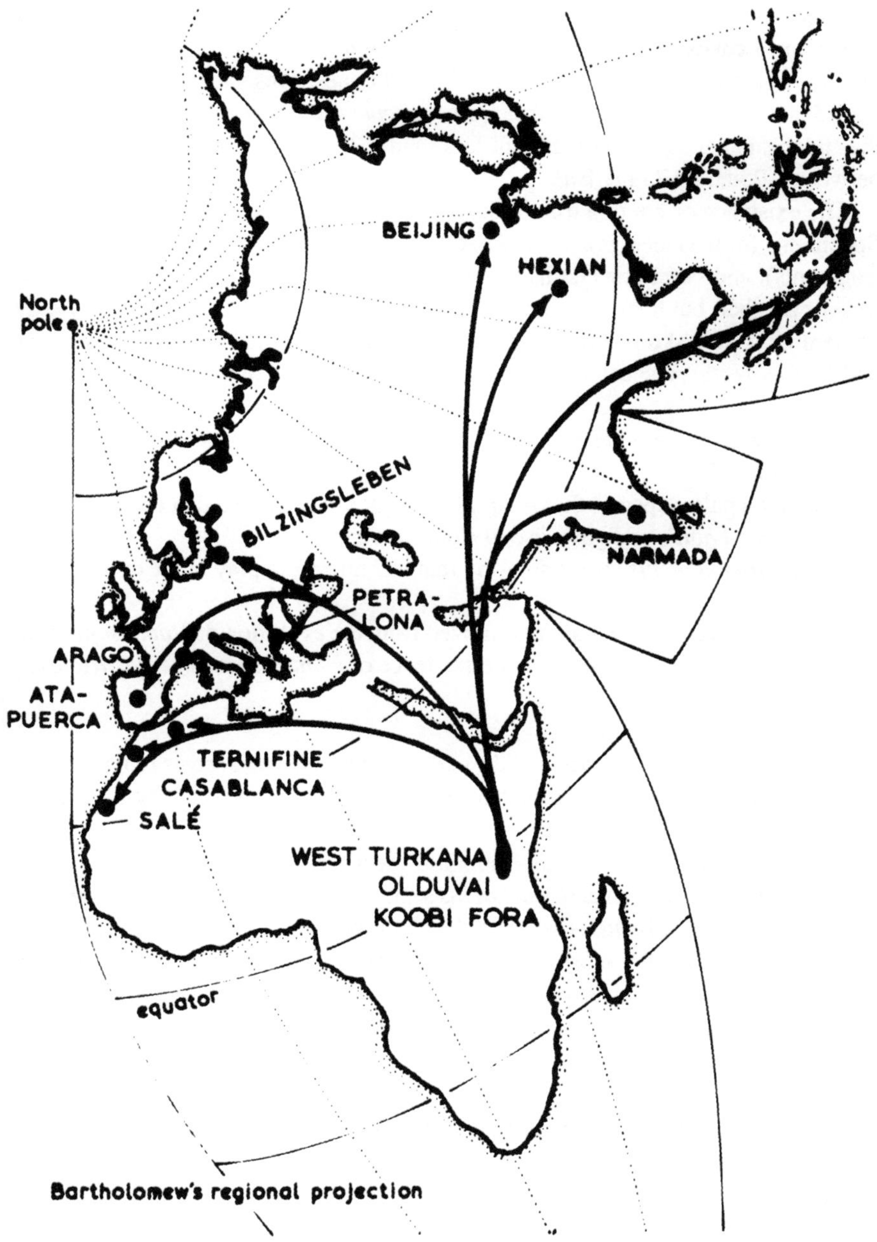

Figure 6. Map of migration of *Homo erectus* out of Africa and into Asia, the Far East and possibly Europe.

belong to the same species. But it was the other side of Lake Turkana that proved to be the most rewarding site. A place called Nariokotome, west of the lake, revealed to Richard Leakey and his team a virtually complete skeleton of a *Homo erectus* boy missing only his feet and most of his hand bones. This site can be dated with confidence to 1.6 my B.P. The other *Homo erectus* sites are dated less securely, but the Indonesian sites are probably all less than 1.3 my B.P. and the Chinese sites between 0.25 and 0.75 my B.P.

The existence or not of a true *Homo erectus* from Europe is a matter of debate. Specimens from Mauer (jaw), Vertésszöllös (occipital bone), Bilzingsleben (skull fragments), and Petralona (skull) have all been attributed to *Homo erectus* at one time or another, but there is no general agreement. Even in Africa two supposedly contemporaneous skulls, Omo I and Omo II, can be aligned with *Homo sapiens* and *Homo erectus*, respectively. Debate centers around all of the following topics:

- Does *Homo erectus* exist as a true taxon, or should it be sunk into *Homo sapiens*?
- Is it a palaeospecies that is simply a segment of the line that emerged from *Homo habilis* and gave rise to *Homo sapiens*?
- Is *Homo erectus* an extinct form that had no part to play in the evolution of *Homo sapiens*?
- Are the Asian forms so far removed from the evolution of *Homo sapiens* in Africa to call into question the existence of *Homo erectus* (*sensu stricto*) in Africa at all?

At present there is evidence that the earliest known *Homo erectus* remains seem to come from Africa and migration from there between 1.6 and 0.25 my B.P. seems to have taken place into Eurasia.

It is always dangerous to equate hominid species with established tool cultures, but in the most general terms *Homo habilis* seems to be associated with the Oldowan culture while *Homo erectus* is associated with the beginnings of hand-axes and their spread through the Old World. Where the boundary between *Homo erectus* and *Homo sapiens* lies within Africa or the Eurasian continent needs consideration of the early evidence of our own species, *Homo sapiens*.

THE EMERGENCE OF HOMO SAPIENS

Africa

One of the first known examples of archaic *Homo sapiens* came from Broken Hill (now known as Kabwe) in what was Rhodesia. The skull was recovered by workmen in 1921 from a lead mine and it had a number of extraordinary features. It is a large skull with large maxillae bearing a set of broken and carious teeth,

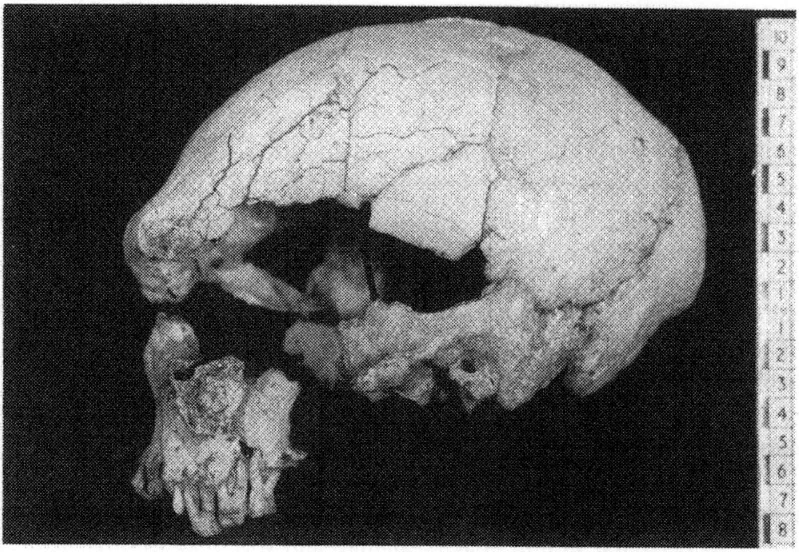

Figure 7. A transitional *erectus/sapiens* fossil from Laetoli, East Africa.

evidence of infected gums and abscesses around the roots of the teeth. The brow ridges are very heavy, but the walls of the cranium are relatively thin and the brain large, (1280 cc), bigger than that of *Homo erectus* by a substantial margin. The heavy brow ridge recalled *Homo erectus*, but clearly the skull was sapient and perhaps an early example of a transitional form. It has been dated to the late Middle Pleistocene perhaps a minimum of 125,000 y B.P. Since the discovery of Kabwe man there have been a number of other finds of transitional "*erectus-sapiens*," human fossils from East Africa (Ndutu and Laetoli 18 from Tanzania, Bodo and Omo II from Ethiopia, Eliye Springs from Kenya) and Klasies River Mouth and Florisbad from South Africa. A category, termed "Archaic *Homo sapiens*," has been used by Bräuer (1989) for those whose dates range from perhaps 350,000 to 100,000 y B.P. They were succeeded by others who were more advanced towards *Homo sapiens* and in whom the *Homo erectus* traits begin to diminish and finally disappear. These include Omo I from Ethiopia and Border Cave from South Africa. The latter site is of particular importance because of the modernity of the anatomy of the remains at a date of between 90,000 and 110,000 y B.P. The Omo I site may be even older, perhaps 125,000 y B.P. The fossil evidence is that *Homo sapiens* were present in Africa at an early stage, and this gives rise to the view that modern man evolved "out of Africa," an idea not supported by all palaeoanthropologists but supported by genetic studies of mitochondrial DNA (Stoneking and Cann, 1989). Those who do not accept the single African origin of *Homo sapiens* draw attention to the claims of other parts of the globe in support of their multiregional view of the evolution of *Homo sapiens* (Wolpoff, 1989).

China, The Far East, and Australia

During the discovery of *Homo erectus* material from Choukoutien additional modern skulls were found from the Upper Cave at this site. These were recognized as modern human, but the cave is now dated at as little as 10,000 y B.P. Other sites in China that have disclosed early *Homo sapiens* include Xujiayao, Dali, Maba and several others (Wu, X. and Wu, M., 1985). This evidence coupled with that of the Ngandong and Sambungmachan skulls from Java can suggest that there is regional continuity down the southeast Asian peninsula and archipelago leading to the peopling of Australia.

Europe and Western Asia

In 1960, an early human skull was unearthed from a cave at Petralona, Greece. The skull was covered in limestone stalagmite and was very well preserved. It showed a mixture of *Homo erectus* and *Homo sapiens* features and it may be as old as 300,000 to 350,000 years B.P. Other examples of early *Homo sapiens* include the Heidelberg jaw (robust like an *erectus* but with small teeth like a *sapiens*), the Steinheim calvaria (small, probably female but with Neanderthal features at an early date), Bilzingsleben (skull fragments that are very *erectus*-like and of an early date, but regarded by some as *Homo sapiens*), Vertésszöllös (an occipital bone from Hungary revealing a large cranial capacity from deposits dated at about 185,000 y B.P.), and the skull and postcranial bones from Arago in southern France that are regarded as pre-Neanderthalers perhaps from the Second Glaciation. A find from Swanscombe, England in 1935, dated at perhaps 250,000 y B.P. is the best known of all these early sapient fossils from Europe, consisting of an occipital bone and two parietals that fit together to form the back portion of a skull that is clearly not that of *Homo erectus*. Without the frontal bone it was difficult to say more, but increasingly it is being regarded as an early sapient with Neanderthal features that are incipient rather than well expressed.

The last site that is of particular interest in this early sapient context is the material from Jebel Qafzeh in Israel; eleven skeletons have been identified from this site and although they do indeed show some archaic features there is no doubt that they belong to *Homo sapiens*. The dating now given to this site is 90,000 y B.P.

It seems clear that there are examples of early members of *Homo sapiens* from a variety of sites in Europe, the Near East, Africa, China, and the Far East. These remains form the basis of the debate that is currently occupying many palaeoanthropologists; that is where, when, and how did *Homo sapiens* evolve? Was it from Africa with subsequent spread to other continents—a single origin with migration?

Or was it a multiregional matter with separate segments of *Homo sapiens* evolving from an earlier form, such as *Homo erectus* and/or Neanderthal man, in separate lines of descent?

The Neanderthalers

Since the mid-nineteenth century it has been known that during the last major glaciation Europe and parts of the Middle East was home to a group of humans whose anatomy was distinct. The first finds were not clearly recognized for what they were, but one in Germany was made famous. It came from the valley of the Neander river and was found at a cave site that was in process of being quarried away. The skeleton was probably complete but only the top of the skull and some limb bones survived for scientific examination in 1856. It was, therefore, two years before the publication of the *Origin of Species* by Charles Darwin and the climate of opinion inclined towards the find being regarded as the pathological or aberrant remains of a hermit or recluse. The limb bones were very powerful and bowed, the vault of the skull sloped backwards and the bones were thick. But it was Huxley, the champion of Darwin, who accepted these remains as human, and King, an anatomist from Galway, who dubbed him *Homo neanderthalensis* (later *Homo sapiens neanderthalensis*). Subsequent finds came from many sites in France, Belgium, Germany, the Middle East (Shanidar, Tabun, Amud, Kebara), and Central Europe from about 70,000 y B.P. to perhaps as recent as 30,000 y B.P.

The features of the Neanderthal skull are that it is long and low but also expanded at the sides, and the cranial capacity is fully up to modern human standards (1300–1400 cc) and in some cases more. The central portion of the Neanderthal face is drawn forward to give a prominent and capacious nose, a degree of prognathism, a low brow, and a heavy, but not continuous, brow ridge. They are short (perhaps 5 feet, 4 inches in stature), thick set, muscular, and big boned with large hands and feet, proportions that conform to those required for living in cold conditions (Bergman and Allen's Rules).

The Neanderthalers were a remarkable group who certainly had the use of fire, clothing, shelter, hunting skills, survival techniques, tool technology, and some ritual behavior in association with death, such as burial with flowers and ritual placement of skulls or animal bones. While thought for many years to have been stooped and incompletely upright, it was later shown that the first reconstruction of a Neanderthal skeleton was faulty. Later finds and reconstructions show that they were fully upright, agile, active, and very strong. In view of their known hunting activities this is not surprising.

The remains of Neanderthals have not been found in Africa or the Far East, neither in America nor Australia. They were a European and West Asian group who survived the cold of the Last Glaciation. Did they die out without issue, or were they absorbed by the burgeoning *Homo sapiens* of modern aspect? This is a question that is central to the debate on the origin of modern man. Those who

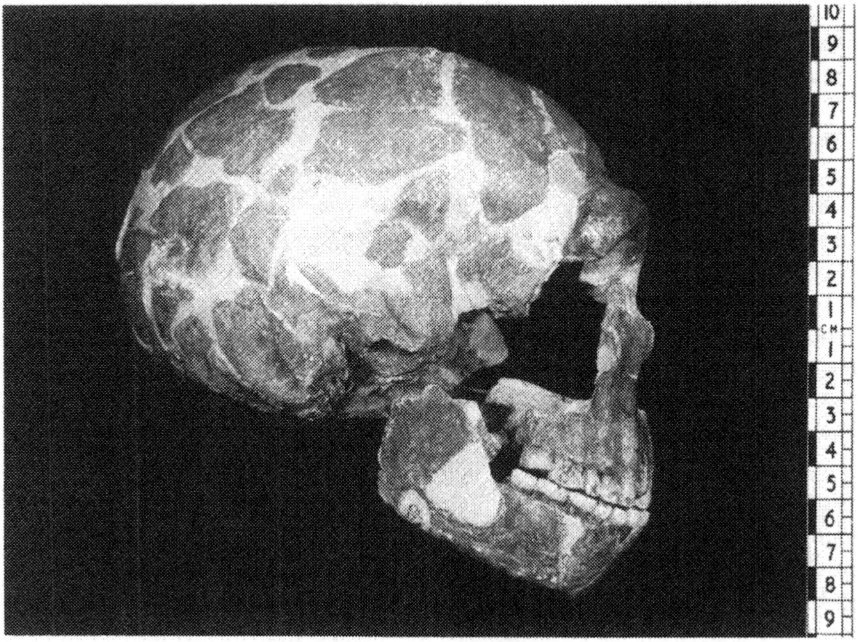

Figure 8. A Neanderthal female skull from Tabun, Israel showing central portion of face drawn forward. (Courtesy of the Natural History Museum).

espouse the multiregional hypothesis would like to see regional continuity in Europe through the Neanderthals to modern man, while others see the Neanderthals as a specialized group adapted to cold conditions who became isolated in the European peninsula during the Last Glaciation, and who died out when that adaptation was no longer appropriate.

Later *Homo sapiens* Worldwide

Today, modern man inhabits all of the continents of the world and many of its remote islands. Only the inhospitable Antarctica does not support a human population of its own. The spread of man to all these areas of the world with their climatic differences is a measure of his increasing ability to control the microenvironment of his immediate surroundings. The use of clothing to conserve heat, the construction of shelter and the use of caves, the ability to make and maintain fire, the ability to store food and water, and finally the construction of rafts to traverse open water have all contributed to the peopling of the planet.

The search for early human remains in North and South America has been undertaken for many years without success; the usual date given for the first wave of migration into North America from Asia is 12,000 y B.P. when the ice sheets of the Last Glaciation were finally retreating to the north. Recent views suggest that the people who made the "Clovis points" stone tool culture may have been as far south as Meadowcroft rock shelter, near Pittsburgh, as early as 14,500 y B.P. Other evidence from South America suggests human occupation as far south as Chile by as early as 12,000 y B.P., but claims for earlier dates have yet to be substantiated.

A similar situation existed in Australia in that the aboriginal people must have entered the island from the southeast Asian archipelago by hopping from one island to another by means of rafts of primitive boats, but no direct evidence of this exists at present. It is possible that the first Australians reached those shores as early as 50,000 y B.P., and it is certain that they were there by 30,000 to 32,000 y B.P. The Australian fossil remains are of two principal types, one gracile and the other robust and this has sparked controversy as to their origins and to the form of the migration that took place during the peopling of Australia (Habgood, 1989).

Europe too had its migrations; it appears that the peoples who brought the technical advances of the Aurignacian culture to Europe came from the east. People of this type were first known from France at a place called Cro-Magnon. The remains from this site, found as early as 1868, were contrasted with the crude and seemingly primitive Neanderthalers and were chosen as the preferred ancestors of modern European man. The Cro-Magnon race were tall, upright, and had slender limbs, large brains and rounded crania, straight foreheads, and jaws with chins—a hallmark of *Homo sapiens*. These people lived in Europe during the latter part of the Last Glaciation and hunted reindeer, bison, mammoth, and horse to great effect. They also produced stone tools, had fine technical skills, and created cave art and sculpture of considerable beauty and refinement.

In anatomical and evolutionary terms, modern man is a single species with a worldwide distribution and with differences of appearance such as hair form, skin color, and nose shape. These are trivial in a biological sense: textbooks of anatomy are universal and not related to race; surgeons trained in one part of the world have a skill that applies to all mankind; and the principles of human medicine apply to all of humanity.

SUMMARY

The scope of this review can be summarized as follows:

- The study of comparative anatomy shows beyond any question that man is an animal, a vertebrate, a mammal, and a primate, and shares with the members of the Order Primates numerous anatomical characteristics and a remarkable genetic closeness in particular to the great apes.
- The history of the Order Primates has been built up from fossil finds in the Earth's crust dating back to 70 million years before the present. The

techniques of dating are becoming more accurate as the resources of geology and geochronology are aided by modern physics.

- Darwin's theory of evolution and its mechanism, natural selection, apply to primate and human evolution and form the theoretical framework for the process.
- Early fossil primates are known from both the Old and the New World, but continental drift and the opening of the Atlantic separated the order into two by the "monkey stage" of evolution. The "ape" stage only developed in Africa and Asia.
- The succeeding "ape-man" stage is confined to Africa where australopithecines occur in rocks from 3 to 5 million years old. The first members of our own genus, *Homo habilis*, are known from East Africa and are dated at about 2 million years before the present. They made simple stone tools. They were succeeded by *Homo erectus*, known from Africa and Asia, who made tools of greater variety and use, who hunted large game, and who probably had the use of fire.
- Out of this stage of human development our own species, *Homo sapiens*, arose, most probably in Africa. From there we spread through Europe and Asia and eventually to Australia and North America peopling these latter continents over a period of about 30,000 years. One other subdivision of *Homo sapiens*, the Neanderthalers, lived in Europe and the Near East and their adaptations to the cold climate gave them advantage until the end of the last ice age.
- The modern races of man are the result of geographic isolation and/or climatic and environmental adaptation over time: all belong to one species, all share the same anatomy, and all interbreed freely and produce fertile offspring.

REFERENCES

Bartsiokas, A. & Day, M. H. (1993). Electron probe energy dispersive X-ray microanalysis (EDXA) in the investigation of fossil bone: The case of Java Man. Proc. Roy. Soc. B. 252, 115–123.

Bräuer, G. (1989). The evolution of modern humans: a comparison of the African and non-African evidence. In: The Human Revolution (Mellars, P. & Stringer, Eds.), pp. 123–154. Edinburgh University Press, Edinburgh.

Dart, R. A. (1925). *Australopithecus africanus*: the man-ape of South Africa. Nature 115, 195–199.

Day, M. H., Leakey, R. E. F., Walker, A. C., & Wood, B. A. (1974). New hominids from East Rudolf, Kenya. Am. J. Phys. Anthrop. 42, 461–476.

Day, M. H. & Napier, J. R. (1964). Hominid fossils from Bed 1, Olduvai Gorge, Tanganyika.

Day, M. H. & Wickens, E. H. (1980). Laetoli Pliocene hominid footprints and bipedalism. Nature 286, 385–387.

Grine, F. E. (1988). Evolutionary History of the "Robust Australopithecines." Aldine de Gruyter, New York.

Habgood, P. J. (1989). The origin of anatomically modern humans in Australia. In: The Human Revolution (Mellars, P. & Stringer, C. B., Eds.), pp. 245–273. Edinburgh University Press, Edinburgh.

Howell, F. C. (1978). Overview of the Pliocene and earlier Pleistocene of the Lower Omo Basin, southern Ethiopia. In: Geological Background to Fossil Man. (Bishop, W.W., Ed.), Scottish Academic Press, Edinburgh.

Leakey, L. S. B. (1959). A new skull from Olduvai, Nature 201, 967–970.

Leakey, L. S. B.,Tobias, P.V., & Napier, J. R. (1964). A new species of the genus *Homo* from Olduvai Gorge. Nature 202, 7–9.

Leakey, M. D. & Hay, R. L. (1979). Pliocene footprints in the Laetolil Beds at Laetoli, northern Tanzania. Nature 278, 317–323.

Leakey, R. E. F. (1973). Further evidence of Lower Pleistocec hominids from East Rudolf, North Kenya, 1972. Nature 242, 170–173.

Linnaeus, C. G. *Systema Naturae.*

Napier, J. R (1962). Fossil hand bones from Olduvai Gorge. Nature 196, 409–411.

Playfair, J. (1802). Illustrations of the Huttonian Theory of the Earth. Dover Publications, New York.

Stoneking, M. & Cann, R. L. (1989). African origin of mitochondrial DNA. In: The Human Revolution. (Mellars, P. & Stringer, C. B., Eds.), pp. 17–30. Edinburgh University Press, Edinburgh.

Tobias, P. V. (1967). Olduvai Gorge, Vol. 2. The Cranium of *Australopithecus*, Cambridge University Press, Cambridge.

Tobias, P. V. (1991). Olduvai Gorge, Vol. 4. The Skull, Endocasts, and Teeth of *Homo habilis*. Cambridge University Press, Cambridge.

Vesalius (1543). *De Humani Corporis Fabrica.*

Walker, A. C. & Leakey, R. E. (1988). The evolution of Australopithecus boisei. In: Evolutionary History of the "Robust Australopithecines" (Grine, F. E., Ed.). Aldine de Gruyter, New York.

Wolpoff, M. H. (1989). Multiregional evolution In: The Evolution of Modern Humans: A Comparison of the African and Non-African Evidence (Mellars, P. & Stringer, C. B., Eds.), pp. 62–108. Edinburgh University Press, Edinburgh.

Wood, B. A. (1991). Koobi Fora Research Project. Hominid Cranial Remains, Vol. 4. Cambridge University Press, Cambridge.

Wu, X. & Wu, M. (1985). Early *Homo sapiens* in China. In: Palaeoanthropology and Palaeolithic Archeology in the People's Republic of China (Wu, K. & Olsen, J. W., Eds.). Academic Press, Orlando.

Zimmerman, M. H. & Angel, J. L. (1986). Dating and age determination of biological materials. Croon Helm, London.

RECOMMENDED READINGS

Day, M. H. (1986). Guide to Fossil Man. Cassell, London.

Desmond, A. & Moore, J. (1991). Darwin. Hutchinson, London.

Fleagle, J. G. (1988). Primate Adaptation and Evolution. Academic Press, San Diego.

Jones, S., Martin, R., & Pilbeam, D. (1992). The Cambridge Encyclopedia of Human Evolution. Cambridge University Press, Cambridge.

Martin, R. D. (1990). Primate Origins and Evolution. Chapman Hall, London.

Rightmire, G. P. (1990). The Evolution of *Homo erectus*. Cambridge University Press, Cambridge.

Smith, F. H. & Spence, F. (1984). The Origin of Modern Humans. Alan Liss, New York.

Wood, B. A. (1992). Origin and evolution of the genus *Homo*. Nature 355.

Wu, R. & Dong, X. (1985). *Homo erectus* in China In: Palaeoanthropology and Palaelithic Archeology in the People's Republic of China (Wu, R. & Olsen, J. W., Eds.), pp. 79–89. Academic Press, Orlando.

Subject Index

www.ingramcontent.com/pod-product-compliance
Ingram Content Group UK Ltd.
Pitfield, Milton Keynes, MK11 3LW, UK
UKHW020233290926
15470UKWH00002B/509